GREAT-ESF Workshop: Stellar Atmospheres in the Gaia Era 2011

AA002802

Journal of Physics: Conference Series Volume 328

Brussels, Belgium
23 – 24 June 2011

Editors:

Alex Lobel
Jean-Pierre De Greve
Walter van Rensbergen

ISBN: 978-1-61839-351-7
ISSN: 1742-6588

Printed from e-media with permission by:

Curran Associates, Inc.
57 Morehouse Lane
Red Hook, NY 12571

Some format issues inherent in the e-media version may also appear in this print version.

Printed by Curran Associates, Inc. (2011)

For permission requests, please contact the Institute of Physics
at the address below.

Institute of Physics
Dirac House, Temple Back
Bristol BS1 6BE UK

Phone: 44 1 17 929 7481
Fax: 44 1 17 920 0979

techtracking@iop.org

Additional copies of this publication are available from:

Curran Associates, Inc.
57 Morehouse Lane
Red Hook, NY 12571 USA
Phone: 845-758-0400
Fax: 845-758-2634
Email: curran@proceedings.com
Web: www.proceedings.com

TABLE OF CONTENTS

QUANTITATIVE SPECTROSCOPY AND COMPARATIVE SPECTRUM MODELING OF COOL STARS

QUANTITATIVE SPECTROSCOPY OF HOT STARS

CONFERENCE REVIEW

Preface – Stellar Atmospheres in the Gaia Era

Volume 328 (2011) of the *Journal of Physics: Conference Series* provides a record of the invited and contributed talks, and of the posters presented at the GREAT-ESF workshop entitled 'Stellar Atmospheres in the Gaia Era: Quantitative Spectroscopy and Comparative Spectrum Modelling' (`http://great-esf.oma.be`). The conference was held 23–24 June 2011 at the Vrije Universiteit Brussel, Belgium. 47 scientists from 11 countries around the world attended the workshop.

The ESA-Gaia satellite (launch mid 2013) will observe a billion stellar objects in the Galaxy and provide spectrophotometric and high-resolution spectra of an unprecedented amount of stars observed with a space-based instrument. The confrontation of these data with theoretical models will drastically advance our understanding of the physics of stellar atmospheres. New stellar populations such as previously unknown emission line stars will be discovered, and fundamental questions such as the basic scenarios of stellar evolution will be addressed with Gaia data.

The 33 presentations and 4 main discussion sessions at the workshop addressed important topics in spectrum synthesis methods and detailed line profile calculations urgently needed for accurate modelling of stellar spectra. It brought together leading scientists and students of the stellar physics communities investigating hot and cool star spectra. The scientific programme of the workshop consisted of 23 oral (6 invited) and 10 poster presentations about cool stars (first day; Comparative Spectrum Modelling and Quantitative Spectroscopy of Cool Stars), and hot stars (second day; Quantitative Spectroscopy of Hot Stars). The hot and cool stars communities use different spectrum modelling codes for determining basic parameters such as the effective temperature, surface gravity, iron abundance, and the chemical composition of stellar atmospheres. The chaired sessions of the first day highlighted new research results with spectral synthesis codes developed for cool stars, while the second day focused on codes applied for modeling the spectra of hot stars.

The workshop addressed five major topics in stellar atmospheres research:

- Spectrum synthesis codes
- Radiation hydrodynamics codes
- Atmospheric parameters, abundance, metallicity, and chemical tagging studies
- Large spectroscopic surveys
- New atomic database

The workshop presentations discussed various important scientific issues by comparing detailed model spectra to identify differences that can influence and bias the resulting atmospheric parameters. Theoretical line-blanketed model spectra were compared in detail to high-resolution spectroscopic observations. Stellar spectra computed (i.e., in the Gaia Radial Velocity Spectrometer wavelength range) with 1-D model atmosphere structures were mutually compared, but also to 3-D models from advanced radiation hydrodynamics codes. Atmospheric parameters derived from spectrum synthesis calculations assuming Local Thermodynamic Equilibrium (LTE) were evaluated against more sophisticated non-LTE models of metal-poor stars and the extended atmospheres of giants and supergiants. The workshop presented an overview of high-resolution synthetic spectral libraries of model spectra computed with the synthesis codes. The spectral model grids will be utilized to derive stellar parameters with the Discrete Source Classifier Algorithms currently under development in the Gaia DPAC consortium (`http://www.rssd.esa.int/index.php?project=GAIA&page=DPAC_Introduction`). They are implemented for training Gaia data analysis algorithms for the classification of a wide variety of hot and cool star types; FGK & M stars, OB stars, white dwarfs, red supergiants, peculiar A and B stars, carbon stars, ultra cool dwarfs, various types of emission line stars, Be stars, Wolf-Rayet stars, etc. A substantial number of oral and poster presentations discussed different techniques for measuring the abundance of various chemical elements from stellar spectra. The presented methods utilize spectra observed with large spectral dispersion, for example for accurately measuring iron, carbon, and nitrogen abundances. These methods are important for ongoing development and testing of automated & supervised algorithms for determining detailed chemical composition in tagging studies of large (chemo-dynamical) spectroscopic surveys planned to complement the Gaia (astrometric and kinematic) census of the Galaxy.

The complete scientific programme is available at `http://http://iopscience.iop.org/1742-6596/328/1/011001`. The workshop website also offers the presentation viewgraphs (in PDF format) and some nice photographs of the talks and poster breaks `http://great-esf.oma.be/program.php`. The papers presented at the workshop and collected here have been edited by A Lobel, J-P De Greve, and W van Rensbergen. The Proceedings essentially follow the order of presentation during the conference program, divided into cool and hot stars (oral papers are followed by poster papers). It also offers a review paper about new research results presented at the conference, including a record of the main discussion sessions. 27 papers passed through the peer review process. The manuscripts were submitted before 15 September 2011 and accepted by the referees before 1 November 2011. We would like to thank the reviewers for their constructive criticism during the preparation of this Volume. It has 225 pages by 27 authors and 135 co-authors, and includes 114 color figures and 21 tables. The articles are freely available at `http://iopscience.iop.org/1742-6596/328/1`.

We would like to express our gratitude for the financial support from the European Science Foundation and the Research Foundation - Flanders. We thank the Vrije Universiteit Brussel for making the meeting and poster rooms freely available during the conference. We specially thank Mrs. Merel Fabré for assistance with the workshop organization and administration at the VUB. We thank the Royal Observatory of Belgium for help with conference grants management, transportation, and administration. We are grateful to all the workshop participants for their valuable contributions and active discussions which made the conference very successful indeed.

Alex Lobel
Royal Observatory of Belgium, Ringlaan 3, B-1180 Brussels, Belgium
E-mail: Alex.Lobel @ oma.be alobel @ sdf.lonestar.org
Web: `http://alobel.freeshell.org`
Jean-Pierre De Greve
Walter van Rensbergen
Vrije Universiteit Brussel, Pleinlaan 2, B-1050 Brussels, Belgium
Editors
Brussels, October 2011.

Journal of Physics: Conference Series **328** (2011) 011001 doi:10.1088/1742-6596/328/1/011001

CONFERENCE POSTER

Journal of Physics: Conference Series **328** (2011) 011001 doi:10.1088/1742-6596/328/1/011001

SCIENTIFIC COMMITTEE
Jose Groh (Germany)
Peter Hauschildt (Germany)
Ulrike Heiter (Sweden)
Alex Lobel (Belgium)
Bertrand Plez (France)
Norbert Przybilla (Germany)
Rosanna Sordo (Italy)

CONFERENCE SPONSORS
European Science Foundation (ESF) `www.esf.org`
Research Foundation - Flanders (FWO-Vlaanderen) `www.fwo.be`
Free University Brussels (VUB) `www.vub.ac.be`
Royal Observatory of Belgium (ROB) `www.astro.oma.be`

Journal of Physics: Conference Series **328** (2011) 011001 doi:10.1088/1742-6596/328/1/011001

CONFERENCE PHOTOGRAPH (23 June 2011)

Journal of Physics: Conference Series **328** (2011) 011001 doi:10.1088/1742-6596/328/1/011001

CONFERENCE PARTICIPANTS

1. Teresa Aparicio Villegas	Instituto de Astrofisica de Andalucia, Granada, Spain
2. Maria Bergemann	Max Planck Institute for Astrophysics, Garching, Germany
3. Ronny Blomme	Royal Observatory of Belgium, Brussels
4. Thomas Dall	European Southern Observatory, Garching, Germany
5. Alex de Koter	University of Amsterdam, The Netherlands
6. Yanping Chen	University of Groningen, The Netherlands
7. Andrea Chiavassa	Université Libre de Bruxelles, Belgium
8. Remo Collet	Max Planck Institute for Astrophysics, Garching, Germany
9. Jean-Pierre De Greve	Vrije Universiteit Brussel, Belgium
10. Nadiya Gorlova	University of Leuven, Belgium
11. Jose Groh	Max Planck Institute for Radioastronomy, Bonn, Germany
12. Peter Hauschildt	Observatory of Hamburg, Germany
13. Artemio Herrero	Instituto de Astrofisica de Canarias, Spain
14. Ulrike Heiter	University of Uppsala, Sweden
15. Anthony Hervé	University of Liège, Belgium
16. Leo Houziaux	Royal Academy of Belgium, Brussels
17. Rene Hudec	Academy of Sciences, Czech Republic
18. Alain Jorissen	Université Libre de Bruxelles, Belgium
19. Andreas Korn	University of Uppsala, Sweden
20. Pavel Koubsky	Astronomical Institute Ondřejov, Czech Republic
21. Alex Lobel	Royal Observatory of Belgium, Brussels
22. Ali Luo	National Astronomical Observatory, Beijing, China
23. Laurent Mahy	University of Liège, Belgium
24. Zazralt Magic	Max Planck Institute for Astrophysics, Garching, Germany
25. Jesús Maldonado	Universidad Autónoma de Madrid, Spain
26. Thomas Masseron	Université Libre de Bruxelles, Belgium
27. Josephina Montalban	University of Liège, Belgium
28. David Montes	Universidad Complutense de Madrid, Spain
29. Thierry Morel	University of Liège, Belgium
30. Pieter Neyskens	Université Libre de Bruxelles, Belgium
31. Fernanda Nieva	Max Planck Institute for Astrophysics, Garching, Germany
32. Giancarlo Pace	University of Porto, Portugal
33. Elena Pancino	INAF – Observatory of Bologna, Italy
34. Bertrand Plez	University of Montpellier, France
35. Norbert Przybilla	Observatory of Bamberg, Germany
36. Andreas Quirrenbach	Observatory of Heidelberg, Germany
37. Alejandra Recio-Blanco	Observatoire de la Côte d'Azur, France
38. Luca Sbordone	Max Planck Institute for Astrophysics, Heidelberg, Germany
39. Sergio Simon Diaz	Instituto de Astrofisica de Canarias, Spain
40. Rosanna Sordo	INAF – Observatory of Padova, Italy
41. Frans Arjen Stap	University of Amsterdam, The Netherlands
42. Marica Valentini	University of Liège, Belgium
43. Griet Van de Steene	Royal Observatory of Belgium, Brussels
44. Sophie Van Eck	Université Libre de Bruxelles, Belgium
45. Walter van Rensbergen	Vrije Universiteit Brussel, Belgium
46. Nicholas Walton	University of Cambridge, United Kingdom
47. Jiannan Zhang	National Astronomical Observatory, Beijing, China

Journal of Physics: Conference Series **328** (2011) 011001 doi:10.1088/1742-6596/328/1/011001

PARTICIPANTS

NAME	EMAIL
A	
Teresa Aparicio Villegas	terenz AT iaa.es
B	
Maria Bergemann	mbergema AT mpa-garching.mpg.de
Ronny Blomme	Ronny.Blomme AT oma.be
C	
Yanping Chen	yanping AT astro.rug.nl
Andrea Chiavassa	achiavas AT ulb.ac.be
Remo Collet	remo AT mpa-garching.mpg.de
D	
Thomas Dall	tdall AT eso.org
Jean-Pierre De Greve	jpdgreve AT vub.ac.be
Alex de Koter	A.deKoter AT uva.nl
G	
Nadya Gorlova	nadya AT ster.kuleuven.be
Jose Groh	jgroh AT mpifr-bonn.mpg.de
H	
Peter Hauschildt	yeti AT hs.uni-hamburg.de
Ulrike Heiter	ulrike.heiter AT physics.uu.se
Artemio Herrero	ahd AT iac.es
Anthony Hervé	herve AT astro.ulg.ac.be
Leo Houziaux	leo.houziaux AT ulg.ac.be
Rene Hudec	rene.hudec AT gmail.com
J	
Alain Jorissen	Alain.Jorissen AT ulb.ac.be
K	
Andreas Korn	andreas.korn AT physics.uu.se

Pavel Koubsky koubsky AT sunstel.asu.cas.cz

L
Alex Lobel alobel AT sdf.lonestar.org
Ali Luo lal AT lamost.org

M
Zazralt Magic magic AT mpa-garching.mpg.de
Laurent Mahy mahy AT astro.ulg.ac.be
Jesus Maldonado jesus.maldonado AT uam.es
Thomas Masseron thomas.masseron AT ulb.ac.be
Josefina Montalban j.montalban AT ulg.ac.be
David Montes dmg AT astrax.fis.ucm.es
Thierry Morel morel AT astro.ulg.ac.be

N
Pieter Neyskens pieter.neyskens AT ulb.ac.be
Fernanda Nieva fnieva AT mpa-garching.mpg.de

P
Giancarlo Pace gpace AT astro.up.pt
Elena Pancino elena.pancino AT oabo.inaf.it
Bertrand Plez bertrand.plez AT univ-montp2.fr
Norbert Przybilla Norbert.Przybilla AT sternwarte.uni-erlangen.de

Q
Andreas Quirrenbach A.Quirrenbach AT lsw.uni-heidelberg.de

R
Alejandra Recio-Blanco arecio AT oca.eu

S
Sergio Simon-Diaz ssimon AT iac.es
Rosanna Sordo rosanna.sordo AT oapd.inaf.it
Frans Arjen Stap frans.stap AT student.uva.nl

V
Marica Valentini valentini AT astro.ulg.ac.be

Griet Van de Steene — gsteene AT oma.be
Sophie Van Eck — svaneck AT astro.ulb.ac.be
Walter van Rensbergen — wvanrens AT vub.ac.be

W
Nicholas Walton — naw AT ast.cam.ac.uk

Z
Jiannan Zhang — jnzhang AT lamost.org

Gaia spectroscopy overview and comparative spectrum modeling for cool giants

U Heiter[1] **and T Lebzelter**[2]

[1]Department of Physics and Astronomy, Uppsala University, Box 516, SE-75120 Uppsala, Sweden

[2]Department of Astronomy, University of Vienna, Türkenschanzstrasse 17, A-1180 Vienna, Austria

E-mail: ulrike.heiter@physics.uu.se

Abstract. The Gaia mission will provide spectroscopic and spectrophotometric observations for a vast number of stars, allowing a characterization in terms of stellar parameters along with the astrometric measurements. The pipeline for astrophysical parameter determination relies on libraries of model spectra for mapping stellar parameters to flux distributions. To maximize the reliability of the parametrization, adequate efforts are required to assess the realism of the model spectra and to calibrate the analysis methods. We describe a spectrum modelling experiment, in which high-resolution optical and infrared spectra of four cool giant stars were analysed by 14 different groups. The resulting variance in derived stellar parameters illustrates the need to be cautious when comparing or combining stellar parameters from different model atmospheres and analysis strategies. The main causes for the differences in parameters derived by different groups seem to lie in the physical input data and in the details of the analysis method.

1. Introduction

The ESA cornerstone mission Gaia[1] is due for launch in 2012. This ambitious astrometric mission will revolutionize our understanding of the formation and evolution of the Milky Way Galaxy. Gaia is a scanning satellite that will repeatedly survey the whole sky to obtain positions, parallaxes and proper motions to μas precision for all of the 10^9 objects brighter than 20 mag. Compared to its predecessor Hipparcos, Gaia will achieve a substantial improvement in terms of astrometric accuracy and completeness limit: parallax and proper-motion accuracy will be 100 times better and the number of stars is increased by a factor 10 000.

The added value of Gaia follows from the non-astrometric instruments on-board. These are the dispersed photometric instruments – the blue photometer at 330–680 nm and the red photometer at 640–1050 nm (RP and BP) with a spectral resolution of about 4–30 nm per pixel, and the medium resolution spectrograph – the RVS or Radial Velocity Spectrometer at 847–874 nm. The RVS wavelength range includes the prominent Ca II IR triplet lines for cooler stars, and the H-Paschen lines for hotter stars, as well as a number of lines of Fe-peak and α elements. The nominal spectral resolution of the RVS is $R = \lambda/\Delta\lambda$=11 500, which will be reduced to about 5 500 for magnitudes fainter than a certain limit around V=12. These instruments will enable the accurate simultaneous measurement of radial velocities as well as the determination of *astrophysical parameters* (APs) – $T_{\rm eff}$, $\log g$, metallicity [Fe/H], and interstellar extinction.

[1] http://sci.esa.int/gaia/

Journal of Physics: Conference Series **328** (2011) 012001 doi:10.1088/1742-6596/328/1/012001

2. Gaia photometric and spectroscopic instruments and data analysis

The building and assembly of the Gaia spacecraft and instruments is in an advanced stage. All instrument parts will be mounted on a torus with about 3 m diameter, made of SiC. Illustrations and images of the Gaia instruments can be found on the image gallery at the Gaia web page maintained by ESA. Useful examples are a schematic layout of the RVS instrument[2], an image of the complete Gaia torus with the 'Folding-Optics Structure' ready to be mounted, which will support among others the RVS[3], and the RVS module with its grating plate and several prismatic lenses[4].

The responsibility for processing and analyzing Gaia data lies with the Gaia Data Processing and Analysis Consortium (DPAC). This consortium of over 300 European scientists and software engineers is developing the software system which will convert the raw data recorded by the Gaia satellite into scientifically meaningful data, for immediate use by the astronomical community. The preparations for data analysis rely on simulations of the data expected from the Gaia instruments. [1] show in their Fig. 3 a number of examples for simulated RP/BP data, for solar metallicity and various combinations of $T_{\rm eff}$ and $\log g$. A large variation with $T_{\rm eff}$ is clearly visible in both bands. Differences in $\log g$ will be more difficult to discern, except at the lowest $T_{\rm eff}$ values around 3000 K. Synthetic spectra calculated for the RVS wavelength range and parameters corresponding to late-type giants are shown in Fig. 1. The calculations were done with the spectrum synthesis code `sme_synth`, which is part of the spectrum analysis package SME [2; 3], using MARCS model atmospheres [4] and line data from VALD [5; 6]. The sensitivity of this spectral range to stellar metallicity is apparent.

The development of the data processing software for the determination of astrophysical parameters has progressed quite far. The software package called 'Apsis' (Astrophysical parameters inference system) has two main tasks: to classify all sources, i.e. to assign probabilities for being a star, galaxy, quasar, etc., and to determine astrophysical parameters for stars: $T_{\rm eff}$, $\log g$, metallicity, extinction, and possibly α-element abundances. The second task will be carried out by two different 'General Stellar Parametrizer' modules – GSP-spec and GSP-phot, specializing on RP/BP and RVS data, respectively. Both of them are based on large grids of model stellar spectra to map stellar parameters to flux distributions.

The GSP-phot module employs several numerical algorithms to determine the set of parameters which best represents the RP/BP flux distribution of an unknown stellar object. One of these (ILIUM) is described in detail in [7]. The performance of the algorithm is assessed by generating a test sample of synthetic flux distributions with random stellar parameters, which are analyzed by GSP-phot. The resulting APs are then compared to the known input APs. The performance depends on the stellar magnitude (i.e. the noise level) and the amount of extinction added to the simulated fluxes. In the best case, for 137 simulated stars at G=15 without extinction, the $T_{\rm eff}$ residuals are within $\pm 0.2\%$, and the $\log g$ residuals are within ± 0.07. However, in general, when introducing variable extinction and including stars with G=15 as well as G=18.5, the $T_{\rm eff}$ residuals range from ± 3 to 13%, the $\log g$ residuals from ± 0.3 to 1.1, and the [Fe/H] residuals from ± 0.5 to 1.3 dex.

The GSP-spec performance for analysis of RVS data depends strongly on the signal-to-noise ratio (SNR) of the spectra. For example, for FGK stars at a SNR of 50, the error in $T_{\rm eff}$ is found to be 100 K on average, i.e. between 2 and 3%, $\log g$ can be determined to 0.2, and [Fe/H] to 0.1 dex. α-element abundance determination might be possible with errors between 0.05 and 0.3 dex for the full range of SNR. More details on the GSP-spec algorithms and their performance can be found in the contribution by A. Recio-Blanco.

The performance tests of Apsis described above are only assessing internal errors stemming

[2] `http://www.rssd.esa.int/index.php?project=GAIA&page=IG_20090923`
[3] `http://www.rssd.esa.int/index.php?project=GAIA&page=IG_20101101b`
[4] `http://www.rssd.esa.int/index.php?project=GAIA&page=IG_20110810a`

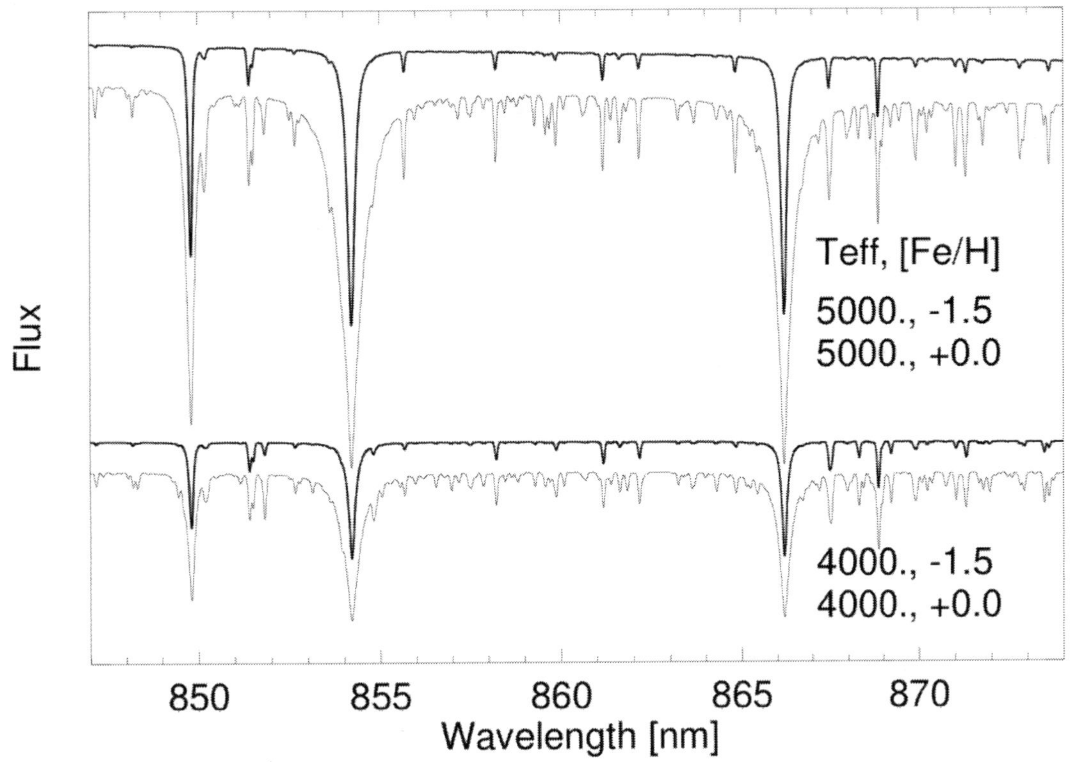

Figure 1. Synthetic spectra calculated with `sme_synth`, MARCS model atmospheres and spectral line data from the VALD database. The spectra are convolved to a resolution of 10 500, corresponding to the Gaia RVS. The parameters correspond to late-type giants with $\log g$=1.5. Besides the prominent Ca II IR triplet lines, a number of weak Fe I, Ti I, and Cr I lines are seen. In the spectrum with higher $T_{\rm eff}$, there are also some Mg I and Si I lines.

from the used algorithm. However, the performance will also depend on the specific stellar spectral library used for training the algorithm. Libraries based on different modeling codes may give different results, and all libraries may to some extent fail to represent spectra of real stars. An example for the change of spectral features with effective temperature within one specific library is shown in Fig. 2. It is doubtful if the exact gradients and the regular appearance of this idealized situation is found in nature. Differences between different libraries of stellar fluxes have been explored by [8]. For a parameter combination of ($T_{\rm eff}$, $\log g$, [Fe/H])=(4000 K, 4.0, 0.0), they point at differences of up to 10% between simulated RP/BP fluxes, and also between line depths in simulated RVS spectra of three different stellar libraries. More details on comparisons of different stellar libraries can be found in the contribution by R. Sordo.

A systematic comparison of synthetic and observed stellar spectra is more difficult, because we do not know the 'true' astrophysical parameters of any observed star (except perhaps for the Sun), and because of the difficulties involved in calibrating observed fluxes. A few attempts can be found in the literature, e.g. [9], who compared MARCS model fluxes with three different observed stellar spectral libraries of low to medium resolution. The comparisons for individual stars and models were not conclusive because of the aforementioned problems, but an average over the differences between models and observations for all stars within each library proved to

Journal of Physics: Conference Series **328** (2011) 012001 doi:10.1088/1742-6596/328/1/012001

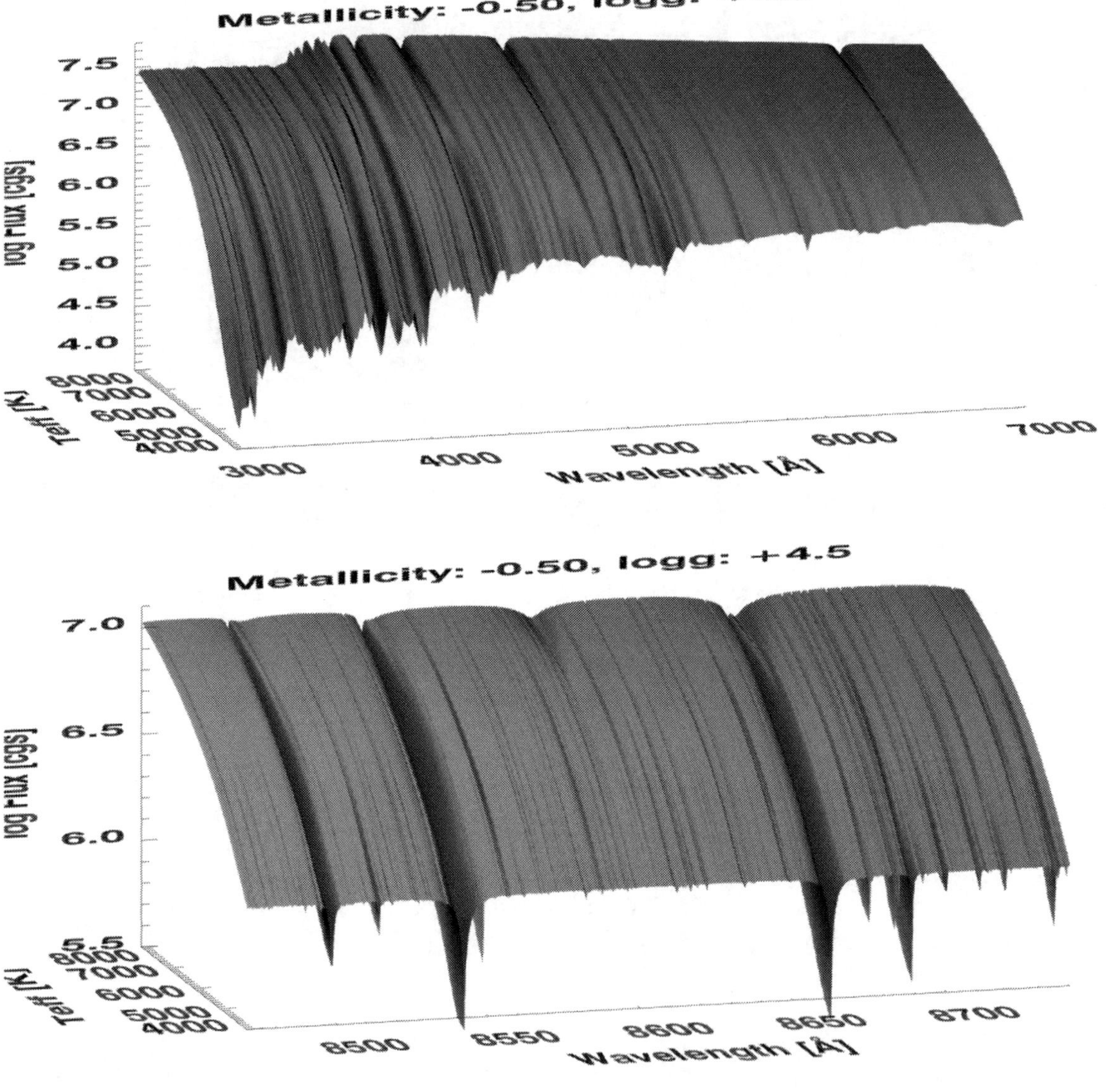

Figure 2. MARCS model atmosphere flux variation with $T_{\rm eff}$. Upper panel: optical wavelength range, lower panel: RVS range.

be useful. It revealed a few distinctive features in the blue and ultraviolet, pointing to missing or erroneous opacities in the models.

For the external calibration of the Gaia AP determination, we are aiming to assess the realism of the synthetic libraries by comparing model spectra and fluxes with those of a dedicated sample of benchmark stars. The APs of these stars should not be based on atmospheric models as far as possible. We are in the process of establishing a sample of benchmark stars with model-independent AP determinations: $T_{\rm eff}$ from an angular diameter measurement and the bolometric flux, and $\log g$ from angular diameter, parallax, and mass. The third atmospheric parameter,

Table 1. Characteristics of the benchmark stars and test spectra. $T_{\rm eff}$ and $\log g$ for α Tau and α Cet are direct determinations from angular diameter, bolometric flux and parallax measurements, and mass determinations, reference [Fe/H] values are taken from the literature (details will be given in Lebzelter et al., in preparation). R is the resolving power, $\Delta\lambda$ the wavelength range, and SNR the signal-to-noise ratio of the spectra.

Object	$T_{\rm eff}$[K]	$\log(g[{\rm cm/s^2}])$	[Fe/H]	R	$\Delta\lambda$ [nm]	SNR
α Tau	3930±40	1.2±0.1	−0.22±0.11	80 000	490–975	> 200
α Cet	3800±60	0.9±0.1	+0.02±0.03	80 000	490–975	> 200
'Star 3'	4257	1.47	−0.4	50 000	1546–1567	125
'Star 4'	3280	0.06	+0.1	50 000	1546–1567	125

metallicity or chemical composition, must be determined from spectroscopy, and thus will always be model-dependent. However, through carefully designed tests we can assess the influence of various modeling assumptions and approximations, as well as the details of the analysis method on the metallicity determination. In the second part of this contribution, we report on a test campaign for two specific benchmark stars.

3. Comparative spectrum modeling for cool giants

The experiment was conducted around a workshop supported by GREAT-ESF in Aug 2010. Observed optical spectra of two benchmark stars and two simulated H-band spectra were analyzed before the workshop by 14 groups using different models and analysis approaches. The analysis details and results were presented and discussed during the workshop. Discussions and follow-up tests continued after the workshop. The participating groups were provided with stellar spectra with R=80 000 and SNR>200 from 490 to 975 nm for two stars of undisclosed identities. The stars were in fact two well-known giant stars with $T_{\rm eff}\lesssim$4000 K (α Tau and α Cet). Spectra for 'Star 3' and 'Star 4' from 1546 to 1567 nm were provided as well. These had been calculated with R=50 000 and artificial noise was added to simulate a SNR of 125. In addition, approximate broad-band colors were given for all four objects. The characteristics of the test stars and spectra are given in Table 1.

The participating groups were asked to calculate model spectra for these objects, matching the observations, and to submit the corresponding best-fit atmospheric parameters to the experiment organizers for comparison. Some information on the groups and their methods is summarized in Table 2. The groups were from institutes all over Europe, as well as from the US, Canada, India, Australia, and Japan. About half of the groups based their analysis on MARCS atmospheric models and related spectrum synthesis codes, a considerable number used the ATLAS suite of codes [10; 11], and the three software packages PHOENIX [12], CODEX [13] and Tsuji08 [14] were employed by one group each. Most of the groups carried out spectrum synthesis at the resolution of the observed spectra or for degraded spectra, while three groups measured equivalent widths from the provided spectra which they compared to calculated equivalent widths.

As an example for the results, we show in Fig. 3 the best-fit spectrum for α Tau in the RVS region from one of the groups using MARCS models. The parameters of the model are ($T_{\rm eff}$, $\log g$, [Fe/H])=(4050 K, 1.5, 0.0), for spherical geometry with one solar mass and a microturbulence of 2 km s^{-1}. These were determined in this case by χ^2-minimization for the RVS region, varying only $T_{\rm eff}$ and $\log g$. Changing the metallicity to −0.25 dex, an equally good fit is obtained for the same $T_{\rm eff}$, but $\log g$=1.0. An inspection of the χ^2-surface shows that it does not vary significantly with $T_{\rm eff}$, and values between 3800 and 4300 K fit the spectrum almost as well. This wavelength region seems to be most sensitive to $\log g$ variations, and somewhat to variations in [Fe/H]. The

Table 2. Groups participating in the spectrum modeling experiment. Model suites employed by the groups are the following: M – MARCS, A – ATLAS, P – PHOENIX, C – CODEX, T – Tsuji08. EW in the Model column indicates that equivalent widths rather than synthetic spectra were used. The numbers in the column headed 'Stars' indicate which of the four test stars were analyzed. † ... symbols used in Figs. 4 and 6.

Contact	Institute	Model	Stars	†	Contact	Institute	Model	Stars	†
Nowotny	Vienna	M1	12	■	Maldonado	Madrid	A2,EW	12	○
Plez	Montpellier	M2	12	●	Neilson	Bonn	A3	34	△
Worley	Nice	M3	12	▲	Peterson	UCO/Lick	A4	1	△
Eriksson	Uppsala	M4	1	▼	Goswami	India	A5,EW	1	▽
Abia	Granada	M5	34	■	Short	Halifax	P	12	◇
Merle	Nice	M6	1	♦	Ireland	Sydney	C	2 4	×
Wahlgren	GSFC	A1	1 3	□	Tsuji	Tokyo	T,EW	4	◇

same group analysed two further wavelength regions, 515–520 nm and 610–640 nm, and found the latter to be most sensitive to $T_{\rm eff}$ variations, while the former was not very sensitive to either parameters. Note that in Fig. 3, even for the best-fit spectrum, differences between observations and calculations reach up to 10%, which is similar to the differences between different synthetic spectral libraries found by [8]. In the blue-most region investigated, the deviations reach up to 30% for the best-fit model.

The group using PHOENIX models used a similar χ^2-minimization approach, but for two larger wavelength regions (490 to 680 nm and 800 to 900 nm). In this case, the bluer region proved to be suited for determining a best-fit $T_{\rm eff}$ value, and to a lesser extent for constraining $\log g$, while the IR region was not sensitive to changes in any parameter. The best-fit model for α Tau had ($T_{\rm eff}$, $\log g$)=(3910 K, 1.75) for solar metallicity, and the same $\log g$, but $T_{\rm eff}$=3800 K for [Fe/H]=−0.5. Here, the relative flux differences in the IR between observed and calculated spectra are up to 20% for any parameter combination around the best fit in the blue region, where the deviations reach up to 50% for certain spectral lines.

Details on the analysis of all participating groups as well as an extensive discussion will be presented in a forthcoming publication (Lebzelter et al.). Here, we present a summary of the results in Fig. 4 for the two benchmark stars α Tau and α Cet. We compare the best-fit parameters determined by all groups and the $T_{\rm eff}$ and $\log g$ values for these stars determined in an independent way (angular diameter, bolometric flux, parallax, mass). The reference [Fe/H] values are previous metallicity determinations based on high-resolution spectroscopy found in the literature (mean of eight references for α Tau and one reference for α Cet).

The overall conclusion is that the spread in the results is larger than the estimated error bars for each point, but most of them agree within 2σ. $T_{\rm eff}$ seems to be better constrained than $\log g$. For α Tau, determinations for all parameters cluster around the reference values, although the $\log g$ results seem to be divided into two groups larger and smaller than the reference value. For α Cet, the optical spectrum fits point towards a lower metallicity than determined before by [15], based on high-resolution spectra in the H-band. We do not see any clear systematic trends with regard to the type of model used, except that groups using ATLAS or PHOENIX models tend to obtain higher $\log g$ values than groups using MARCS models.

In an attempt to pin down the factors determining the differences in the results, we did a direct comparison of synthetic optical spectra calculated for a fixed set of parameters by eight groups, using six different codes or code versions and sets of input line data. Considerable differences in the depths of numerous spectral lines between the different synthetic spectra became apparent over the whole wavelength range. They are largest in the blue region and decrease towards the

Journal of Physics: Conference Series **328** (2011) 012001 doi:10.1088/1742-6596/328/1/012001

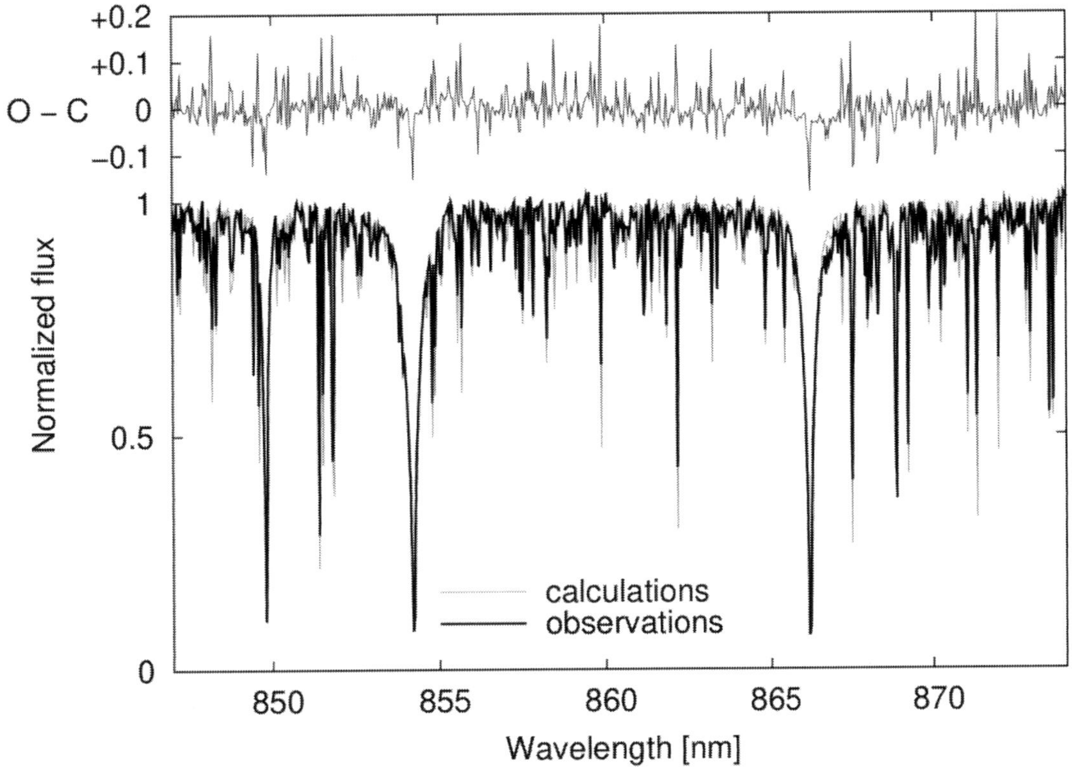

Figure 3. One of the best fit spectra for α Tau using MARCS model atmospheres for the Gaia RVS region. For the purpose of presentation, the spectra are sampled at 0.04 nm. Solar abundances, $T_{\rm eff}$=4050 K, $\log g$=1.5. Bottom: Scaled flux, observed and calculated; top: observed minus calculated flux.

near-IR. For example, in the interval 511 to 512 nm we see typical differences in normalized line depth of 0.4. In the interval 656 to 657 nm differences are mostly below 0.1, with a few outliers of up to 0.4. The main cause for these differences presumably lies in the input line data, both oscillator strengths and broadening parameters, as well as in the methods and data used for calculating the continuum flux.

Besides the model differences, the results of the experiment are clearly affected by different methods and strategies employed for parameter determination. Half of the groups used the provided color information for an initial guess of the parameters or as an additional constraint. Each group selected subsections of the available wavelength range to work with. For the optical spectra, Fig. 5 shows the wavelength ranges used by the 11 groups who analysed these spectra. Two of the groups using the same model and similar wavelength ranges arrived at similar $T_{\rm eff}$ values, but different $\log g$ and [Fe/H]. The differences may be caused by using wavelength ranges, which are not sensitive to changes in these parameters. One of the groups ran their analysis method twice, for different wavelength regions and degrading the spectra to different resolutions. Using the RVS region at low resolution resulted in higher $T_{\rm eff}$ values than using the region 490 – 670 nm at higher resolution.

The infrared spectra of 'Star 3' and 'Star 4' were only analysed by three and four groups,

Journal of Physics: Conference Series **328** (2011) 012001 doi:10.1088/1742-6596/328/1/012001

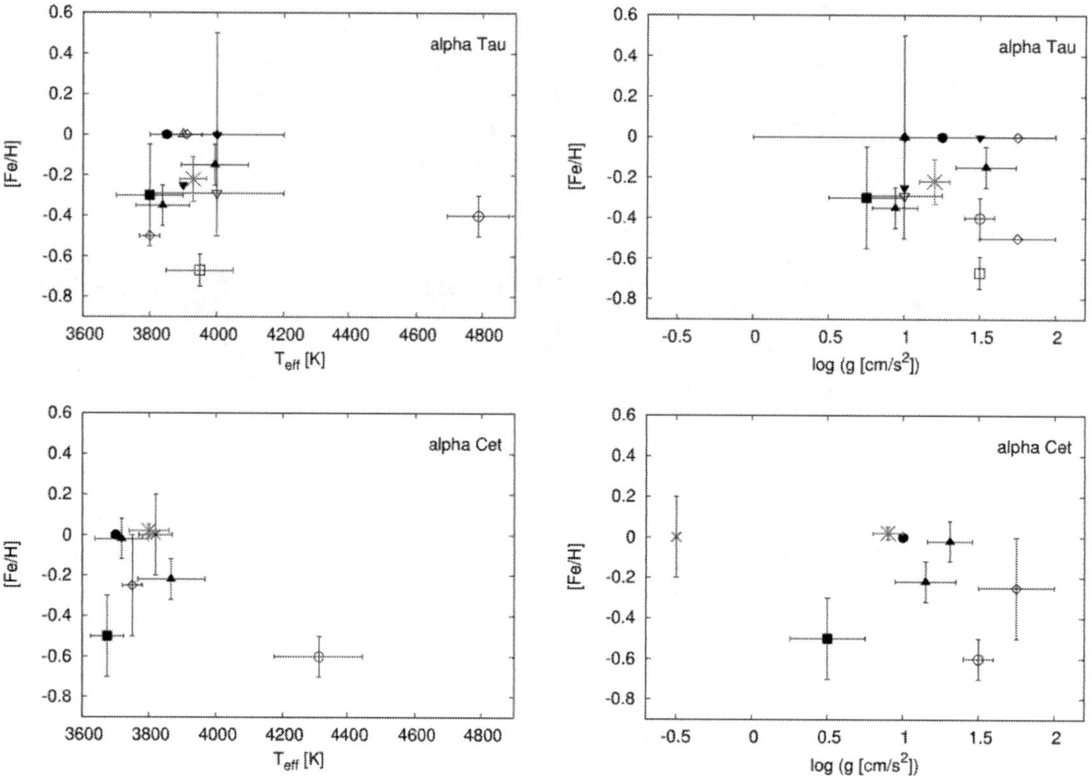

Figure 4. Best-fit parameters determined by 14 different groups for α Tau (top) and α Cet (bottom). Black symbols represent different groups as defined in Table 2. Not all groups provided error estimates for all parameters. M3 provided two results for different wavelength and resolution set-ups. M4 and P provided two results each for α Tau, for two different adopted metallicity values. The large cross-symbol in red represents atmospheric-model-independent $T_{\rm eff}$ and $\log g$ determinations and literature [Fe/H] values.

respectively. The resulting best-fit values for $T_{\rm eff}$ and $\log g$ are shown in Fig. 6, compared to the input parameters used to generate the artificial observations. For Star 3, the derived $T_{\rm eff}$ values are in reasonable agreement with the reference value. The resulting $\log g$ and [Fe/H] values from all three groups are higher than the reference values. It is unclear if this is caused by a systematic difference between the employed models and the model used to generate the test spectra, or if it is due to the analysis methods used. For Star 4, all parameters are close to the input values.

4. Conclusions

The Gaia mission promises great advances for stellar astrophysics, and the hardware and software are nearing completion. To ensure the best possible use of these resources, we have to put adequate efforts into calibrating our data analysis methods.

The spectrum modeling experiment described in Section 3 illustrates the need to be cautious when comparing or combining stellar parameters from different model atmospheres and analysis strategies. As no systematic differences between 'model families' (MARCS, ATLAS, etc.) are apparent, we conclude that the main causes for the differences in parameters derived by different

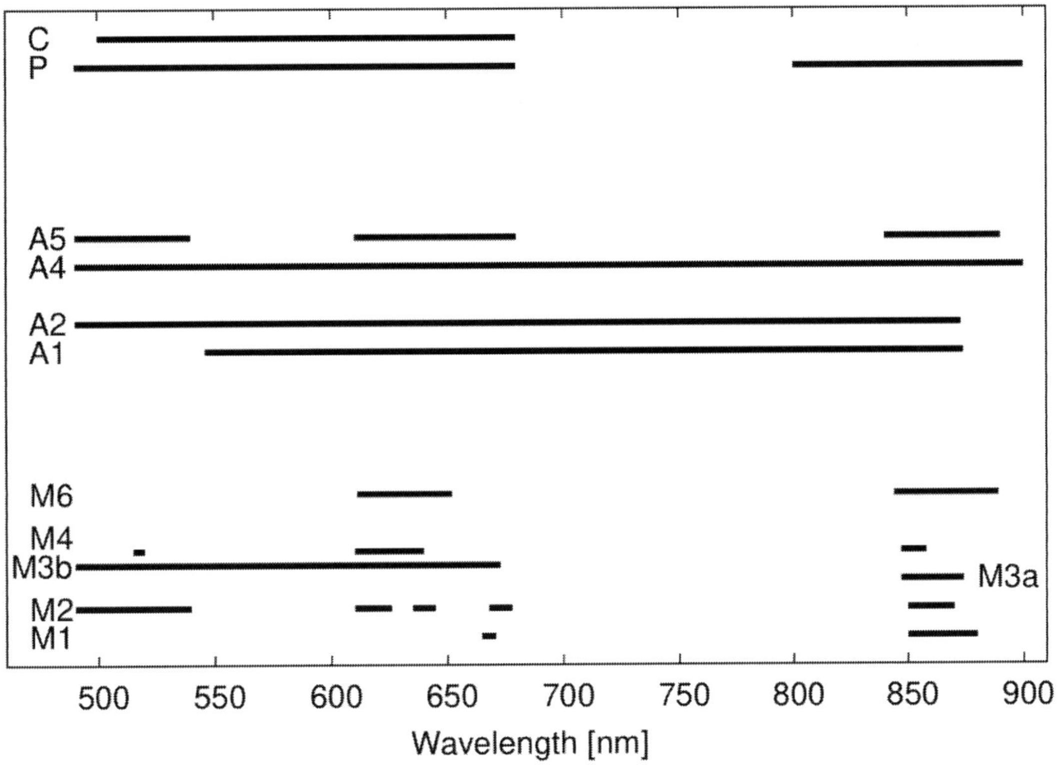

Figure 5. Wavelength ranges used by the different groups analyzing the optical spectra.

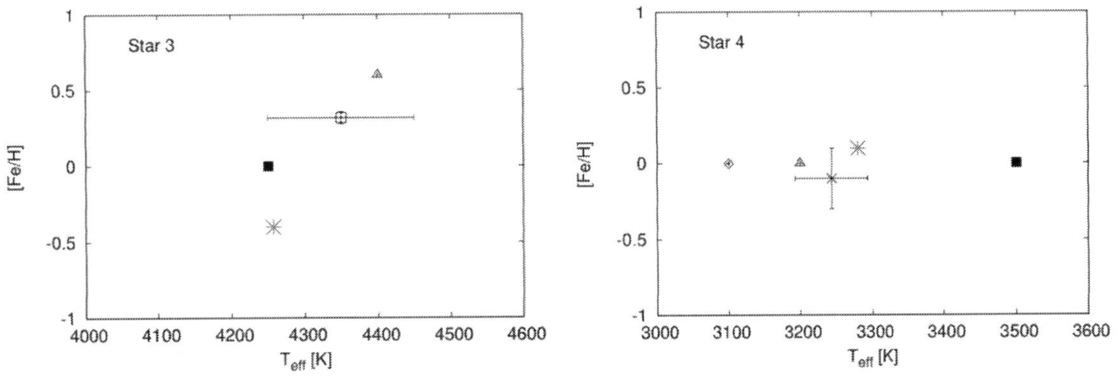

Figure 6. Best-fit parameters for 'Star 3' and 'Star 4'. Black symbols represent different groups as defined in Table 2. Not all groups provided error estimates. Asterisk-symbols in red represent the input parameters used to generate the artificial observations.

groups lie in the physical input data (e.g. oscillator strengths) and in the details of the analysis method (e.g. selected wavelength regions). However, due to complexity of problem, we could not derive clear trends for the effect of any given assumption. We note that the experiment represents a typical situation in abundance analysis and provides an overall assessment of the current status of this field, at least for the case of cool giant stars.

Acknowledgments

UH acknowledges support from the Swedish National Space Board, and from the ESF. TL acknowledges support from the Austrian Science Fund (FWF). We thank the groups participating in the spectrum modeling experiment for their work.

References

[1] Jordi C, Gebran M, Carrasco J M, de Bruijne J, Voss H, Fabricius C, Knude J, Vallenari A, Kohley R and Mora A 2010 *A&A* **523** A48 (*Preprint* 1008.0815)

[2] Valenti J A and Fischer D A 2005 *ApJS* **159** 141–166

[3] Valenti J A and Piskunov N 1996 *A&AS* **118** 595–603

[4] Gustafsson B, Edvardsson B, Eriksson K, Jørgensen U G, Nordlund Å and Plez B 2008 *A&A* **486** 951–970 (*Preprint* 0805.0554)

[5] Heiter U, Barklem P, Fossati L, Kildiyarova R, Kochukhov O, Kupka F, Obbrugger M, Piskunov N, Plez B, Ryabchikova T, Stempels H C, Stütz C and Weiss W W 2008 *Journal of Physics Conference Series* **130** 012011

[6] Kupka F, Piskunov N, Ryabchikova T A, Stempels H C and Weiss W W 1999 *A&AS* **138** 119–133

[7] Bailer-Jones C A L 2010 *MNRAS* **403** 96–116 (*Preprint* 0911.5242)

[8] Vallenari A and Sordo R 2009 Impact on GOG simulations of different input stellar libraries gaia DPAC Technical Note AV-009 URL http://www.rssd.esa.int/llink/livelink/open/2867451

[9] Edvardsson B 2008 *Physica Scripta Volume T* **133** 014011

[10] Kurucz R 1993 *ATLAS9 Stellar Atmosphere Programs and 2 km/s grid. Kurucz CD-ROM No. 13. Cambridge, Mass.: Smithsonian Astrophysical Observatory, 1993.* **13**

[11] Kurucz R 1993 *SYNTHE Spectrum Synthesis Programs and Line Data. Kurucz CD-ROM No. 18. Cambridge, Mass.: Smithsonian Astrophysical Observatory*

[12] Hauschildt P and Baron E 1999 *J. Comp. App. Math.* **109** 41

[13] Ireland M J, Scholz M and Wood P R 2008 *MNRAS* **391** 1994–2002

[14] Tsuji T 2008 *A&A* **489** 1271–1289 (*Preprint* 0807.4387)

[15] Meléndez J, Asplund M, Alves-Brito A, Cunha K, Barbuy B, Bessell M S, Chiappini C, Freeman K C, Ramírez I, Smith V V and Yong D 2008 *A&A* **484** L21–L25 (*Preprint* 0804.4124)

Journal of Physics: Conference Series **328** (2011) 012002 doi:10.1088/1742-6596/328/1/012002

NLTE effects on Fe I/II in the atmospheres of FGK stars and application to the abundance analysis of their spectra

Maria Bergemann[1], **Karin Lind**[1], **Remo Collet**[1] **and Martin Asplund**[1]

[1] Max-Planck Institute for Astrophysics, Karl-Schwarzschild Str. 1, 85741, Garching, Germany

E-mail: mbergema@mpa-garching.mpg.de

Abstract. We describe the first results from our project aimed at large-scale calculations of NLTE abundance corrections for important astrophysical atoms and ions. In this paper, the focus is on Fe which is a proxy of stellar metallicity and is commonly used to derive effective temperature and surface gravity. We present a small grid of NLTE abundance corrections for a sample of Fe I lines and discuss how the NLTE effects influence the determination of effective temperature, surface gravity, and metallicity for late-type stars.

1. Introduction

Iron is a key element in stellar astrophysics. The very complex atomic structure of its lowest ionization stages, Fe I and Fe II, gives rise to a wealth of spectral lines all across the spectrum of a typical late-type star. This atomic property, coupled to a relatively large abundance makes Fe a reference for spectroscopic estimates of stellar parameters using the method of excitation-ionization balance.

The classical implementation of this method in spectrum analysis codes involves three assumptions: local thermodynamic equilibrium (LTE), hydrostatic equilibrium, and 1-D geometry. These approximations strongly reduce the complexity of the problem, thus permitting analysis of very large stellar samples in short timescales. Yet, in conditions when the breakdown of 1-D static LTE models occurs the inferred stellar parameters suffer from large systematic biases. To assess the latter, more physically realistic modeling is necessary.

Studies of NLTE effects on the Fe I/Fe II level populations for FGK stars trace back to Athay & Lites (1972). Since then, vast amounts of work have been performed in the field, demonstrating that NLTE effects on the excitation-ionization balance of Fe I/Fe II are significant and cannot be ignored even for analyses of solar-type stars (Mashonkina et al. 2011 and references therein). Yet, despite major efforts aimed at understanding how non-local thermodynamic equilibrium affects the line formation of Fe, there has never been an attempt to accurately quantify these deviations in a systematic way for a wide range of stellar parameters, and to apply them to a large stellar sample.

We present a new NLTE model atom for Fe I/Fe II applied in large-scale calculations of NLTE abundance corrections of late-type stars. We discuss NLTE effects influencing atomic level populations at typical conditions in the atmospheres of late-type stars and provide a small grid of the NLTE corrections for Fe I lines. The model has been tested on a number of well-studied

Journal of Physics: Conference Series **328** (2011) 012002 doi:10.1088/1742-6596/328/1/012002

stars with independently-determined parameters, including metal-poor giants and turnoff stars. These tests performed with classical 1-D hydro-static model atmospheres and averages of 3-D hydrodynamical simulations of stellar convection will be presented in Bergemann et al. (in preparation). A complete grid of NLTE abundance corrections computed with the MULTI2.3 statistical equilibrium code will be presented in Lind et al. (in preparation).

2. Methods

2.1. Model atmospheres and codes

The calculations presented here were performed with the 1-D LTE plane-parallel MAFAGS-ODF models (Fuhrmann et al. 1997, Grupp 2004). In these models convective energy transport is accounted for using the mixing-length theory of Böhm-Vitense (1958), with the mixing length parameter $\alpha_{\rm mlt}$ set equal to 0.5. Line opacity is represented for all chemical elements up to Ni, and for various diatomic molecules (H_2, CH, CO, MgH, etc). This accounts for nearly 20 million atomic and molecular lines. The reference solar abundances were compiled from various literature sources, giving preference to the determinations by the Munich group. The model atmospheres provide the partition functions and the partial pressures of all relevant atoms and molecules.

The NLTE statistical equilibrium of Fe was computed with a revised version of the DETAIL code (Butler & Giddings 1985), which solves the multi-level NLTE radiative transfer for a given static 1-D model atmosphere. The latest version of the code is based on the Accelerated Lambda Iteration (ALI) method. LTE and NLTE line formation calculations with the departure coefficients from DETAIL were performed with a revised version of the spectrum synthesis code SIU (Reetz 1999). The major update in the code relates to the automated computation of NLTE abundance corrections[1], which generally proceeds by constructing the grids of LTE and NLTE line equivalent widths for a range of stellar parameters and interpolating the corresponding curves-of-growth for the input observed equivalent widths or abundance values.

2.2. Model atom

The NLTE model atom of Fe was constructed using all available atomic data from the Kurucz[2] database, which also includes laboratory data from the NIST compilation[3]. All predicted energy levels of Fe I with the same parity above $E_{\rm low} \geq 5.1$ eV were grouped into super-levels. Transitions between the components of the super-levels were also grouped. The total transition probability of a super-line is a weighted average of the $\log gf$s of individual transitions. Thus, in the final atomic model, the number of energy levels is 296 for Fe I and 112 for Fe II, with the uppermost excited levels located at 0.03 eV and 2.72 eV below the respective ionization limits of 7.9 eV and 16.19 eV. The model is closed by the Fe III ground state. The total number of radiatively-allowed transitions is 16 207 (13 888 Fe I and 2 316 Fe II).

The photoionization cross-sections for 136 states of Fe I were taken from Bautista (1997) and the hydrogenic approximation was adopted for all other levels. The rates of transitions induced by inelastic collisions with electrons (e^-) and H I atoms were computed using different recipes. The electron cross-sections from states with allowed bound-bound and bound-free transitions were computed with the formulas of van Regemorter (1962), respectively, Seaton (1962). These states were also coupled by inelastic H I collisions using Drawin's formula (Drawin 1969) in the version of Steenbock & Holweger (1984). All the other states are connected by forbidden transitions induced by inelastic collisions with e^- and H I atoms. These are computed using the

[1] the NLTE abundance correction is defined to be the difference in the abundances required to match NLTE and LTE line profiles or equivalent widths

[2] http://kurucz.harvard.edu/atoms.html

[3] http://www.nist.gov/pml/data/asd.cfm

Journal of Physics: Conference Series **328** (2011) 012002 doi:10.1088/1742-6596/328/1/012002

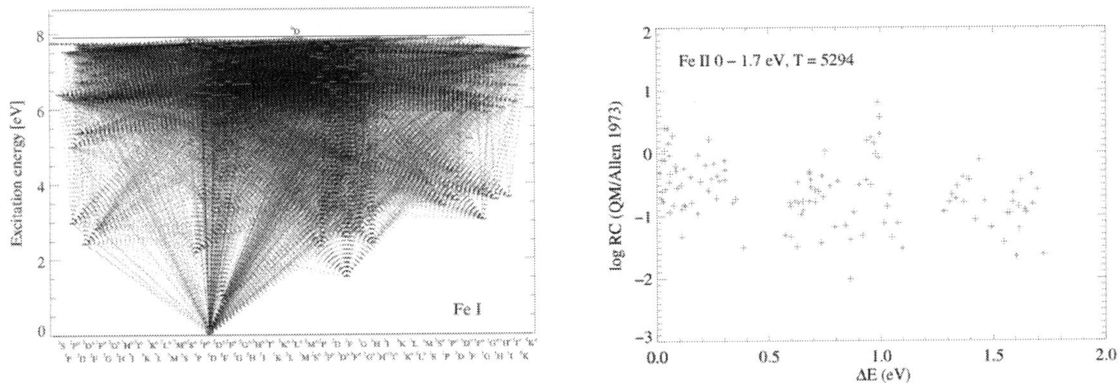

Figure 1. Left-hand panel: Grotrian diagram of the Fe I atom. Right-hand panel: comparison of the quantum-mechanical rate coefficients for e^--induced forbidden transitions between the lowest energy levels of Fe II (Ramsbottom et al. 2007) with the classical Allen's (1973) recipe.

formulae of Allen (1973) and Takeda (1994). Quantum-mechanical calculations exist only for e^--induced forbidden transitions between the 16 lowest Fe II energy levels (Ramsbottom et al. 2007). These data are compared with Allen's (1973) recipe in Fig. 1.

The influence of e^- and H I collisions on the statistical equilibrium of Fe was carefully investigated by performing test calculations with various scaling factors to the above-mentioned formulae. Furthermore, the NLTE synthetic profiles were compared with observed spectra of well-studied stars to check how well the test model atoms satisfy the constraint of ionization-excitation balance of Fe I/Fe II under the restriction of different stellar parameters. The final choice of the scaling factors will be discussed in Bergemann et al. (in preparation).

3. Statistical equilibrium of Fe

3.1. NLTE effects

Figure 2 shows level departure coefficients[4], b_i, as a function of continuum optical depth $\log \tau_{5000}$ for a number of model atmospheres computed with different stellar parameters. The latter are representative of stars typically used in Galactic chemical evolution studies. Only selected energy levels of Fe I and Fe II typical for their depth-dependence are included in the plot: $a\,^5$D (the ground state of Fe I), $z\,^7$D° (2.4 eV), $e\,^5$D (5.4 eV), $t\,^3$H$_\circ$, $a\,^6$D (the ground state of Fe II), and $z\,^4$D° (5.5 eV). All these levels give rise to radiatively-permitted transitions, which are used in the detailed abundance analysis of the Sun and of a number of metal-poor stars in Bergemann et al. (in preparation).

The common result for all the models shown in Fig. 2 is that in the optically thin atmospheric layers the majority of the Fe I levels are underpopulated compared to LTE, $b_i < 1$. The Fe II number densities remain close to LTE values, $b_i \approx 1$, and a minor overpopulation of the Fe II ground state develops only close to the outer atmospheric boundary.

Deviations from LTE in the distribution of atomic level populations arise because the mean radiation field, J_ν, at different depths and frequencies is not equal to the Planck function, $B_\nu[T_e(\tau)]$, which is defined by the local temperature $T_e(\tau)$ at each depth. For Fe I, an excess of the mean intensity over the Planck function in the UV continua leads to over-ionization, which sets in as soon as the optical depth in the photoionization continua of the low-excitation Fe I

[4] Departure coefficient is defined to be the ratio of NLTE to LTE atomic level populations, $b_i = n_i^{\rm NLTE}/n_i^{\rm LTE}$

Journal of Physics: Conference Series **328** (2011) 012002 doi:10.1088/1742-6596/328/1/012002

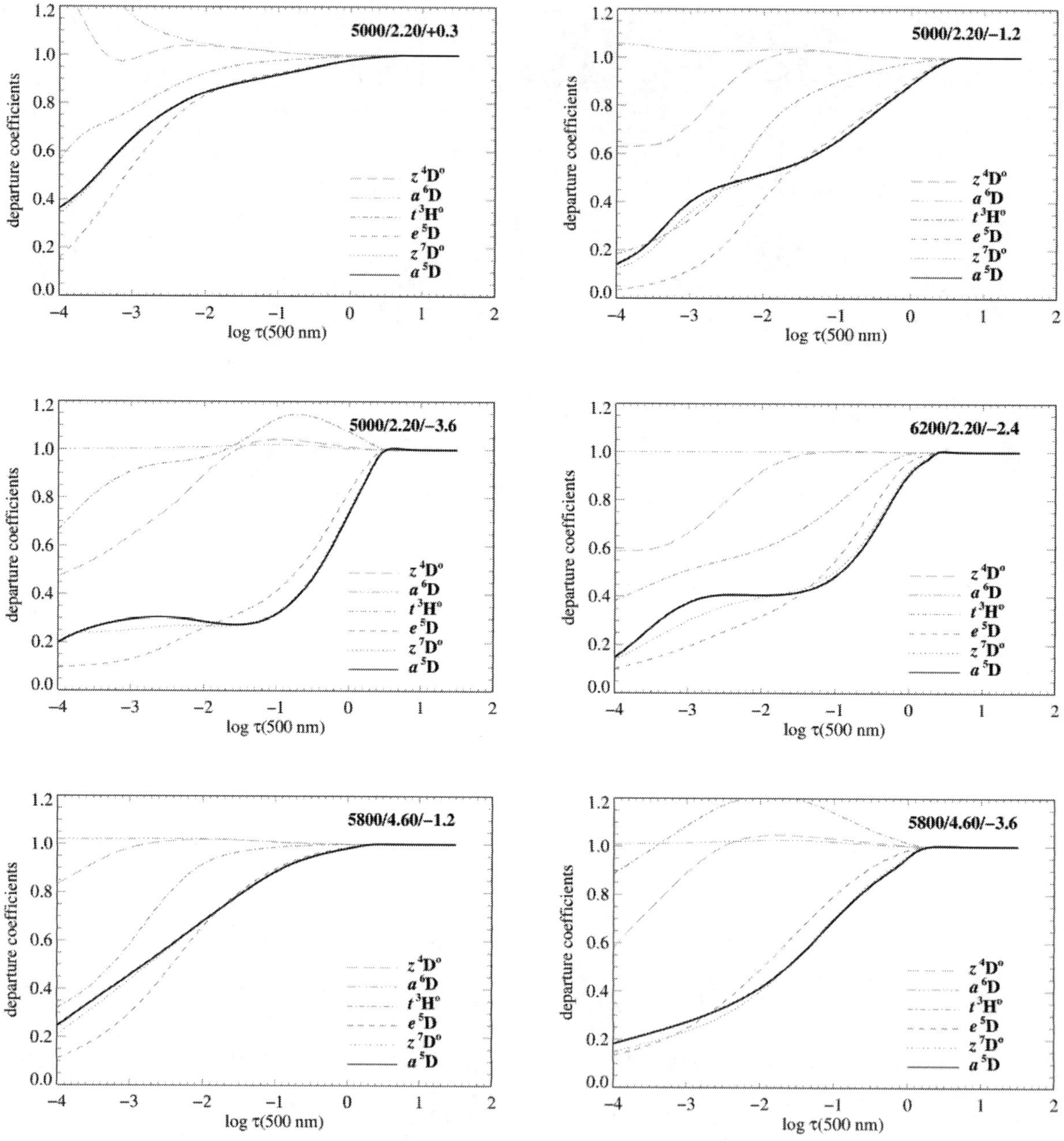

Figure 2. The departure coefficients of selected Fe I and Fe II levels for a number of model atmospheres from the grid; stellar parameters are indicated in each figure.

levels falls below unity, i.e. $\log \tau_{5000} \sim 0$. In the outer layers, the ionization balance is dominated by the Fe I levels with excitation energies of $4-5$ eV. In the infra-red continuum, $J_\nu < B_\nu$ leads to over-recombination, which is very efficient in our atomic model with only a 0.03 eV energy gap between the upper Fe I levels and the ground state of Fe II.

Due to an extremely tight radiative and collisional coupling of the atomic energy levels, the excitation balance of Fe I is also affected by radiative imbalances in the line transitions. These include radiative pumping by a super-thermal UV radiation field of non-local origin at the line frequencies, as well as photon suction driven by photon losses in the large-probability transitions between highly-excited levels. These processes leave a characteristic imprint on the behavior of b_i-factors in the outer atmospheric layers. In Fig. 2, this is seen as the onset of deviations from the relative thermal equilibrium between the Fe I levels, which occurs, depending on the model metallicity, at $\log \tau_{5000} \sim -1 \ldots -2$. The NLTE effects on the line formation are different for the strong and weak Fe I lines. The former are mostly shaped by deviations of their source function from the Planck function, which yields a clear effect in the line cores. Weak lines are pre-dominantly controlled by their opacity, thus the major change is due to the shift of their optical depth scale in NLTE.

3.2. NLTE abundance corrections

We computed NLTE abundance corrections, hereafter $\Delta_{\rm NLTE}$, to Fe I lines for a small grid of model atmospheres covering the metallicity range $-3.6 \leq$ [Fe/H] $\leq +0.3$. The results are presented in Fig. 3 for parameters representative of giants (*left-hand panel*) and dwarfs (*right-hand panel*). Our sample of Fe I lines includes the transitions between 0 to 5 eV levels, and line equivalent widths W_λ that range from 5 mÅ to a few Å depending on the stellar parameters. We used 30 to 130 lines per model.

In the plot, each symbol corresponds to a mean $\Delta_{\rm NLTE}$ obtained by averaging individual NLTE corrections for all Fe I lines in the sample, and the errorbars correspond to the line-to-line variations in Δ for each model atmosphere. The former quantity roughly demonstrates how much the Fe I/Fe II ionization balance, and thus, the determination of the surface gravity, is affected by NLTE, whereas the latter helps to assess the influence of NLTE on the excitation balance and the determination of the effective temperature.

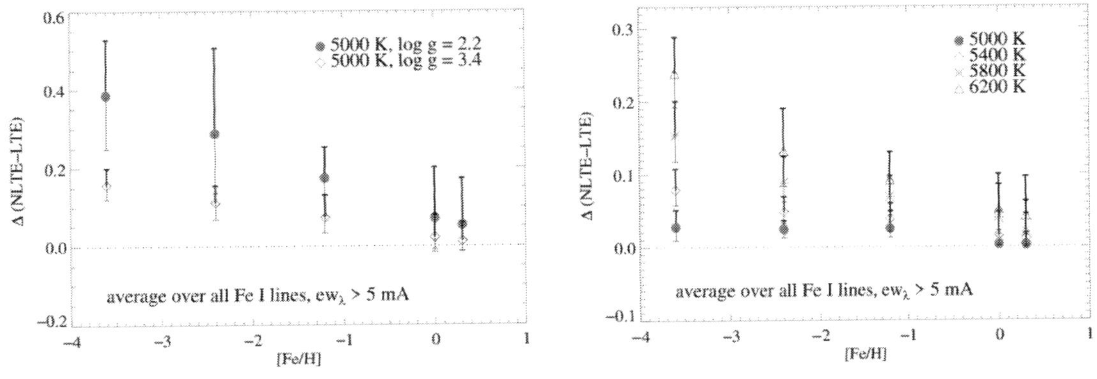

Figure 3. NLTE abundance corrections for different stellar parameters. Right-hand panel: all model atmospheres adopt $\log g = 4.6$.

The NLTE corrections increase with decreasing [Fe/H] and $\log g$, thus the ionization balance achieved assuming LTE leads to progressively underestimated gravities and metallicities. In particular, at $T_{\rm eff} = 5000$ K (Fig. 3, left-hand panel), the mean NLTE corrections are twice as

Journal of Physics: Conference Series **328** (2011) 012002 doi:10.1088/1742-6596/328/1/012002

large in the $\log g = 2.2$ model compared to the $\log g = 3.4$ model, reaching $+0.4$ dex at [Fe/H] ~ -3.5. It is also interesting that the NLTE corrections for the dwarf models critically depend on $T_{\rm eff}$. Figure 3 (*right-hand panel*) shows that for the coolest models, $T_{\rm eff} \sim 5000$ K, one may safely adopt the LTE approximation down to the lowest metallicities. However, LTE leads to severe errors in surface gravity estimates for warm turn-off stars of $T > 6000$ K.

On the other side, the non-negligible errorbars indicate that there are line-to-line variations of the NLTE abundance corrections for the same model atmosphere. The spread of $\Delta_{\rm NLTE}$ among Fe I lines with different equivalent widths W_λ and excitation potentials $E_{\rm low}$ strongly increases with decreasing metallicity for giants, and less so for dwarfs. Furthermore, this scatter is not random, but corresponds to a trend of $\Delta_{\rm NLTE}$ with $E_{\rm low}$, as we show in Fig. 4. The figure demonstrates the variation of NLTE corrections with the lower level excitation potential for models with parameters: $T_{\rm eff} = 5000$, $\log g = 2.2$, $[{\rm Fe/H}] = -2.4$ (*left-hand panel*) and $T_{\rm eff} = 5800$, $\log g = 4.6$, $[{\rm Fe/H}] = -2.4$ (*right-hand panel*). The stronger lines, $W_\lambda > 60$ mÅ, tend to exhibit smaller departures from LTE compared to the weaker lines, but they also demonstrate systematically increasing $\Delta_{\rm NLTE}$-values with increasing $E_{\rm low}$. No such effect is observed for the weak lines of $W_\lambda \leq 60$ mÅ, which, with some exceptions, have very similar NLTE corrections independent of their excitation potential.

Thus, our results suggest that LTE $T_{\rm eff}$ determinations become very inaccurate at [Fe/H] < -2 and there are systematic effects in the $T_{\rm eff}$-scale of dwarfs and giants.

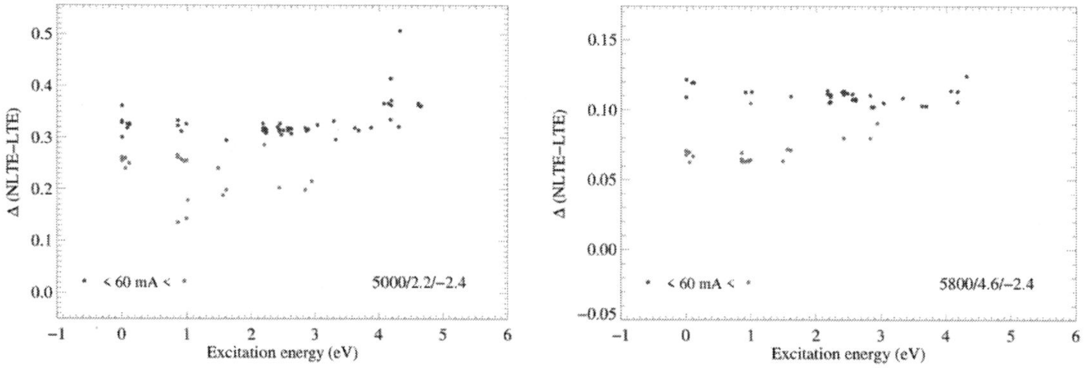

Figure 4. NLTE abundance corrections for the Fe I lines as a function of the lower level excitation potential. Red and blue colors separate the spectral lines with W_λ greater and less than 60 mÅ, respectively. Stellar parameters are indicated in the plots.

4. Conclusions

We have constructed a new NLTE Fe model atom using the most up-to-date theoretical and experimental atomic data. The model has been tested on a number of well-studied late-type stars with the parameters determined by other independent methods (Bergemann et al., in preparation).

Our preliminary results for the NLTE effects on Fe I and Fe II in the atmospheres of late-type stars are qualitatively consistent with other studies. Kinetic equilibrium of Fe, which is a minority ion at the typical conditions of these cool and dense atmospheres, favors lower number densities of Fe I compared to LTE. The number densities of Fe II are hardly affected by NLTE. In general, this leads to a weakening of Fe I lines compared to LTE, which, in turn, requires a larger Fe abundance to fit a given observed spectral line. The magnitude of the departures and the NLTE corrections critically depend on the stellar parameters. Whereas NLTE abundance corrections for the Fe I lines do not exceed ~ 0.05 dex for solar metallicity dwarfs, they can be as large as 0.6 dex for metal-poor giants.

We find that both the excitation and ionization balance are affected by NLTE effects on Fe I. The NLTE corrections to the Fe I lines increase with decreasing [Fe/H] and $\log g$, thus the ionization balance achieved assuming LTE leads to progressively underestimated surface gravities and metallicities. On the other side, Fe I levels are neither in thermal equilibrium with respect to each other, and for the stronger lines the NLTE abundance corrections show trends with the line excitation potential. In particular, we find that LTE $T_{\rm eff}$ determinations become very inaccurate at low metallicities and significant systematic effects in the $T_{\rm eff}$-scale of dwarfs and giants can be expected.

The NLTE abundance corrections will become publicly available through an interactive online database, which is currently under construction. Users will be able to retrieve NLTE abundances and/or NLTE abundance corrections for the input, e.g., observed, equivalent line widths. We also consider an option to provide a functional dependency of the NLTE corrections on spectral line parameters for a grid of stellar parameters ($T_{\rm eff}$, $\log g$, [Fe/H]). This is a more convenient solution for the implementation in automated codes used in large-scale spectroscopic studies, such as follow-up spectroscopy of Gaia targets.

Acknowledgments
We would like to thank Luca Sbordone for the help with the revision of SIU.

References

Allen C W 1973, *Astrophysical Quantities* **3rd ed** Athlone Press, London
Athay R G & Lites B W 1972 *ApJ* **176** 809
Bautista M A 1997 *A&AS* **122** 167
Böhm-Vitense E 1958 *Zeitschrift für Astrophysik* **46** 108
Butler K and Giddings J 1985 *Newsletter on Analysis of Astronomical Spectra* University of London **9**
Drawin H W 1969 *Z. Physik* **225** 470
Fuhrmann K, Pfeiffer M, Frank C, Reetz J, and Gehren T 1997 *A&A* **323** 909
Grupp F 2004 *A&A* **420** 289
Mashonkina L, Gehren T, Shi J R, Korn A J, and Grupp F 2011 *A&A* **528** 87
Ramsbottom C A, Hudson C E, Norrington P H, and Scott, M P 2007 *A&A* **475** 765
Reetz J K 1999 *PhD thesis, LMU Munich*
Steenbock W and Holweger H 1984 *A&A* **130** 319
Takeda Y 1994 *PASJ* **46** 53
van Regemorter H 1962 *ApJ* **136** 906

Journal of Physics: Conference Series **328** (2011) 012003 doi:10.1088/1742-6596/328/1/012003

The StaggerGrid project: a grid of 3-D model atmospheres for high-precision spectroscopy

Remo Collet[1,2,3], Zazralt Magic[1] and Martin Asplund[1,4]

[1] Max-Planck-Institut für Astrophysik, Karl-Schwarzschild-Str. 1, D–85741 Garching b. München, Germany

[2] Centre for Star and Planet Formation, Natural History Museum of Denmark University of Copenhagen, Øster Voldgade 5-7, DK–1350 Copenhagen, Denmark

[3] Astronomical Observatory/Niels Bohr Institute, Juliane Maries Vej 30, DK–2100 Copenhagen, Denmark

[4] Research School of Astronomy & Astrophysics, Cotter Road, Weston ACT 2611, Australia

E-mail: `[remo,magic,asplund]@mpa-garching.mpg.de`

Abstract. In this contribution, we present the StaggerGrid, a collaborative project for the construction of a comprehensive grid of time-dependent, three-dimensional (3-D), hydrodynamic model atmospheres of solar- and late-type stars with different effective temperatures, surface gravities, and chemical compositions. We illustrate the main characteristics of these 3-D models and their effects on the predicted strengths, wavelength-shifts, and shapes of spectral lines, highlighting the differences with respect to calculations based on classical, one-dimensional, hydrostatic models, and discuss some of their possible applications to elemental abundance analysis of stellar spectra in the context of large observational surveys.

1. Introduction

The Gaia mission [1, 2] will measure high-precision parallaxes and proper motions for 10^9 galactic stars down to apparent magnitude $V = 20$, as well as radial velocities for 10^8 stars with $V \leq 13$. In connection with the mission, a number of ground-based surveys are being planned that will carry out observations to complement the Gaia data. Over the next five years, the upcoming Gaia-ESO public survey, for instance, will acquire high-resolution spectra for more than 10^5 galactic stars using the VLT/FLAMES multi-fibre spectrograph. The goal of the survey is to homogeneously derive chemical abundances for all these stars; combined with astrometry from Gaia, this information will allow to trace the most detailed and extensive chemo-dynamical map of the stellar components of the Milky Way.

Processing the enormous amount of data from this and similar surveys and extracting from them stellar elemental abundances will require fast access to grids of model stellar atmospheres and synthetic spectra covering the relevant range in terms of stellar parameters and chemical compositions. Grids of traditional one-dimensional (1-D), stationary, hydrostatic model stellar atmospheres such as MARCS [3, 4] or ATLAS [5, 6] are already available or are being re-computed together with atlases of synthetic spectra to match the specific needs of the various surveys. The main advantage of classical model atmospheres is that the simplifying assumption of a 1-D stratification allows to invest the computational resources on the solution of the radiative transfer equation for a very large number (typically 10^5 or more) of wavelength points. However,

1-D models can only treat convective energy transport in an approximate manner via recipes such as the mixing-length theory (MLT) [7] or the full-spectrum-of-turbulence (FTS) model [8], which are all dependent on a number of free parameters. In late-type stars, the convective flows reach the stellar surface, affecting the atmospheric layers and, consequently, the actual spectral energy distribution in the emergent radiative flux. Furthermore, bulk gas flows in stellar atmospheres and associated Doppler shifts also affect the broadening, shapes, wavelength shifts, and strengths of spectral lines. It is therefore important to properly account for their effects in order to extract accurate and precise elemental abundances from the analysis of stellar spectra. Because traditional stationary, 1-D, hydrostatic model atmospheres of late-type stars lack a consistent description of atmospheric velocity fields, Doppler broadening of spectral lines is modelled in 1-D analyses by introducing additional free parameters such as micro-turbulence and macro-turbulence that generally need adjusting and tuning on an individual star basis.

In more recent years, on the other hand, a lot of efforts have been invested in the development of realistic, time-dependent, three-dimensional (3-D), radiation-hydrodynamic simulations of stellar surface convection that can be directly applied as 3-D model stellar atmospheres for spectral synthesis calculations (see [9] for a recent review). Three-dimensional simulations of stellar surface convection successfully reproduce important observational constraints such as the spatial properties and temporal evolution of the solar granulation pattern [10, 11, 12], centre-to-limb intensity variations at the surface of the Sun [13, 14] as well as other nearby stars [15, 16], as well as the detailed shapes and wavelength shifts of spectral lines in solar- and late-type stars [17, 18, 19]. At present, a number of codes suitable for constructing realistic 3-D hydrodynamic model atmospheres are available and actively developed. A few prominent examples in this respect are the STAGGER [20], BIFROST [21], CO^5BOLD [22], and MURAM [23] codes. Three-dimensional models have recently started to be employed for spectroscopic stellar abundance analyses [24, 25, 26, 27, 28]. Similarly as in the case of 1-D models, organized grids of 3-D model atmospheres are currently being developed and becoming available for this purpose. The recently presented CIFIST collection of CO^5BOLD model atmospheres [29], for instance, is a pioneering example of such a grid. In this contribution, we will present the general outlines of the STAGGERGRID, an alternative project for the construction of a grid of 3-D model stellar atmospheres of late-type stars with the STAGGER-CODE, and we will discuss some of its possible applications in the context of spectral line formation and elemental abundance analysis.

2. The StaggerGrid

We are using a custom version of the MPI-parallel STAGGER-CODE to carry out time-dependent, three-dimensional, radiation-hydrodynamic simulations of convection at the surface of late-type stars for a range of effective temperatures, surface gravities, and metallicities. The general outline of the grid in stellar parameter space is shown in Fig. 1.

The simulations are of the so-called *star-in-the-box* kind: each simulation's domain is a rectangular, three-dimensional volume located at the surface of the star, with periodic boundaries horizontally and open boundaries vertically. The mass, momentum, and energy conservation equations for a compressible, viscous flow are discretized and solved on a Cartesian numerical mesh with $240\times240\times240$ grid-points as default numerical resolution. The code uses artificial viscosity operators to deal with the effects of numerical diffusion. The parameters for these operators have been tuned using solar surface convection simulations to minimize such effects without causing undesired or excessive smoothing of shocks. Once tuned, these parameters are kept fixed for all models. Changing these parameters does affect the sharpness of, e.g., temperature and density inhomogeneities but essentially does not alter the main features of the simulations, such as the basic morphology of the convective flows or the average temperature and density stratifications.

We choose the physical size of the domains so that the simulations typically host about ten

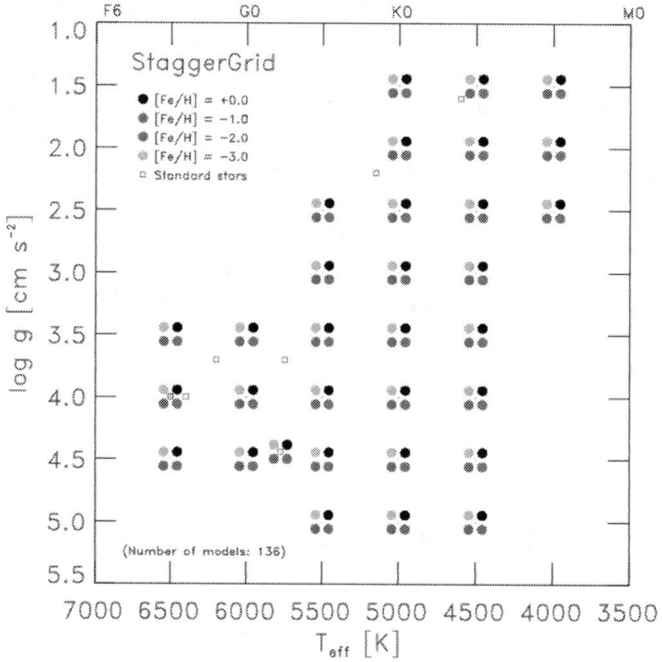

Figure 1. Overview of the StaggerGrid: the dots indicate the effective temperatures ($T_{\rm eff}$) and surface gravities ($\log g$) of the 3-D surface convection simulations being computed for the project. The coloured circles represent the different compositions (scaled solar metallicity [14] with [Fe/H] ranging from +0 to −3, with a α-enhancement value of $[\alpha/{\rm Fe}]= +0.4$ for the metal-poor models). The empty squares indicate the $T_{\rm eff}$- and $\log g$-values adopted for the surface convection simulations of the Sun and other standard stars important for stellar spectroscopy (see also Table 1).

convective granules at any given time and cover about twelve pressure scale heights vertically, extending from $\log \tau_{\rm Ross} \leq -4$ to $\log \tau_{\rm Ross} \geq +6$ in terms of Rosseland optical depth. In the setup currently used for the StaggerGrid models, we assume a constant vertical gravitational acceleration in the simulation box and neglect sphericity effects. The physical size of the computational domain of low-surface-gravity StaggerGrid models is, however, still sufficiently small compared with the stellar radius, implying that these approximations are acceptable. At the bottom boundary, located deep below the surface, in the convectively unstable layers, we impose constant pressure across the whole layer and require the inflowing gas to have constant entropy per unit mass.

Our goal is to construct model atmospheres that can be used for accurate abundance determinations, so we try to include as realistic input physics as possible in the simulations: we implement a state-of-the-art equation-of-state [30] and up-to-date continuous [3] (see also Trampedach et al., in prep.) and line opacities [4].

In order to model the temperature stratification in the outer stellar layers properly, it is paramount to account for the energy exchange between gas and radiation. At each time-step during the simulation, we solve the radiative transfer equation along the vertical as well as eight other inclined directions (two θ- and four ϕ-angles) using a Feautrier-like method [31] and compute the necessary radiative heating rates for the energy conservation equation. The choice of a Feautrier-like method implies that each direction is effectively counted twice, once for outgoing rays and once for incoming rays.

Journal of Physics: Conference Series **328** (2011) 012003 doi:10.1088/1742-6596/328/1/012003

Temperature gradients at the surface of red giant models can become very steep. In general, a fixed geometrical depth scale does not allow to properly resolve these gradients, which can give rise to numerical artifacts in the outgoing intensity pattern at the surface. In our version of the STAGGER-CODE, we have therefore implemented an adaptive depth scale to solve this issue (Collet et al., in prep.).

In order to reduce the computational burden for the solution of the radiative transfer equation, we approximate the source function with the Planck function at the local temperature ($S_\nu = B_\nu(T)$) and neglect the contribution of scattering to the total extinction in the optically thin layers [32]. Furthermore, we use a multi-group or opacity-binning approximation [33, 34] to account for the dependence of opacity on wavelength: we sort the monochromatic opacities into groups (or *bins*) according to wavelength range and opacity strength, then solve the radiative transfer equation for the individual group mean opacities and the integrated group source functions [32]. We calibrate the opacity-binning method for each specific choice of stellar parameters in order to achieve an as accurate as possible representation of the heating rates. We normally do that by finely adjusting the criteria for opacity-bin-membership for each individual simulation until the difference between the heating rates computed with opacity-binning and with the full monochromatic solution for the average stratification from the 3-D model is minimized.

Table 1. Stellar parameters of some standard stars for which 3-D hydrodynamic STAGGER-CODE models have been computed as part of the grid. For all stars, we have assumed a standard solar composition with the abundances of metals scaled proportionally to the relevant [Fe/H], with α-enhancement of $[\alpha/\mathrm{Fe}] = +0.4$ for the metal-poor models. In addition, for HE1327−2326 and HE0107−5240, we have also accounted for the peculiar CNO-enhancement of these stars.

Star	$T_{\mathrm{eff}}/[\mathrm{K}]$	$\log g/[\mathrm{cms}^{-2}]$	[Fe/H]
Sun	5780	4.44	+0.0
Procyon	6500	4.00	+0.0
HD 140283	5750	3.70	−2.5
HD 84937	6400	4.00	−2.0
G64-12	6500	4.00	−3.0
HD 122563	4600	1.60	−3.0
HE1327−2326	6200	4.00	−5.0
HE0107−5240	5200	2.20	−5.0

At the present time, a grid of models using a radiative transfer solution with six opacity bins is nearing completion. In addition, we have also computed models for some reference stars (see Table 1) using a more refined opacity-binning scheme with twelve opacity bins. We plan to eventually extend the twelve-bin opacity binning scheme to the computation of all STAGGERGRID models.

To produce the initial 3-D snapshot of a simulation for a given set of targeted stellar parameters, we take the physical structure from another simulation snapshot previously computed for other stellar parameters and scale it appropriately. We do that by looking at the ratios of spatial scales and of various important physical quantities such as temperature, density, and pressure from 1-D model envelopes corresponding to the same stellar parameters. The scaled model is then allowed to adjust and relax. Spurious p-mode-like oscillations are then damped and filtered out to rid the simulation from excess energy caused by the imperfections of the scaling procedure and to allow only the natural modes of oscillation to survive.

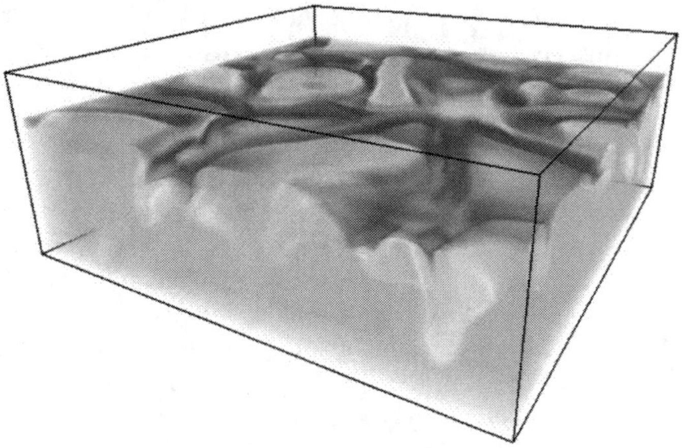

Figure 2. Specific entropy per unit mass for a metal-poor red giant simulation: red-orange hues indicate high-entropy regions, purple-blue-green hues low-entropy ones. The physical size of the box is about 1200×1200 Mm^2 horizontally and 450 Mm vertically. The warm inflowing gas from the bottom of the simulation domain has constant, high, specific entropy per unit mass. As the ascending gas reaches the optically thin layers at the surface, it rapidly cools via radiation losses, lowering its entropy. It eventually becomes denser than the surrounding material and falls back toward the interior of the star in narrow downdrafts.

3. Results and applications

As an illustration of the typical physical structures resulting from 3-D simulations, in Fig. 2, we show the gas entropy per unit mass in a representative snapshot of a STAGGER-CODE red-giant surface convection simulation. The apparent characteristic surface convective pattern with large granules with warm, ascending gas surrounded by an intergranular network of cooler, denser, downflowing material emerges naturally from the simulations without the need for adjustable parameters. The first fundamental difference between 3-D and 1-D model stellar atmospheres is therefore that the 3-D models self-consistently predict the emergence of density and temperature inhomogeneities at the stellar surface and their correlation to macroscopic velocity fields. The non-linear dependence of the populations of energy levels of atoms and molecules as well as of ionization balance and molecular equilibrium on such inhomogeneities ultimately leads to appreciable differences between the profiles and strengths of spectral lines generated using 3-D and 1-D models, even in those cases where the 1-D and the average 3-D stratifications are not too dissimilar from each other [35].

Another difference, especially important in the context of spectral line formation calculations, is that 1-D model atmospheres of late-type stars generally predict a steeper temperature stratification as a function of optical depth than 3-D simulations at the optical surface and in the layers immediately below it. The effects of this are apparent, for instance, in the different predictions of centre-to-limb variations (CLV) in the outgoing continuum radiation intensity. In the Sun's case, in particular, the predicted CLV can be tested directly against observations, showing that the 3-D simulations, contrary to 1-D models, can successfully reproduce such variations across the solar spectrum [14].

Finally, a third important difference is that 3-D stellar surface convection simulations of metal-poor late-type stars predict significantly cooler upper-atmospheric temperature stratifications than 1-D models [24, 35] (Fig. 3, left-hand panel). In the 3-D hydrodynamic case, the temperature in these layers is essentially regulated by two mechanisms: radiative heating due to reabsorption of continuum-radiation by spectral lines and adiabatic cooling associated

Journal of Physics: Conference Series **328** (2011) 012003 doi:10.1088/1742-6596/328/1/012003

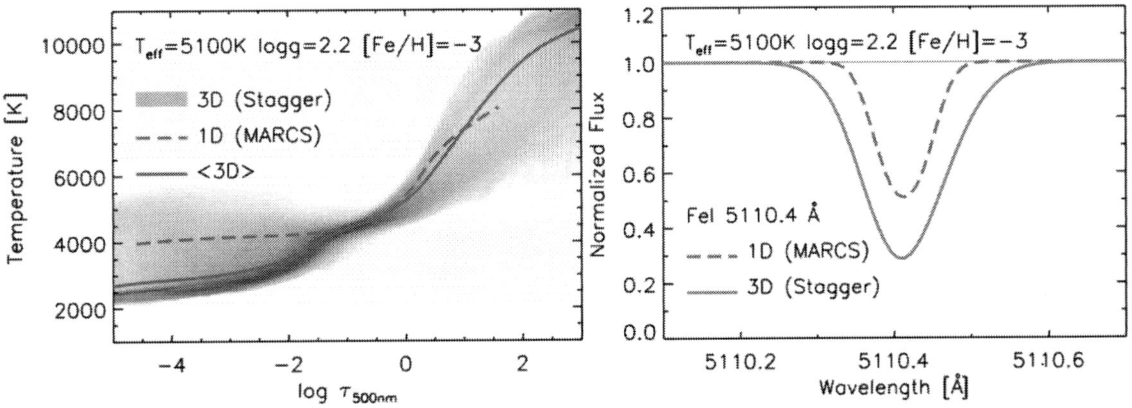

Figure 3. *Left-hand panel, grey shaded area:* Temperature distribution as a function of optical depth in a representative snapshot from a 3-D STAGGER-CODE simulation of metal-poor red giant; *blue line:* mean temperature stratification; *red dashed line:* temperature stratification from corresponding 1-D MARCS model. *Right-hand panel:* synthetic profiles for the Fe I (neutral iron) spectral line at 5110.4 Å computed in local thermodynamic equilibrium (LTE) with the 3-D (*blue line*) and 1-D (*red dashed line*) models of red giant stellar atmospheres shown in the left-hand panel, assuming the same Fe abundance.

with diverging flows above granules. At low metallicities, the contribution of spectral lines to the total opacity in the upper atmosphere is relatively weak: the significance of radiative heating in these layers is therefore reduced relative to adiabatic cooling and the temperature balance is shifted to lower temperatures. Stationary, 1-D, hydrostatic model atmospheres do not account for such adiabatic cooling component associated with gas expansion. The thermal balance in such models is regulated purely via heating and cooling by radiation, ultimately resulting in artificially high temperatures compared with the 3-D case.

Differences between the temperature stratifications in the upper atmosphere of 1-D and 3-D models can amount to ~1000 K in some cases, severely affecting the excitation, ionization, and molecular equilibria in those layers. Such temperature differences can have a particularly large impact on the strengths of synthetic line profiles from temperature-sensitive species and, consequently, on the elemental abundances that can be derived by comparing theoretical and observed profiles. As an example, Fig. 3, right-hand panel, shows the predicted profile of a neutral Fe line, computed in local thermodynamic equilibrium (LTE) using a 3-D model atmosphere of a very metal-poor red giant star and its corresponding 1-D counterpart, assuming the same Fe abundance in the two cases. Because of the line's temperature-sensitivity, the resulting 3-D line profile is significantly stronger than the 1-D one. This also implies that a 3-D LTE analysis of this line would require in this case a significantly lower Fe abundance than a 1-D LTE analysis to reproduce the strength of a given, observed, line profile. Differences between the derived 3-D and 1-D abundances can be of the order of −0.5 dex or even larger in magnitude for lines from neutral atoms and molecules [35, 36]. Based on the experience with previous simulations carried out by our group [37, 35] with a predecessor of the STAGGER-CODE [10] and from preliminary analyses with the present STAGGERGRID models, we see that the 3D−1D LTE abundance differences derived from lines from molecules or neutral atoms typically tend to increase in magnitude (i.e., become more negative) when the metallicity decreases, or when the effective temperature increases, or, also, when the surface gravity decreases. At the moment, however, these results are still based on a limited number of tests restricted to some parts of the grid. We are therefore planning to carry out a more systematic comparison of the results of abundances determinations with 3-D and 1-D models and study in greater detail the trends of the 3D−1D abundance differences with stellar parameters.

4. Comparison with other models

The STAGGER and CIFIST grids are conceptually similar in terms of basic structure and purposes. Both grids also rely on essentially the same opacity sources and use up-to-date, realistic equation-of-state packages and input physics. The main differences are in the adopted numerical methods (codes), basic resolution, and physical extension of the models (current box-in-the-star STAGGERGRID models use a higher numerical resolution and generally extend down to deeper layers) and in the implementations of radiative transfer and opacity binning. The latter, in particular, may be responsible for the apparent differences between the resulting temperature stratifications from simulations of metal-poor stars. The STAGGERGRID metal-poor simulations predict a cooler temperature stratification in the upper-photospheric layers compared with analogous CIFIST models computed for the same stellar parameters [38, 32, 39]. A systematic comparison of the two grids has not been carried out yet, but it is being planned. However, it is important to mention that, in spite of the differences between the two grids at low metallicity, the STAGGERGRID and CIFIST solar surface convection simulations are actually in very good agreement with each other, as well as with the current solar simulation by the MURAM group [40].

5. Summary and outlook

In light of the results we have presented here, accounting for the differences between 3-D and 1-D models is paramount in order to accurately determine elemental abundances and other stellar properties from the analysis of stellar spectra. Once the grid will be completed, we will therefore compute synthetic spectra for all 3-D models and their 1-D counterparts, covering the range from ultraviolet to infrared wavelengths. These spectra will be used for computing synthetic colours, deriving stellar parameters and abundances, studying the properties of stellar surface convection across the H-R diagram, and for many other applications (see, e.g., Chiavassa et al., these proceedings).

We will also construct and make publicly available average 3-D stratifications from the STAGGERGRID models. The information from the full 3-D structures will also be used to provide physical constraints to free parameters used in 1-D analyses such as micro- and macro-turbulence. This will facilitate the implementation of the main results from 3-D modelling in existing, commonly used, 1-D spectral line formation packages. This would be particularly useful in order to systematically and consistently study the combined effects of granulation and departures from local thermodynamic equilibrium in late-type stellar atmospheres (see Bergemann et al., these proceedings).

In conclusion, the STAGGERGRID will offer a powerful and flexible tool for progressing toward precise and accurate analyses of stellar spectra and elemental abundances determinations, and, when combined with Gaia-related follow-up surveys, it will provide a significant leap forward in our understanding of Galactic chemical evolution.

6. Acknowledgments

The STAGGERGRID project is a collaboration involving scientists from several institutes including the Max Planck Institute for Astrophysics (MPA, Garching), the Niels Bohr Institute (NBI, Copenhagen), the Observatoire de la Côte d'Azur (OCA) in Nice, the Institute of Astronomy and Astrophysics (IAA) in Brussels, and the Joint Institute for Laboratory Astrophysics (JILA, Boulder).

References

[1] Perryman M A C, de Boer K S, Gilmore G, Høg E, Lattanzi M G, Lindegren L, Luri X, Mignard F, Pace O and de Zeeuw P T 2001 *A&A* **369** 339–363 (*Preprint* arXiv:astro-ph/0101235)
[2] Prusti T 2011 *EAS Publications Series* (*EAS Publications Series* vol 45) pp 9–14
[3] Gustafsson B, Bell R A, Eriksson K and Nordlund Å 1975 *A&A* **42** 407–432
[4] Gustafsson B, Edvardsson B, Eriksson K, Jørgensen U G, Nordlund Å and Plez B 2008 *A&A* **486** 951–970 (*Preprint* 0805.0554)
[5] Kurucz R L 1993 *Opacities for Stellar Atmospheres* (*Kurucz CD-ROMs* vol 2–12) (Cambridge, Mass.: SAO)
[6] Kurucz R L 2005 *Memorie della Societa Astronomica Italiana Supplementi* **8** 14–24
[7] Böhm-Vitense E 1958 *Zeitschrift fur Astrophysik* **46** 108–143
[8] Canuto V M and Mazzitelli I 1991 *ApJ* **370** 295–311
[9] Nordlund Å, Stein R F and Asplund M 2009 *Living Reviews in Solar Physics* **6** 2–117
[10] Stein R F and Nordlund Å 1998 *ApJ* **499** 914–933
[11] Danilovic S, Gandorfer A, Lagg A, Schüssler M, Solanki S K, Vögler A, Katsukawa Y and Tsuneta S 2008 *A&A* **484** L17–L20 (*Preprint* 0804.4230)
[12] Wedemeyer-Böhm S and Rouppe van der Voort L 2009 *A&A* **503** 225–239 (*Preprint* 0905.0705)
[13] Koesterke L, Allende Prieto C and Lambert D L 2008 *ApJ* **680** 764–773 (*Preprint* 0802.2177)
[14] Asplund M, Grevesse N, Sauval A J and Scott P 2009 *ARA&A* **47** 481–522 (*Preprint* 0909.0948)
[15] Aufdenberg J P, Ludwig H and Kervella P 2005 *ApJ* **633** 424–439 (*Preprint* arXiv:astro-ph/0507336)
[16] Bigot L, Kervella P, Thévenin F and Ségransan D 2006 *A&A* **446** 635–641
[17] Nordlund Å and Dravins D 1990 *A&A* **228** 155–217
[18] Allende Prieto C, Asplund M, López R J G and Lambert D L 2002 *ApJ* **567** 544–565 (*Preprint* astro-ph/0111055)
[19] Ramírez I, Collet R, Lambert D L, Allende Prieto C and Asplund M 2010 *ApJ* **725** L223–L227 (*Preprint* 1011.4077)
[20] Nordlund Å and Galsgaard K 1995 A 3D MHD code for Parallel Computers Tech. rep. Niels Bohr Institute, University of Copenhagen URL http://www.astro.ku.dk/~kg/Papers/MHD_code.ps.gz
[21] Gudiksen B V, Carlsson M, Hansteen V H, Hayek W, Leenaarts J and Martínez-Sykora J 2011 (*Preprint* 1105.6306v1) URL http://arxiv.org/abs/1105.6306v1
[22] Freytag B, Steffen M and Dorch B 2002 *Astronomische Nachrichten* **323** 213–219
[23] Vögler A, Shelyag S, Schüssler M, Cattaneo F, Emonet T and Linde T 2005 *A&A* **429** 335–351
[24] Asplund M, Nordlund Å, Trampedach R and Stein R F 1999 *A&A* **346** L17–L20 (*Preprint* astro-ph/9905059)
[25] Asplund M, Carlsson M and Botnen A V 2003 *A&A* **399** L31–L34 (*Preprint* astro-ph/0302406)
[26] Collet R, Asplund M and Trampedach R 2006 *ApJ* **644** L121–L124 (*Preprint* astro-ph/0605219)
[27] Caffau E, Ludwig H, Steffen M, Freytag B and Bonifacio P 2011 *Sol. Phys.* **268** 255–+ (*Preprint* 1003.1190)
[28] González Hernández J I, Bonifacio P, Ludwig H, Caffau E, Behara N T and Freytag B 2010 *A&A* **519** A46+ (*Preprint* 1005.3754)
[29] Ludwig H, Caffau E, Steffen M, Freytag B, Bonifacio P and Kučinskas A 2009 *Mem. Soc. Astron. Italiana* **80** 711–+ (*Preprint* 0908.4496)
[30] Mihalas D, Däppen W and Hummer D G 1988 *ApJ* **331** 815–825
[31] Feautrier P 1964 *SAO Special Report* **167** 80–+
[32] Collet R, Hayek W, Asplund M, Nordlund Å, Trampedach R and Gudiksen B 2011 *A&A* **528** A32+ (*Preprint* 1101.3265)
[33] Nordlund Å 1982 *A&A* **107** 1–10
[34] Skartlien R 2000 *ApJ* **536** 465–480
[35] Collet R, Asplund M and Trampedach R 2007 *A&A* **469** 687–706 (*Preprint* arXiv:astro-ph/0703652)
[36] Collet R, Nordlund Å, Asplund M, Hayek W and Trampedach R 2009 *Mem. Soc. Astron. Italiana* **80** 719–+ (*Preprint* 0909.0690)
[37] Asplund M and García Pérez A E 2001 *A&A* **372** 601–615 (*Preprint* astro-ph/0104071)
[38] Bonifacio P, Spite M, Cayrel R, Hill V, Spite F, François P, Plez B, Ludwig H, Caffau E, Molaro P, Depagne E, Andersen J, Barbuy B, Beers T C, Nordström B and Primas F 2009 *A&A* **501** 519–530 (*Preprint* 0903.4174)
[39] Ludwig H and Steffen M 2011 *Proceedings of Rome Conference*
[40] Beeck B, Collet R, Steffen M, Asplund M, Cameron R H, Freytag B, Hayek W, Ludwig H and Schüssler M 2011 *A&A* Submitted

Journal of Physics: Conference Series **328** (2011) 012004 doi:10.1088/1742-6596/328/1/012004

Stellar physics with the ALHAMBRA photometric system

T Aparicio Villegas[1], E J Alfaro[1], J Cabrera-Caño[2,1], M Moles[1,3], N Benítez[1], J Perea[1], A del Olmo[1], A Fernández-Soto[4], D Cristóbal-Hornillos[1,3], J A L Aguerri[5], T Broadhurst[6], F J Castander[7], J Cepa[5,8], M Cervio[1], R M González Delgado[1], L Infante[9], I Márquez[1], J Masegosa[1], V J Martínez[10,11], F Prada[1], J M Quintana[1] and S F Sánchez[12]

[1] Instituto de Astrofísica de Andalucía (CSIC), E-18008, Granada, Spain
[2] Facultad de Física, Departamento de Física Atómica, Molecular y Nuclear. Universidad de Sevilla, Sevilla, Spain
[3] Centro de Estudios de Física del Cosmos de Aragón (CEFCA), C/ General Pizarro, 1, 44001 Teruel, Spain
[4] Instituto de Física de Cantabria (CSIC-UC), 39005, Santander, Spain
[5] Instituto de Astrofísica de Canarias, La Laguna, Tenerife, Spain
[6] School of Physics and Astronomy, Tel Aviv University, Israel
[7] Institut de Ciencies de l'Espai, IEEC-CSIC, Barcelona, Spain
[8] Departamento de Astrofísica, Facultad de Física, Universidad de la Laguna, Spain
[9] Departamento de Astronomía, Pontificia Universidad Católica, Santiago, Chile
[10] Departament d'Astronomía i Astrofísica, Universitat de Valencia, Valencia, Spain
[11] Observatori Astronómic de la Universitat de Valencia, Valencia, Spain
[12] Centro Astronómico Hispano-Alemán, Almería, Spain

E-mail: `terenz@iaa.es`, `emilio@iaa.es`

Abstract. The ALHAMBRA photometric system was specifically designed to perform a tomography of the Universe in some selected areas. Although mainly designed for extragalactic purposes, its 20 contiguous, equal-width, medium-band photometric system in the optical wavelength range, shows a great capacity for stellar classification. In this contribution we propose a methodology for stellar classification and physical parameter estimation ($T_{\rm eff}$, log g, [Fe/H], and color excess E$(B-V)$) based on 18 independent reddening-free Q-values from the ALHAMBRA photometry. Based on the theoretical Spectral library BaSeL 2.2, and applied to 288 stars from the Next Generation spectral Library (NGSL), we discuss the reliability of the method and its dependence on the extinction law used.

1. Introduction

The ALHAMBRA (Advanced Large, Homogeneous Area Medium Band Redshift Astronomical) survey is a project aimed at getting a photometric data super-cube, which samples a cosmological fraction of the Universe (almost 4 square degrees in 8 discontinuous regions of the sky) with sufficient precision to draw an evolutionary track of its content and properties (see [1], for a more detailed description of the scientific objectives of the project).

The number, width and position of the filters composing an optimal filter set to accomplish

Journal of Physics: Conference Series **328** (2011) 012004 doi:10.1088/1742-6596/328/1/012004

these objectives was discussed and evaluated by [2], yielding a filter set formed by a uniform system of 20 constant-width, non-overlapping, medium-band filters in the optical range [3] plus the three standard JHK_s near-infrared (NIR) bands. The capabilities of the optical ALHAMBRA photometric system to perform an accurate spectral classification and physical parameters estimation is examined making use of synthetic photometry on the theoretical spectra from BaSeL 2.2 [4], and 288 observed spectra from the Next Generation Spectral Library (NGSL) [5]. The methodology developed, Q Fit Algorithm (QFA), leads to estimates of $T_{\rm eff}$, log g, [Fe/H] and E($B-V$) for a great variety of spectral and luminosity classes, from O- to M-type stars, and including hot white dwarfs and chemically peculiar stars. The method is based on reddening-free Q-parameters, as has been done with other photometric systems, such as [6] for the UPXYZVS Vilnius photometric system [7], or [8] for the UBV system.

2. Stellar classification and estimating physical parameters: *Q* Fit Algorithm

The methodology is mainly based on reddening-free Q-parameters, which do not depend on the distance or the reddening of the objects, but lie on the choice of an interstellar extinction law. This parameter was originally defined by [8] for the UBV system, and in the same way, for the ALHAMBRA system the definition would be:

$$Q_{ijkl} = (m_i - m_j) - \frac{E_{ij}}{E_{kl}}(m_k - m_l)\,, \tag{1}$$

where m_i is the AB-magnitude in the ALHAMBRA band i estimated as:

$$m_i = -2.5\log\frac{\int f_\nu S_\nu\, d(\log\nu)}{\int S_\nu\, d(\log\nu)} - 48.6\,. \tag{2}$$

S_ν is the response function of the system corresponding to the atmosphere-telescope-filter-detector combination, and f_ν is the flux of an object per unit frequency in $\mathrm{erg\,cm^{-2}\,s^{-1}\,Hz^{-1}}$. E_{ij} is the color excess of bands i and j, where

$$E_{ij} = A_i - A_j\,, \tag{3}$$

and A_i the extinction in band i.

2.1. Methodology

The Q Fit Algorithm has the following main steps:

(i) Generation of 18 independent Q-parameters from ALHAMBRA photometry: theoretical spectra and observed spectra.

(ii) Estimation of the main physical parameters, $T_{\rm eff}$, log g, and [Fe/H], by comparing observations to predictions from models.

(iii) Estimation of E($B-V$) from 19 correlated ALHAMBRA colors.

However, other factors have to be taken into account to obtain the most accurate estimation of the parameters possible with this methodology, such as zero point corrections for the synthetic ALHAMBRA photometry of BaSeL models, or checking the dependence of the method on the interstellar extinction law used to generate the Q-values and to estimate the color excess of stars.

We present below a more detailed description of each step.

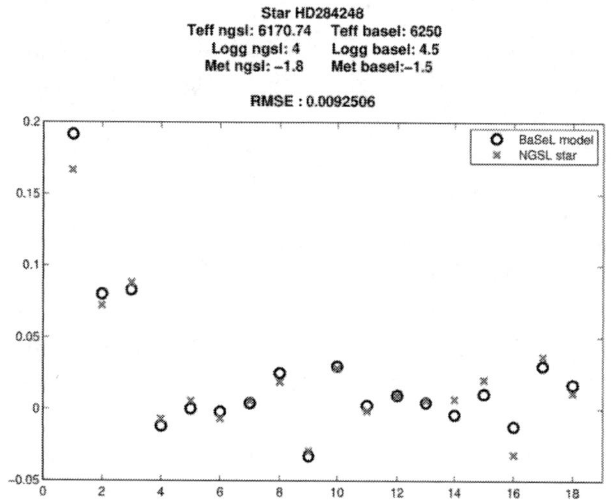

Figure 1. Example of Q fit. The graph shows the 18 Q-values of the star HD 284248 with red crosses, and the values of the model (black circles) from BaSeL that minimizes the euclidean distance. The physical parameters of the object are also shown. The top left-hand side labels are parameters from the literature, while at the right-hand side ones are from the BaSeL model. The Root Mean Square Error (RMSE) of the fit is also shown.

2.2. Generation of 18 independent Q-parameters
Since the optical photometric system is formed by 20 bands, we can generate 18 independent Q-parameters adopting for their definition three correlated bands instead of four different bands. Therefore, if we enumerate the ALHAMBRA magnitudes with the 20 different bands as m_i, for $i = 1 : 20$, the Q-values are defined by:

$$Q_{ijk} = (m_i - m_j) - \frac{E_{ij}}{E_{jk}}(m_j - m_k), i = 1 : 18, j = i + 1, k = i + 2 . \quad (4)$$

We generate these values for theoretical spectra from the BaSeL 2.2 library. Since the expected number of stars with metallicity below −3 in the ALHAMBRA Project is very small, we select BaSeL models with [Fe/H] above or equal to −3 to optimize the algorithm. For the white dwarfs, we compute the Q-values from synthetic ALHAMBRA colors of the white dwarf atmosphere models of [9].

On the other hand, the same 18 Q-parameters are generated for a subset of primary standard stars of the ALHAMBRA photometric system formed by 288 stars from the NGSL [5]. They are observed spectral energy distributions (SEDs) with synthetic AB ALHAMBRA magnitudes that form the basis of the zero-point calibration of the photometric system [3]. The subset of stars presents a wide range of $T_{\rm eff}$-, log g-, and metallicity values (3440 K$\leq T_{\rm eff} \leq$44500 K, 0.45 $\leq$ log $g \leq$ 7.5, and $-3 \leq$ [Fe/H] $\leq$ 0.7), and a range of E$(B-V)$-values between 0 and 0.6. They include several hot white dwarfs, and a large number of peculiar stars of different spectral types. The physical parameters of the 288 stars are obtained from the literature. Indeed, most of them are well-known stars and so one can easily find several values of their $T_{\rm eff}$, log g, [Fe/H], and color excess. We choose the mean of these distributions of parameters to compare with the values we estimate in the algorithm.

2.3. Estimates of main physical parameters, $T_{\rm eff}$, log g, and [Fe/H]
Considering the 18-dimensional euclidean space formed by 18 independent Q-parameters for each object, we then search for the model that minimizes the euclidean distance between both the star and the model. The physical parameters of the model ($T_{\rm eff}$, log g, and [Fe/H] in case

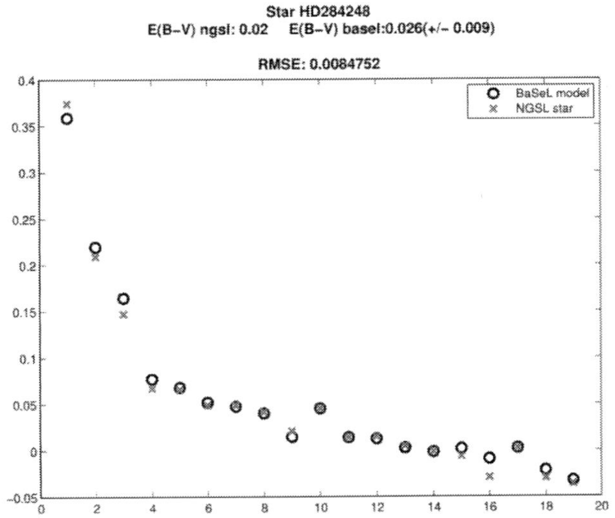

Figure 2. Example of colors fit. The graph shows the 19 ALHAMBRA colors of HD 284248 dereddened with E($B-V$) obtained from the method (red crosses), together with the colors of the model from BaSeL (black circles). The top labels list the literature color excess (left-hand label) and the E($B-V$) obtained from the fit with the error (right-hand label). The RMSE of the fit is also listed.

of BaSeL models, and $T_{\rm eff}$ and log g in case of the white dwarf models) are directly associated to the star.

Figure 1 shows an example of the fit. The star we try to classify is HD 284248. The figure shows the 18 Q-values of the star with red crosses, and the Q-values of the model in black circles. The values are generated using the extinction law of [10]. The x-axis shows the number of the Q-parameters in increasing order. The top labels of the graph list the physical parameters of the star from the literature at the left-hand side. The right-hand side labels provide the model parameters associated to the star by the algorithm. The root mean square error (RMSE) of the fit is also listed.

2.4. Estimates of E(B-V)

For the estimation of E($B-V$) we proceed as follows. From the synthetic photometry of the model we find 19 unreddened ALHAMBRA colors:

$$m_{0i} - m_{0(i+1)}, i = 1:19\,. \tag{5}$$

Hence, with the reddened colors of the star, we can determine the color excess in each band as the difference between the color of the star and the unreddened color of the model:

$$E(m_i - m_{i+1}) = (m_{1i} - m_{1(i+1)}) - (m_{0i} - m_{0(i+1)}), i = 1:19\,. \tag{6}$$

We can then generate 19 values of E($B-V$) from each different color excess by adopting an extinction law. The E($B-V$) associated to the star would be the median of these 19 values:

$$E(B-V) = median_i(\alpha_i \cdot E(m_i - m_{i+1})), i = 1:19\,. \tag{7}$$

Returning to the former example, Fig. 2 shows the colors fit for the same star, HD 284248. The graph shows with red crosses the color 19 values of the star already dereddened with the color excess (obtained as we explain above). The black circles show the unreddened 19 colors of the associated model. The top label also lists the value of E($B-V$) from the literature, and the E($B-V$)-value obtained with our method plus the error. The latter error value is the MAD of

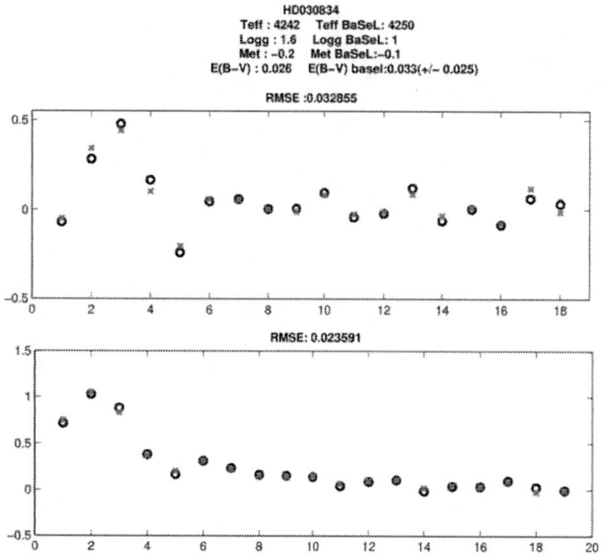

Figure 3. Example of Qs fit (upper panel) and colors fit (lower panel) of the K-type star HD 030834. Symbols are as in Fig. 2. The RMSE-values of both fits are larger compared to HD 284248, although the spectral classification of HD 030834 appears to be precise. The larger RMSEs result from the complexity of the Q-distribution and color values.

the distribution of color excesses, divided by the square root of the number of elements (19 in this case). The RMSE of the fit is also shown.

2.5. Zero-point corrections

To improve the parameter estimation method, we correct the synthetic photometry of the theoretical spectra from BaSeL with synthetic photometry of the primary standard stars. We run the algorithm for the 288 NGSL stars and BaSeL models to compute the offsets. Next, we only select stars with RMSE-values in the Q fit of less or equal than 0.03, thereby rejecting white dwarfs. The resulting average of 190 selected stars depends on the extinction law used by the algorithm. In particular, 191 stars are selected by using [10], 193 stars with [11], and 187 stars with [12]. For each of these stars and their associated model, and for each of the 19 colors, we then determine the differences between the colors of both. The zero-point corrections are the mean of these differences. The parameter determination of the algorithm becomes, in most cases, more accurate with the corrections, and yields smaller average RMSE-values for fits to all different spectral types. In most cases we obtain more accurate physical parameters compared to the literature values (they generally improve after correcting), which directly results from our estimates of the color excess. In particular, we also compute smaller RMSE-values for fits to the coolest stars (of G- to M-type) we expect to observe in the ALHAMBRA fields. The RMSE-values provide important information about the fit quality for estimating the parameters, which also depend on the grid density in the employed library of models. The RMSEs-values are also related to the stellar spectral type. Cool stars have on average larger RMSEs because they show more complex Q distributions or SEDs. Their range of Q-values or colors is also wider than for the hottest stars (see Figure 3).

3. Choice of the interstellar extinction law and results

The Q-parameters are very sensitive to the employed extinction law. Our fit method is suited to assess how the results vary with different extinction laws. In particular, we checked the validity of the method with three extinction laws of [11], [12], and [10]. For all extinction laws we use the value of $R_V = 3.1$ for Vega. The results of the Q Fit Algorithm using different extinction laws

Table 1. Statistical indicators computed with the QFA results for 288 NGSL stars and BaSeL models, corrected for the photometric zero-points, and using three different extinction laws of [10], [11], and [12]. The errors are the standard deviations of the differences divided by the square root of the number of elements (see text).

	Nandy *et al*	Cardelli *et al*	Fitzpatrick
$Mean(MAD(\mathrm{E}(B-V))/\sqrt{19})$	0.018	0.019	0.019
$Std(MAD(\mathrm{E}(B-V))/\sqrt{19})$	0.016	0.016	0.016
$\overline{\lvert\delta(\mathrm{E}(B-V))\rvert}$	0.069±0.005	0.062±0.007	0.062±0.006
$\overline{\delta(\mathrm{E}(B-V))}$	−0.043±0.006	−0.037±0.008	−0.038±0.006
$\overline{\lvert\delta(T_{\mathrm{eff}})\rvert}$	583.20±93	454.63±61	511.74±69
$\overline{\delta(T_{\mathrm{eff}})}$	−505.32±94	−386.20±63	−422.23±71
$\overline{\lvert\delta(\log g)\rvert}$	0.57±0.03	0.51±0.03	0.51±0.03
$\overline{\delta(\log g)}$	−0.17±0.05	−0.09±0.04	−0.10±0.04
$\overline{\lvert\delta([\mathrm{Fe/H}])\rvert}$	0.49±0.02	0.44±0.03	0.49±0.03
$\overline{\delta([\mathrm{Fe/H}])}$	−0.32±0.03	−0.27±0.03	−0.25±0.04

and BaSeL photometry, and corrected for the corresponding zero-points, are listed in Table 1 for various statistical indicators. The color excesses determined by the algorithm are the median of the 19 E$(B-V)$-values from the colors fit, which allows us to also calculate the MAD of the distribution. The mean of the 288 MADs is given in the first row, and the standard deviation in the second row. The first row is an estimate of the internal precision of the extinction law for computing the color excess. Based on this parameter we conclude that any of the three extinction laws can be adopted by the algorithm using our data because we find that the Nandy *et al* extinction law has the smallest value for this estimator, while the differences with Fitzpatrick and Cardelli *et al* are not significant. The third row lists the mean differences (using absolute values) between the color excess from the literature and our algorithm. The next row provides the same mean differences, however without using absolute values. Rows 5 and 6 are the same as the former two, but computed using the effective temperature-, log g-, and [Fe/H]-values. The differences using these values are not significant and do not allow us to conclude if one extinction law should be preferred over the other. Table 2 lists row Nos. 4, 6, 8, and 10 of Table 1 separately for hot, intermediate, and cool stars[1].

4. Conclusions

We examine the capability of the ALHAMBRA photometric system for stellar classification and for estimating atmospheric parameters. With the methodology we develop, the Q Fit Algorithm

[1] The minimum RMSE fit with the Q-parameters yield in some cases negative E$(B-V)$-values. In the current version of the algorithm when E$(B-V)$ is negative, we adopt the next best fit providing a positive E$(B-V)$-value. It produces a bias in the E$(B-V)$-estimates that directly affects the value of temperature parameter.

Journal of Physics: Conference Series **328** (2011) 012004 doi:10.1088/1742-6596/328/1/012004

Table 2. Statistical indicators of the QFA results for 288 NGSL stars and BaSeL models corrected for the photometric zero-points, and using three different extinction laws of [10], [11], and [12], for hot, intermediate, and cool stars. The errors are the standard deviations of the differences divided by the square root of the number of elements (see text).

	Nandy *et al*		
	O- to B-	A- to F-	G- to M-
$\overline{\delta(\mathrm{E}(B-V))}$	0.048±0.03	−0.012±0.007	−0.071±0.008
$\overline{\delta(T_{\mathrm{eff}})}$	−2505±1149	−257.4±62	−378.9±27
$\overline{\delta(\log g)}$	−0.46±0.14	−0.21±0.05	−0.09±0.09
$\overline{\delta([\mathrm{Fe/H}])}$	0.44±0.26	0.23±0.10	−0.46±0.04
	Cardelli *et al*		
	O- to B-	A- to F-	G- to M-
$\overline{\delta(\mathrm{E}(B-V))}$	0.048±0.03	−0.007±0.007	−0.055±0.01
$\overline{\delta(T_{\mathrm{eff}})}$	−1008±821	−239.9±66	−342.8±61
$\overline{\delta(\log g)}$	−0.36±0.18	−0.17±0.04	−0.002±0.06
$\overline{\delta([\mathrm{Fe/H}])}$	1.02±0.31	0.57±0.09	−0.42±0.04
	Fitzpatrick		
	O- to B-	A- to F-	G- to M-
$\overline{\delta(\mathrm{E}(B-V))}$	0.041±0.03	−0.007±0.007	−0.056±0.008
$\overline{\delta(T_{\mathrm{eff}})}$	−1365±812	−254.5±68	−326.9±36
$\overline{\delta(\log g)}$	−0.41±0.18	−0.17±0.04	−0.02±0.06
$\overline{\delta([\mathrm{Fe/H}])}$	0.92±0.27	0.54±0.10	−0.44±0.04

(QFA), it is possible to reliably estimate the T_{eff}-, log g-, [Fe/H]-, and E($B-V$)-values of stars for a wide variety of spectral and luminosity classes, from O- to M-type, including hot white dwarfs and chemically peculiar stars. The QFA is based on 18 independent reddening-free Q-parameters that allow us to the perform parameter estimates without considering the reddening or distances of the objects. To test our method we use a set of primary standard stars of the ALHAMBRA system, consisting of 288 stars from the NGSL. The parameters of these stars are obtained from the literature and compared to the parameters computed with the algorithm. While the T_{eff}-, log g-, and [Fe/H]-values are directly assigned by the BaSeL models, the determination of the color excess (or the value of E($B-V$)) is based on the 19 correlated ALHAMBRA colors, which therefore depends on the employed extinction law. Finally, we test the validity of our method for estimating physical parameters with a number of statistical indicators. We also investigate the dependence of the algorithm on the adopted extinction law. We find that the computed differences are not significant and do not allow us to conclude if one extinction law should be preferred over the other. An advanced analysis of our method and a detailed description of the physical parameters of the NGSL stars will be presented in a forthcoming paper.

References

[1] Moles M *et al* 2008 *Astron. Astrophys.* **136** 1325
[2] Benítez N *et al* 2009 *Astrophys. J.* **692** L5
[3] Aparicio Villegas T *et al* 2010 *Astron. J.* **139** 1242
[4] Lejeune T, Cuisinier F and Buser R 1998 *Astron. Astrophys.* **130** 65
[5] Gregg M D *et al* 2004 *Bull. Am. Astron. Soc.* **36** 1496
[6] Straizys V and Sviderskiene Z 1972 *Astron. Astrophys.* **17** 312
[7] Straizys V 1970 *Vil. Astrono. Observ. Biul.* **28** 6
[8] Johnson H L and Morgan W W 1953 *Astrophys. J.* **117** 313
[9] Holberg J B and Bergeron P 2006 *Astron. J.* **132** 1221
[10] Nandy K, Thompson G I, Jamar C, Monfils A and Wilson R 1975 *Astron. Astrophys.* **44** 195
[11] Cardelli J A, Clayton G C and Mathis J S 1989 *Astrophys. J.* **345** 245
[12] Fitzpatrick E L 1999 *Publ. Astron. Soc. Pac.* **111** 63

Journal of Physics: Conference Series **328** (2011) 012005 doi:10.1088/1742-6596/328/1/012005

Cool star model atmospheres for Gaia : ATLAS, MARCS, and PHOENIX

Bertrand Plez

Laboratoire Univers et Particules de Montpellier, CNRS UMR 5299, Université Montpellier II, cc072, 34095 Montpellier, France

E-mail: `bertrand.plez@univ-montp2.fr`

Abstract. I present the widely used model atmosphere codes ATLAS, MARCS, and PHOENIX, and I compare their output model structures and spectra for cool stars of FGKM-types. While model atmosphere stratifications agree closely with each other in the 1-D approximation, this is not the case for spectra. Differences between model spectra from different codes are largest in the blue-UV, but smaller differences appear in all regions, especially in the molecular features of cooler model spectra. I recommend the groups to try to solve these discrepancies together. In the meantime, users must be careful when using these spectra in regimes where they differ.

I discuss here only comparisons of spectra at solar metallicity, and this should be extended to other metallicities. Detailed comparisons with carefully calibrated spectrophotometric data, and high resolution spectra for stars with well known parameters are also of prime importance. It appears that we still need better line positions for molecules. Finally we should remember that 1-D models are only a step towards a better representation of reality, and we should keep developing, and carefully test 3-D, NLTE models.

1. Introduction

Lots of information we have on stars comes from our comparison of observed spectra with spectra we compute. Synthetic spectra are calculated using a variety of physical data (e.g., atomic and molecular line data, partition functions, dissociation energies), and a physical description of the environment in which spectral lines and continuum occur, the stellar atmosphere. Models of stellar atmospheres are thus a key ingredient in our quest of information on stars. Real star atmospheres display a wealth of dynamical phenomena, such as convection and mass-loss, are structured by magnetic field, with frequent departures from equilibrium in the population of atomic levels. Some recent attempts to include this complexity in stellar atmosphere modeling have widened our capacity to interpret observations, at a considerable computing cost (see, e.g. [1], and Collet in this Volume). On the other hand, it is possible to achieve a quite satisfactory description of stellar atmospheres and spectra, at a much lower cost, by imposing a few, although quite restrictive approximations: local thermodynamic equilibrium (LTE, actually not strictly, as scattering in the continuum is properly treated in 1-D, an issue only recently tackled in 3-D models [2],[3]), hydrostatic equilibrium, 1-D geometry (plane-parallel, PP, or spherically symmetric, Sph), and an energy balance that includes convection treated with the mixing length approximation (MLT). It is surprising how well these approximations work, as long as one is not concerned with extreme cases, or one is not looking at the few percent accuracy level. I will

Journal of Physics: Conference Series **328** (2011) 012005 doi:10.1088/1742-6596/328/1/012005

discuss only such classical modeling, and the main three codes available to compute cool star atmospheres. One should, however, keep in mind the limitations of these models discussed at length in numerous papers, and in this Volume by Collet, Przybilla, and Groh.

2. Model atmosphere codes

2.1. ATLAS

The ATLAS code was developed by R. Kurucz in the 1970s [4], and later updated by him and colleagues, see e.g. [5], [6]. The currently used versions are ATLAS9, using Opacity distribution functions (ODF), and ATLAS12, using OS, with 30 000 points. A major contribution of R. Kurucz was to compute, and provide to the community with an extensive database of atomic and molecular line data (see kurucz.harvard.edu, for data, and codes, including the synthetic spectrum code SYNTHE). They are in wide use in the community, and the atomic data is included in the VALD database [7]. A comparison of ODF and OS models of FGK stars shows very little differences, for the same chemical composition and line data. It was shown, however, by [8] that ODF is a poor approximation when different sources of opacity uncorrelated in wavelength occur at different atmospheric depths. This is the case in, e.g., cool carbon stars, with CO and CN at depth, and HCN and C_2H_2 in the surface layers. Opacity sampling offers a better description of the radiation field, provided a sufficiently large number of wavelength points are used (of the order of 10^5, [9]). Comparisons of ATLAS9 and ATLAS12 models where made by U. Heiter (`camd08.ast.cam.ac.uk/Greatwiki/WGB4StellarAtmospheres/Workshop2010`), who also gives useful information on the code and its history, as well as a number of links to grids of models (`wwwuser.oat.ts.astro.it/castelli/`, and `www.univie.ac.at/nemo/cgi-bin/dive.cgi`). There is a spherical version of ATLAS by J. Lester and H. Neilson [10], on `www.astro.utoronto.ca/ lester/Programs`. A workshop was dedicated to ATLAS12, and the various tools and data associated with it in 2005 (see `sait.oat.ts.astro.it/MSAIS/8/index.html`)

2.2. MARCS2008

The MARCS model atmosphere code has been in use since the mid-1970s [11]. A detailed history of its development may be found in [9], and in [12]. The models are computed in PP or Sph 1-D geometry, LTE, hydrostatic equilibrium, with a detailed account of opacity through opacity sampling (OS) with more than 10^5 wavelength points. A major update of the input physical data, esp. of opacities, resulted in the publication of a new grid [12], available online (www.marcs.astro.uu.se). The inclusion of a huge number of molecular lines allows the computation of cool stars, including M, S, and C-types. The temperature range is limited downwards to M dwarfs, due to the lack of dust opacities, and to the use of the electron pressure as a primary variable. The hot end is early F or late A-type stars, due to limits in the number of ions that are included. The details of the implementation of the code and of the input data are discussed in [12]. Sphericity effects were discussed by [13], [14], and [12]. There is a synthetic spectrum code, turbospectrum [15], available upon request to B. Plez, that includes the same physical data, and routines than the MARCS code.

2.3. PHOENIX

The PHOENIX code was born in the 1990s see, e.g. [16], and references therein. It is much more general than MARCS and ATLAS, as it can handle relativistic dynamical media, and massive NLTE, in 1-D. Models published through the years cover supernovae to brown dwarfs and planetary atmospheres. A 3-D version has been developed recently [17].

Journal of Physics: Conference Series **328** (2011) 012005 doi:10.1088/1742-6596/328/1/012005

2.4. Limitations of the models

We may improve opacities, input data, as it has been extensively done in the past decade, but there are limitations intrinsic to the methods : sampling of the opacities, LTE, 1-D. The relaxation of the LTE hypotheses is outside the scope of this paper, but discussions can be found elsewhere in this volume. Opacity sampling is now routinely done with a number of wavelengths in excess of 10^5, which is a big advantage of simple classical models over more sophisticated 3-D, or NLTE models. A 1-D model can be computed within minutes on a laptop, whereas it is a major computational effort to increase the number of frequency bins in 3-D hydrodynamical models. The impact of a coarse sampling of opacities on the thermal structure, and the emergent sampled spectral energy distributions (SED) was discussed by [9]. The conclusion is that 10^5 sampling points between 100 and 20 000 nm are sufficient to compute cool star (FGKM) model atmospheres with errors on the thermal stratification not exceeding a few degrees Kelvin in any layer. One exception is the optically thin layers of metal-poor stars due to the fact that only a few lines dominate their thermal equilibrium. However, other effects (hydrodynamical, and NLTE) are likely to have a stronger impact on the temperature of these layers. Sampled SEDs are more a problem, as many lines affect the emergent flux, without effects on the thermal structure. A sampling density sufficient for an accurate thermal structure may not allow the computation of a sampled SED to better than 10% in some spectral regions. Detailed spectra can, and should of course be computed afterwards with much larger spectral resolution ($> 10^6$ points), based on the thermal stratification calculated with much less points.

3. Comparison of model structures

Before looking at emergent spectra, it is important to verify if the three codes, using different numerical methods and schemes, as well as different input physical data, produce similar model stratifications. [12] show comparisons between MARCS, ATLAS, and PHOENIX thermal structures for solar and metal-poor composition, for dwarfs, giants, and supergiants models of effective temperatures, $T_{\rm eff}$ between 3000 K and 7000 K. Large differences are not found, except with previous generation PHOENIX models (NextGen).

4. Comparison of spectra

4.1. The SED

A comparison of ATLAS and previous generation PHOENIX (NextGen) models was made by [18], by fitting the SEDs of a sample of target stars. Here, I wish to compare models between each other. They should of course also be compared to observations. This was done for MARCS models, in a preliminary way, by [19]. One simple way to look at the differences in the overall shape of spectra is color-color, and color-$T_{\rm eff}$ diagrams. I therefore computed Johnson-Cousins UBVRIJHK photometry for a number of MARCS and PHOENIX (kindly provided by P. Hauschildt) model spectra using the filters of [20]. For ATLAS models the photometry computed with the same filters is available on `wwwuser.oat.ts.astro.it/castelli/`. This was done for solar metallicity models only. Figure 1 shows the $T_{\rm eff}$ - V$-$K relation for giants and dwarfs. Small differences may be seen around 3500 K and below (note that there are no ATLAS models cooler than 3500 K). With its large leverage on the SED of cool stars, the V$-$K index is a very powerful tool for temperature determination, quite insensitive to gravity (except for M-type stars), and metallicity, and thus not so sensitive to modeling details. It is, however, affected by reddening. Figure 2 shows that there seems to be a problem in the near-IR flux (too blue) of the dwarf ATLAS models for $T_{\rm eff} \leq 4000$ K. The $T_{\rm eff}$ - B$-$V relation of Figure 3 is more affected by metallicity and gravity. The calculations of the three codes agree well above 4000 K for both giants and dwarfs, which indicates that opacities are accounted for in very similar ways in the blue-green part of the spectrum. This is not so surprising as they all use the VALD or Kurucz atomic line data, which is part of VALD. Below 4000 K large differences show up. The

Journal of Physics: Conference Series **328** (2011) 012005 doi:10.1088/1742-6596/328/1/012005

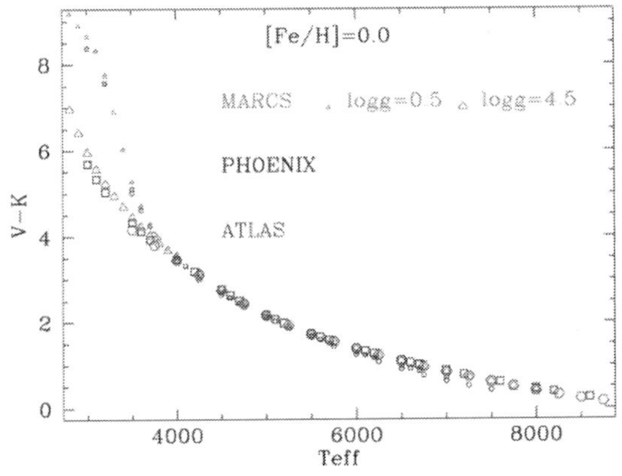

Figure 1. $T_{\rm eff}$ - V−K relation for dwarf (large symbols) and giant (small symbols) star models (log g=4.5 and 0.5 cgs respectively. Blue circles are ATLAS models, red triangles are MARCS models, and black squares are PHOENIX models.

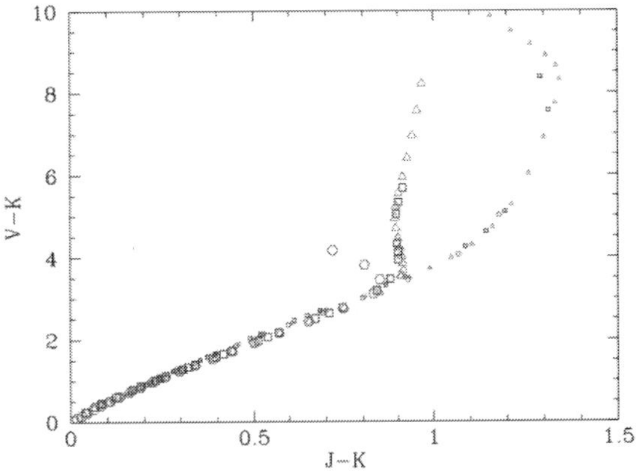

Figure 2. J−K - V−K relation. Symbols as in Figure 1.

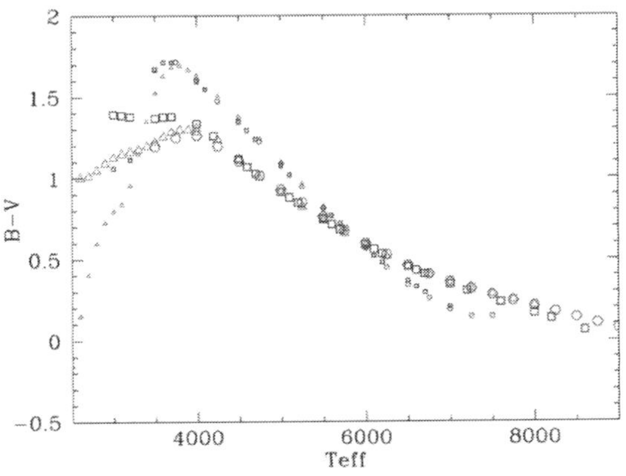

Figure 3. $T_{\rm eff}$ - B−V relation. Symbols as in Figure 1.

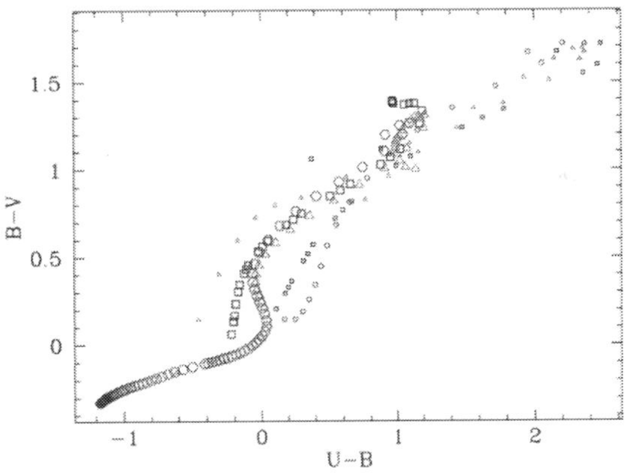

Figure 4. U−B - B−V relation. Symbols as in Figure 1.

U−B - B−V plot of Figure 4 displays differences of up to half a magnitude. It appears that the blue-violet part of the spectrum is the most problematic, and these differences should be investigated in detail, in order to pinpoint their origin, and find a cure.

4.2. the RVS domain

The next comparison we can make of direct interest to Gaia is high-resolution spectra in the RVS domain. To that end I computed high-resolution ($R = \lambda/\Delta\lambda = 500\,000$) spectra for a number of MARCS models using the Turbospectrum code. The input atomic line data comes from VALD, and the molecular lines from the MARCS database. I collected PHOENIX spectra (courtesy of P. Hauschildt) for similar models, that use a partly different set of line data. All spectra were degraded to the resolution of the RVS by convolution with a Gaussian profile of 26 $\mathrm{km\,s^{-1}}$. This comparison thus gives an estimate of remaining discrepancies due to both differences in the model structures, and in the line data used to compute the spectra in different groups. Figure 5

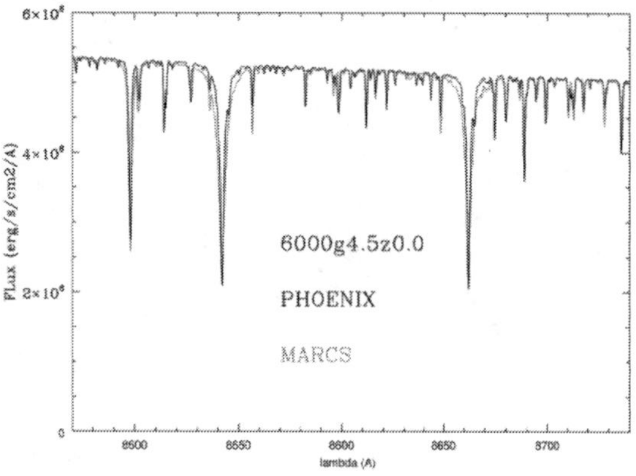

Figure 5. Synthetic spectra in the Gaia/RVS range for stellar parameters typical of a solar composition G dwarf. The black solid line is for a PHOENIX model, and the red solid line for a MARCS calculation. Spectra were degraded to the resolution of the Gaia/RVS.

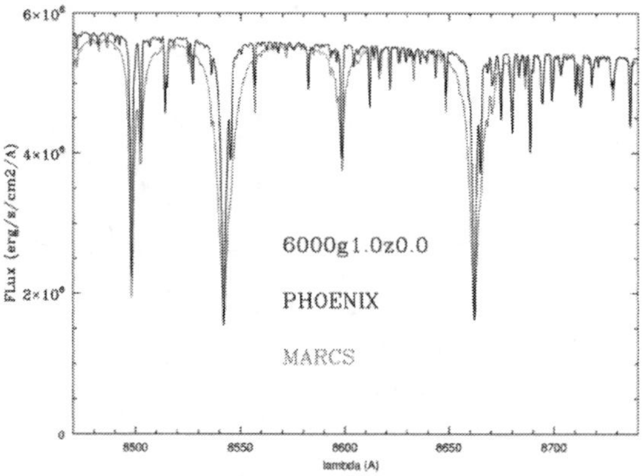

Figure 6. Synthetic spectra in the Gaia/RVS range for stellar parameters typical of a solar composition G supergiant. Symbols as in Figure 5.

to Figure 10 show comparisons at this resolution. Note that to really understand where the differences stem from, it is necessary to carry out an analysis at much higher resolution, as most features seen at the RVS resolution are blends of lines, both atomic and molecular, especially at lower $T_{\rm eff}$. My purpose here is only to estimate at which level two independent calculations (codes, and physical and line data) differ. This gives us an idea of the magnitude of errors we may expect from the current models. Figure 5 and Figure 6 display model spectra representative of a dwarf, and a supergiant star slightly hotter than the Sun. The most prominent difference between MARCS and PHOENIX spectra is that the latter have markedly narrower Ca II lines. In MARCS/Turbospectrum collisional line broadening is treated according to the recipes of [21], [22]. This is obviously not the case in PHOENIX, but this implementation is under way. There are a number of other differences in the details of fainter features. An interesting fact is that MARCS predicts broader Ca II lines in the lower gravity model, with lower pressure. Naively one would expect the opposite. This is a well-known effect of contrast between line and continuum opacity. In the supergiant model, the electron pressure is 30 times lower than in the dwarf, leading to a 30 times lower H^- opacity, and thus to stronger lines. In both cases Ca is almost fully ionized. Figure 7 and Figure 8 show a similar comparison for a dwarf and a supergiant model of K-type. In the supergiant model, molecular lines begin to show up, and they become much more prominent in M-type stars, like in the M dwarf model of Figure 9. The enlarged plot of the same spectra in Figure 10 demonstrates that there is work to be done on molecular line lists (TiO here), at least on the line positions. This demands careful comparisons with observed spectra of standard stars. It is also obvious from Figure 7 that the situation is much worse for FeH (with band head at 8690 Å), with a large uncertainty in the line strengths.

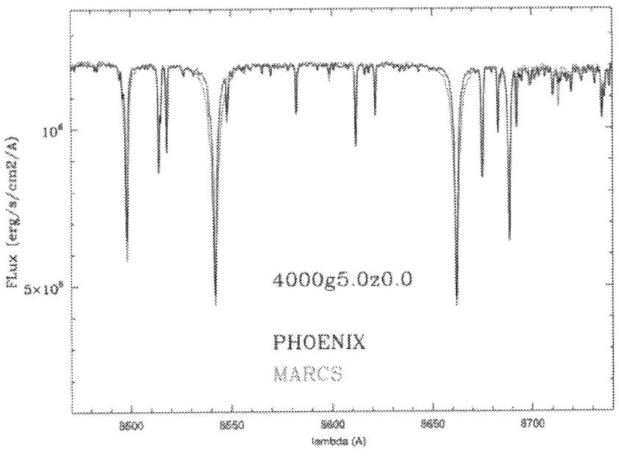

Figure 7. Synthetic spectra in the Gaia/RVS range for stellar parameters typical of a solar composition K dwarf. Symbols as in Figure 5.

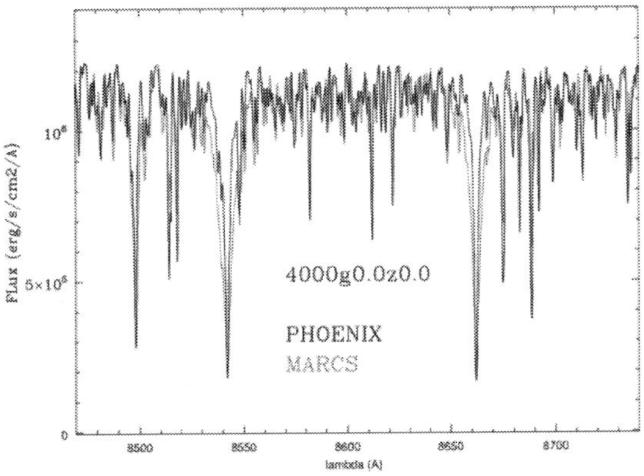

Figure 8. Synthetic spectra in the Gaia/RVS range for stellar parameters typical of a solar composition K supergiant. Symbols as in Figure 5.

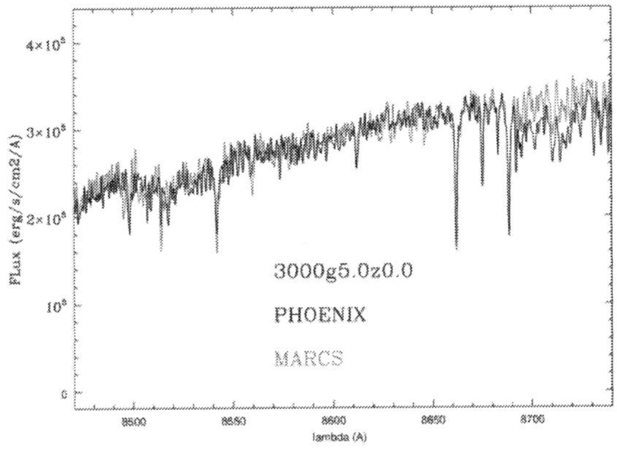

Figure 9. Synthetic spectra in the Gaia/RVS range for stellar parameters typical of a solar composition M dwarf. Symbols as in Figure 5.

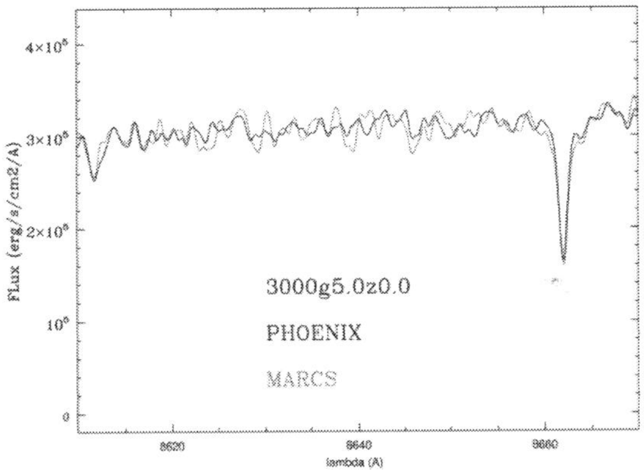

Figure 10. Zoom in on the 8610-8670 Å window of the spectra of Figure 9. Symbols as in Figure 5.

5. Conclusions and recommendations

The non-exhaustive comparisons presented above show that there are still significant differences between ATLAS, MARCS, and PHOENIX model spectra, although model structures are quite similar. Judging from color comparisons, most large differences are in the UV, and their origin should be investigated. ATLAS model spectra deviate in the near-IR, below 4000 K. PHOENIX model spectra deviate in B−V below 4000 K. In the RVS domain, from the comparison of MARCS and PHOENIX model spectra, it appears that a number of faint features do not match each other. In PHOENIX calculations, collisional broadening is too approximate, and should be improved following the recipe of [21], [22]. This is under way. M dwarf model spectra reveal important differences in the details of TiO bands, and in the strength of FeH. The MARCS and PHOENIX groups should compare in detail their line lists, and settle this problem. I discuss here only comparisons of spectra at solar metallicity, and this should be extended to other metallicities. While model atmosphere stratifications agree closely with each other in the 1-D approximation, this is not the case for spectra. Detailed comparisons with carefully calibrated spectrophotometric data, and high resolution spectra for stars with well-known parameters are of prime importance to test the models. Large efforts are indeed being devoted within the Work Package "Provide calibration training data" of CU8 to this end. It appears that we still need better line positions for molecules. Close collaboration with physicists is very productive, and must be encouraged. Progress is being made for, e.g., CN, C_2, CH, C_2H_2. Finally we should remember that 1-D models are only a step towards a better representation of reality, and we should keep developing, and carefully test 3-D, NLTE models.

Acknowledgements

Ulrike Heiter, and Peter Hauschildt are thanked for sending material used in this work. The organizers are thanked for providing financial support. The constructive comments from the referee, Dr. F. Castelli, Osservatorio Astronomico di Trieste, contributed to improve this paper.

References

[1] Chiavassa A, Haubois X, Young J S, Plez B, Josselin E, Perrin G and Freytag B 2010 *A&A* **515** A12+ (*Preprint* 1003.1407)
[2] Collet R, Hayek W, Asplund M, Nordlund Å, Trampedach R and Gudiksen B 2011 *A&A* **528** A32+ (*Preprint* 1101.3265)
[3] Hayek W, Asplund M, Collet R and Nordlund Å 2011 *A&A* **529** A158+ (*Preprint* 1108.3366)
[4] Kurucz R L 1970 *Atlas: A computer program for calculating model stellar atmospheres* (SAO Special Report 309)
[5] Castelli F and Kurucz R L 2003 *Modelling of Stellar Atmospheres* (*IAU Symposium* vol 210) ed N Piskunov, W W Weiss, & D F Gray pp 20P–+
[6] Sbordone L, Bonifacio P and Castelli F 2007 *IAU Symposium* (*IAU Symposium* vol 239) ed T Kuroda, H Sugama, R Kanno, & M Okamoto pp 71–73
[7] Heiter U, Barklem P, Fossati L, Kildiyarova R, Kochukhov O, Kupka F, Obbrugger M, Piskunov N, Plez B, Ryabchikova T, Stempels H C, Stütz C and Weiss W W 2008 *Journal of Physics Conference Series* **130** 012011–+
[8] Ekberg U, Eriksson K and Gustafsson B 1986 *A&A* **167** 304–310
[9] Plez B 2008 *Physica Scripta Volume T* **133** 014003–+ (*Preprint* 0810.2375)
[10] Lester J B and Neilson H R 2008 *A&A* **491** 633–641 (*Preprint* 0809.1870)
[11] Gustafsson B, Bell R A, Eriksson K and Nordlund A 1975 *A&A* **42** 407–432
[12] Gustafsson B, Edvardsson B, Eriksson K, Jørgensen U G, Nordlund Å and Plez B 2008 *A&A* **486** 951–970 (*Preprint* arXiv:0805.0554)
[13] Plez B, Brett J M and Nordlund A 1992 *A&A* **256** 551–571
[14] Heiter U and Eriksson K 2006 *A&A* **452** 1039–1048 (*Preprint* arXiv:astro-ph/0603273)
[15] Alvarez R and Plez B 1998 *A&A* **330** 1109–1119 (*Preprint* arXiv:astro-ph/9710157)
[16] Hauschildt P H, Baron E and Allard F 1997 *ApJ* **483** 390–+ (*Preprint* arXiv:astro-ph/9607087)
[17] Baron E, Chen B and Hauschildt P H 2009 *American Institute of Physics Conference Series* (*American Institute of Physics Conference Series* vol 1171) ed I Hubeny, J M Stone, K MacGregor, & K Werner pp 148–160 (*Preprint* 0908.4273)
[18] Bertone E, Buzzoni A, Chávez M and Rodríguez-Merino L H 2004 *AJ* **128** 829–841 (*Preprint* arXiv:astro-ph/0406215)
[19] Edvardsson B 2008 *Physica Scripta Volume T* **133** 014011–+
[20] Bessell M S, Castelli F and Plez B 1998 *A&A* **337** 321–+
[21] Anstee S D and O'Mara B J 1995 *MNRAS* **276** 859–866
[22] Barklem P S, Piskunov N and O'Mara B J 2000 *A&AS* **142** 467–473

Journal of Physics: Conference Series **328** (2011) 012006 doi:10.1088/1742-6596/328/1/012006

Stellar libraries for Gaia

R Sordo [1], A Vallenari [1], R Tantalo [2], C Liu [3], K Smith [3], F Allard [4], R Blomme [5], J-C Bouret [6], I Brott [7], P de Laverny [8], B Edvardsson [9], Y Frémat [5], U Heber [10], E Josselin [11], O Kochukhov [12], A Korn [12], A Lanzafame [13], C Martayan [14,15], F Martins [11], B Plez [11], A Schweitzer [16], F Thévenin [8] and J Zorec [17]

[1] INAF-Osservatorio Astronomico, Vic. dell'Osservatorio 5 I-35122 Padova (Italy)
[2] Astronomy Dept., Padova Univ., Vic. dell'Osservatorio 3 I-35122 Padova (Italy)
[3] MPIA, Knigstuhl 17, Heidelberg 69117 (Germany)
[4] CRAL-ENS 46, Allée d'Italie 69364 Lyon Cedex 07 (France)
[5] Royal Observatory, Ringlaan 3, 1180, Brussel (Belgium)
[6] Lab. d'Astroph. de Marseille, 38 rue Joliot-Curie, 13388 Marseille Cedex 13 (France)
[7] Astronomical Inst., Utrecht Univ., Princetonplein 5, 3584 CC Utrecht (The Netherlands)
[8] OCA, B.P. 4229, 06304 Nice Cedex 4 (France)
[9] Dept. of Physics and Astronomy, Uppsala Astron. Obs., Box 515, 751 20 Uppsala (Sweden)
[10] Dr. Remeis-Sternwarte, Astronomisches Inst. der Universität Erlangen-Nürnberg, Sternwartstr. 7, 96049 Bamberg (Germany)
[11] GRAAL, CNRS, Universitè Montpellier II, Pl. Bataillon, 34095 Montpellier Cedex (France)
[12] Dept. of Physics and Astronomy, Uppsala Univ., BOX 515, 75120 Uppsala (Sweden)
[13] Dip. di Fisica e Astronomia, Univ. di Catania, Via S. Sofia 78, 95123 Catania (Italy)
[14] ESO, Alonso de Cordova, 3107, Vitacura, Santiago (Chile)
[15] GEPI, Observatoire de Paris, CNRS, Universit Paris Diderot, (France)
[16] Hamburger Sternwarte, Gojenbergsweg 112, 21029 Hamburg (Germany)
[17] Institut d'Astrophysique de Paris, UMR 7095 du CNRS, Univ.P&MC (France)

E-mail: `rosanna.sordo@oapd.inaf.it`

Abstract. Gaia will observe up to a billion stellar sources. Automated algorithms are under development to derive the atmospheric parameters of all observed spectra, from low resolution optical spectra alone or in synergy with high resolution spectra in the near-IR Ca II triplet region. To do so, a large database of state-of-the-art stellar libraries has been produced for the Gaia community, computed using different codes optimized for specific purposes. The choice to use different spectral codes in different regions of the H-R diagram raises the problem of the coherence of the different spectra, specifically in the transition zones. We present a comparison between the libraries from the point of view of spectra simulations for training the Gaia algorithms. We also present the implementation of these libraries into a Simple Stellar Population code.

1. Gaia

ESA's Gaia mission, to be launched in 2013, will obtain accurate positions, parallaxes, and proper motions for 10^9 objects all over the sky to magnitude G=20 (V=20–22). It will provide an incredibly accurate map of a large portion of our Galaxy. The amount of astrophysical problems that will be addressed with this wealth of data will be enormous considering the large impact of the Hipparcos mission ([1]) on all fields of astronomy. The impact of Gaia data is strongly related to the precision and accuracy of the astrometric (μ-arcsec) and photometric (millimag) catalog, both connected to the spectrophotometric data Gaia will observe combined with the astrometry. For all observed sources Gaia will obtain low dispersion spectra in the optical range that will not only be used for correcting the astrometry for chromaticity effects, but (possibly) also to characterize the objects. Since no input catalog is foreseen or even possible for Gaia, it would be of lesser impact to create a map of the Galaxy without at least some physical information about the objects. Follow-up observations of a billion of objects is simply impossible. Gaia *must* therefore be able to characterize the sources with *a)* a discrete classification of sources (e.g., single stars, galaxies, QSOs, asteroids), and possibly *b)* their extinction values and Astrophysical Parameters (APs), i.e., $T_{\rm eff}$, $\log g$, etc. for stars, Equivalent Width for Quasars, the Star Formation Rate for Galaxies, etc. Automated algorithms are currently developed using all knowledge for modeling astrophysical objects. Large and up-to-date libraries of stellar spectra are therefore of fundamental importance. A massive effort is ongoing to cover, as far as is currently possible, the H-R diagram for all stars to the best of our present knowledge, producing synthetic datasets of stars from the coolest to the faster evolutionary stages, and to the most evolved stars. Although we will only focus on stars in this paper, the coupling of a stellar spectrum (for all possible evolutionary stages) with stellar evolutionary models allows us to also derive global properties of stellar populations (such as unresolved galaxies) from their integrated light (we quote among others [2, 3, 4, 5]).

2. Spectroscopy on-board Gaia

Gaia will have aboard two different spectrophotometric instruments:

- The Blue and Red Photometer (BP/RP) will measure the spectral energy distribution of all objects, from 300 to 1100 nm with very low resolution (3–30 nm/pix).
- The Radial Velocity Spectrometer (RVS) will take higher resolution spectra of all objects to $V{\sim}16$, in a limited wavelength region around the near-IR Ca II triplet for deriving radial velocity information of the majority of stars (i.e. for cool stars see [6]). The RAVE survey observes the same spectral region with large diagnostic capability ([7]).

The Gaia instruments will convolve the stellar source spectra with their photometric bands. The Gaia community of providers has agreed to compute a database of spectra for all types of sources (we will only discuss stars) with a significantly larger resolving power than the final spectra. The spectra available for Gaia cover with a 3-pixels resolution:

- for BP/RP: 300–1100 nm wavelength range, 0.1 nm step (low resolution);
- for RVS: 840–890 nm wavelength range, 0.001 nm step (high resolution).

The currently available libraries are listed in Table 1. The parameter space ($T_{\rm eff}$, $\log g$, metallicity, etc.) is sampled with discrete steps, providing a sufficiently fine grid for reliable linear interpolation. The steps are chosen based on common knowledge about stellar classification for sampling each stellar subtype; from 250 K for cool stars to 5000 K for very hot stars. Steps of 0.5 dex in $\log g$ or metallicity are sufficient for the interpolation ([24]). Further tests are presented in Section 3.1. We present in Fig. 1 the coverage in the $T_{\rm eff}$-$\log g$ plane for all libraries with $T_{\rm eff}$<10 000 K compared to the widely used BASEL library ([25]). The comparison with two overplotted isochrones, for 10 Gyr (blue dotted line) and 6 Myr (black dotted line), underscores our efforts to cover the complete parameter space. These spectral libraries have an unprecedented uniformity (same steps and same metallicity) in parameter space coverage and resolution (see Table 1 for the references). Different codes are used to produce them; the BASEL and Munari libraries cover the entire parameter space. The other codes are strongly optimized for specific purposes or $T_{\rm eff}$-regimes, and provide a more adequate treatment of all physically important

Journal of Physics: Conference Series **328** (2011) 012006 doi:10.1088/1742-6596/328/1/012006

Table 1. Synthetic stellar libraries computed for Gaia. The separation into two sets is explained in Section 6. UCD indicates Ultra-Cool Dwarfs, while RSG means Red Supergiants.

Name	$T_{\rm eff}$ (K)	$\log g$	[Fe/H]	References
Munari	3500–47500	0.0–5.0	−2.5–1.0	[8]
O, B stars	15000–50000	1.0–5.0	−5.0–1.0	[9]
wind, mass loss				
Ap/Bp stars	7000–16000	4.0	+0.0	[10]
abundance				
A stars	6000–16000	2.5–4.5	+0.0	[11]
MARCS	2800–8000	−0.5–5.5	−5.0–1.0	[12]
PHOENIX	3000–10000	−0.5–5.5	−2.5–0.5	[13]
UCD	400–3000	−0.5–5.5	−2.5–0.5	[14]
C stars	4000–8000	0.0–5.0	−5.0–0.0	[12, 15]
Be	15000–25000	4.0	+0.0	[16]
radius considered;				
WR	25000–51000	2.8–4.0	+0.0	[16]
rate considered;				
WD	6000–90000	7.0–9.0	-	[17]
sdOB	26000–100000	4.8–6.4	+0.0	[18]
$\Delta \log g = 0.2$				
NLTE Cool stars	4000–6000	4.5–5.5	+0.0	[19]
Fast Rotators	9000–25000	3.6–4.2	+0.0	[20]
RSG	3000–4300	−1.0–0.5	+0.0	[21]
MARCS RVS	2800–8000	−0.5–5.5	−5.0–1.0	[22]

processes in stellar atmospheres; NLTE effects, dust models for very cool stars, effects of mass-loss, emission from circumstellar envelopes, variations of single element chemical abundances, magnetic field, and α-enhancement. These libraries are described in further detail in [26]. It is important to emphasize that, although many codes can compute 3-D stellar atmospheres with very high complexity and accuracy, the time required to calculate a single stellar spectrum is currently prohibitive even for modern computers. For this reason the PHOENIX library for Gaia is for example computed in the 1-D approximation. In the following section we will mainly focus on the low-resolution database. It is however important to emphasize the coherence between the two databases at the different spectral resolutions. Each spectrum in the BP/RP database corresponds to one in the RVS database computed with the same code. It allows us to simultaneously simulate a *source* in both the BP/RP and the RVS domain, providing consistent spectra.

3. AP algorithms for Gaia

As we explain above, automated algorithms are currently under development for characterizing the observed sources. The algorithms are developed for processing both a given dataset and to exploit the synergy with astrometry. Figure 2 shows BP/RP spectra computed by the Gaia Object Generator (GOG, [27]) after convolving them with the response function of the instrument. The sequences in temperature and gravity are shown for the BASEL library. The impact of using other libraries is shown below.

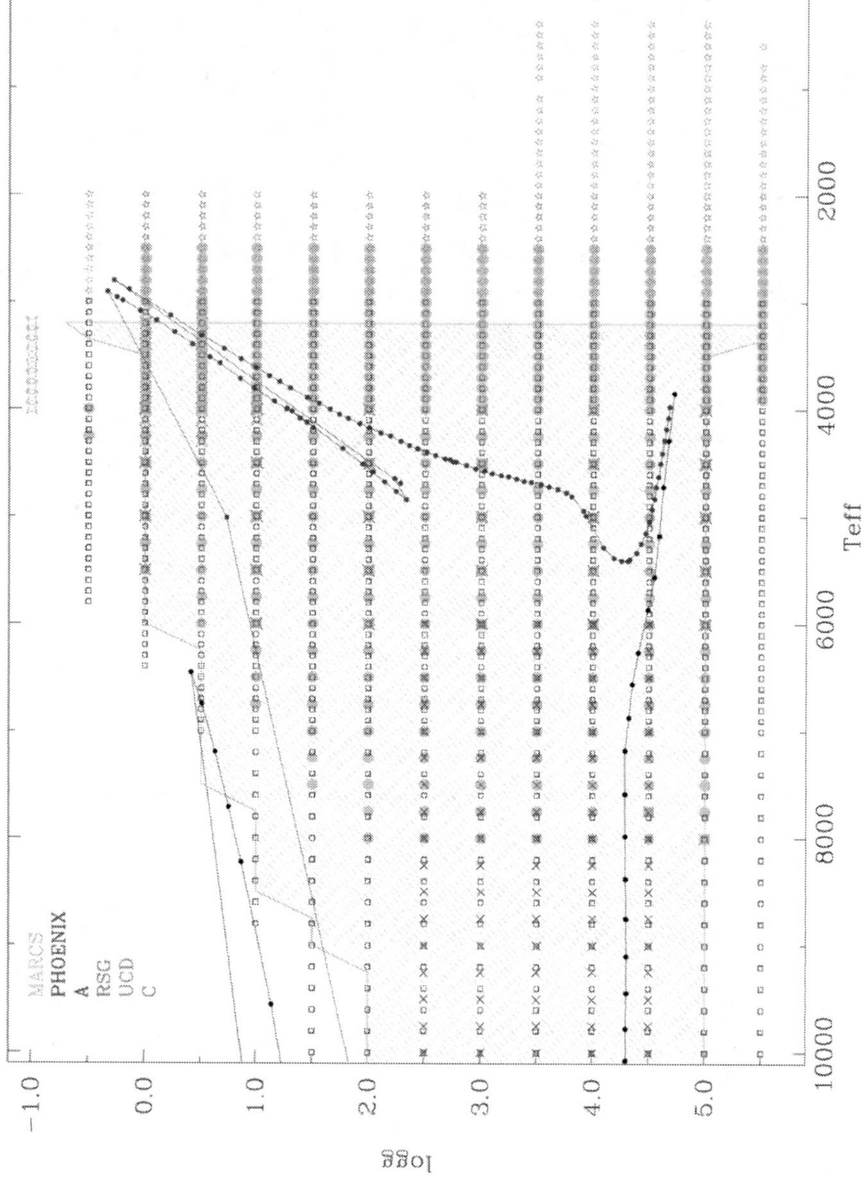

Figure 1. $T_{\rm eff}$-log g plane for a sample of libraries presented in Table 1 for $T_{\rm eff}$<10 000 K with solar metallicity. We do not show NLTE Cool Stars and peculiar A-type stars because they cover a very narrow region. The coverage by the BASEL library is shown shaded to facilitate the comparison. The [8] library is not reported since it covers the same region of the BASEL library for $T_{\rm eff}$>3500 K. The isochrone of an old population (blue dotted line) and of a young population (black dotted line) for solar metallicity from the Padova family ([23]) is overplotted.

Journal of Physics: Conference Series **328** (2011) 012006 doi:10.1088/1742-6596/328/1/012006

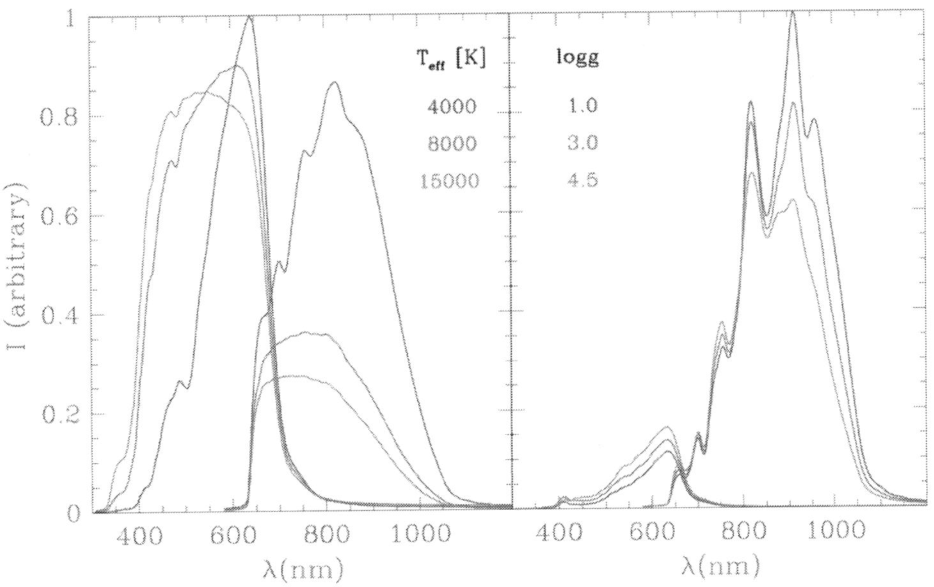

Figure 2. GOG simulations in the BP/RP range. Left-hand panel: temperature sequence: $T_{\rm eff}$=4000, 8000, 15 000 K. Left-hand panel: gravity sequence at $T_{\rm eff}$=3000 K, from supergiant to dwarf stars.

The *Coordination Unit 8* of the Gaia Data Processing and Analysis Consortium is working on a set of algorithms to exploit the different measurements Gaia will provide. The algorithms are both supervised and unsupervised, and will perform physical classification or provide astrophysical parameters. To summarize them:

- the Discrete Source Classifier (DSC) is used to assign an astrophysical class (QSO, star, galaxy) to an object using BP/RP spectra, magnitude, and astrometry. DSC has a modular design with subclassifiers using only photometric data, or in combination with astrometric data or astrophysical priors. The algorithm has been recently used to photometrically classify BHB stars (see [28] for details).
- GSP-Phot: is responsible for the determination of astrophysical parameters from BP/RP spectra and astrometry. Different algorithms are under evaluation for testing the effect of the inclusion of *priors* in the Bayesian treatment (see for example [29]).
- GSP-Spec is responsible for the analysis of the RVS spectra of FGK stars. Different algorithms are under development, including distance minimization, projection, and pattern recognition.
- the Extended Stellar Parametrizer (ESP) considers hot stars and *peculiar* objects, such as Emission Line stars, Ultra Cool stars, cool active stars, etc. Hot stars suffer from a poor diagnostic potential at this $T_{\rm eff}$ of the RVS wavelength region (see Blomme, this Volume).

All the algorithms are developed/trained with synthetic datasets. The available databases of synthetic spectra are used to couple the spectra to predictions from evolutionary track models (yielding mass, radius, age, etc.) for building an 'emitting source'. The spectrum is then simulated using the *Gaia Object Generator* (GOG, [27]) to obtain spectroscopic, astrometric,

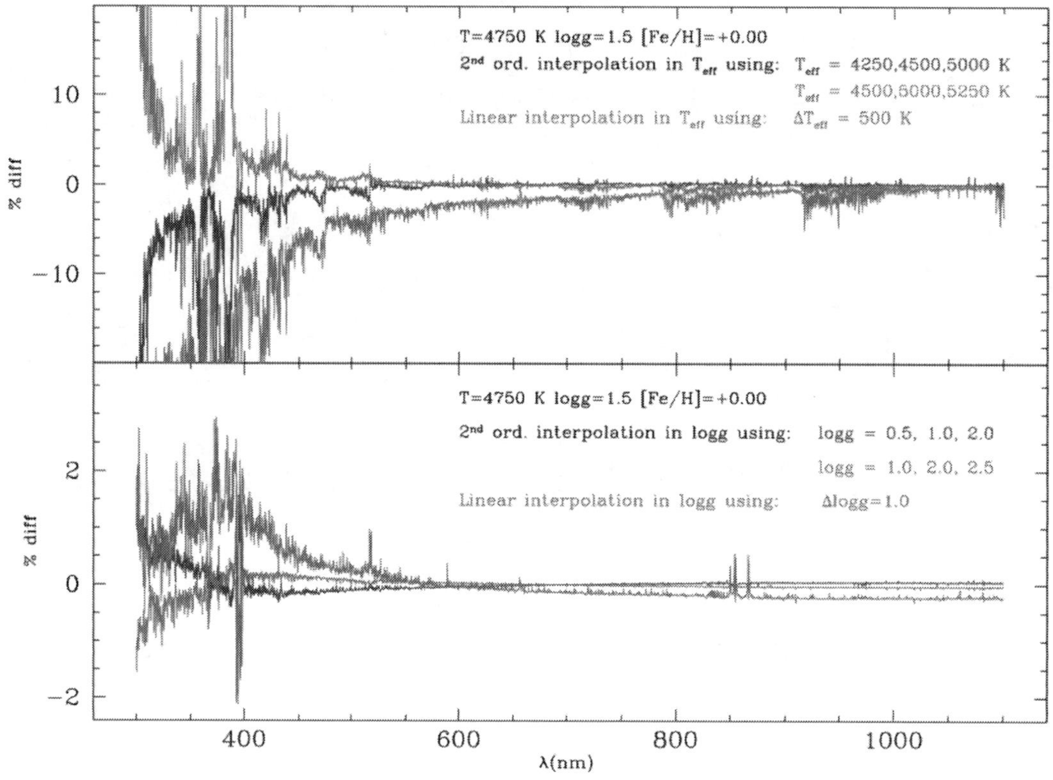

Figure 3. Results of interpolation tests (see Section 3.1). The rebuilding test for one of the MARCS grid nodes with 1 nm sampling is shown for different interpolators (in 1-D, linear or cubic) and compared to the original spectrum. The interpolation is performed in $T_{\rm eff}$ (top panel) or $\log g$ (bottom panel).

and photometric measurements. Extinction is applied to the spectra being one of the *desiderata* of the final Gaia catalog.

3.1. Interpolation procedure

The preparation of training data involves interpolation. The available libraries are computed with fine, but discrete, steps. Each library is simulated by adding the extinction at discrete steps to test the behavior of the simulated spectrum for various parameters. No preferred choice of libraries is made, i.e., all are simulated even in overlapping regions. Next, the libraries are interpolated inside their domain to create a continuous coverage of the AP space. We neither extrapolate nor mix different libraries and linearly interpolation them. As mentioned before, [24] provided an upper error limit of ~10% introduced by this method. However, the tests were limited to the RVS domain for larger resolution. We can therefore consider the error value valid for the RVS wavelength region. To test the interpolation procedure for libraries with 0.1 nm sampling, we computed a subgrid of synthetic spectra with the Linux porting of the Kurucz's codes ([30, 31] both at the nodes and at half of the nodes for $T_{\rm eff}$, $\log g$, and [Fe/H], where

$T_{\rm eff}$=3500, 5000, 8000, 15 000 K, $\log g$=3.0, 4.5, and [Fe/H]=−2.0, +0.0. Next, the semi-nodes spectra are reproduced by linear interpolation and compared to the properly computed ones. The interpolation of [Fe/H] is only tested around solar metallicity. The interpolation introduces an additional uncertainty of at most ∼10% around 3500 K, which rapidly drops to ∼4% around 5000 K, and even less at larger $T_{\rm eff}$.

A problem with the MARCS library results from missing nodes due to problems with the convergence of the model. It considerably limits the final AP space of the simulations, especially at small $T_{\rm eff}$ and/or very low gravity, for which model convergence are a problem. We tested the possibility to compute by interpolation at least some of the missing nodes. The test is shown in Fig. 3 for an existing node of $T_{\rm eff}$=4750 K, $\log g$=1.5, and [Fe/H]=+0.0, computed with 1-D linear interpolation (from 2 adjacent existing nodes) or by cubic interpolation (with a parabola using 3 existing nodes). The tests show the very limited error introduced by cubic interpolation in $T_{\rm eff}$ *or* $\log g$. If we consider $\lambda > 400$ (relevant for Gaia) the error is less than 10%. Tests at lower temperatures (in the Ultra cool dwarfs regime, see Table 1) instead favor linear interpolation. The non-linear effects become too large by mixing spectra with a temperature difference larger than 2 steps (equal to 200 K), which occurs with cubic interpolation.

4. Comparison of the libraries

An inspection of Fig. 1 shows the simultaneous coverage of some regions of the parameter space by different spectral libraries. Different codes adopt different approaches for handling the same physical process in the same physical conditions. Moreover, they can use different atomic or molecular line lists (see Plez, this Volume) or different abundance scales. The BASEL and Munari libraries use Kurucz's 1-D codes with very basic assumptions (even if NLTE effects are introduced in the calculations). Their advantages and drawbacks are well-known in the literature (see among others [8, 32]). Other codes focus on specific AP regions and introduce a more advanced treatment of specific processes, as described above. It is difficult to correctly compare spectra from different libraries. It is generally known that for an observed spectrum an ad hoc tuning of the computation parameters is possible that allows each code to produce a model in good agreement with the observation. A set of reference stars, critically compiled and with well-determined parameters for comparison, is missing. The definition of 'well-determined parameters' is ambiguous, also discussed in Heiter's contribution (this Volume). 'Who decides which ones are *the* parameters?' was asked at this workshop. Binary stars can be used to pinpoint the model atmospheres with some caveats (see [33, 34, 35]), but the question is still open and perhaps awaiting interferometry and asteroseismology.

The problems with the differences between different libraries, not only in the Gaia context, have been addressed in various papers, both for details about line reproduction and in terms of broad-band colors. We quote among others: [8] where Main Sequence stars are reproduced using spectra from different libraries and their colors are compared with photometric calibrations. They find that discrepancies are present for all libraries, including peculiar behaviors. [32] test colors and Lick indexes for some of the libraries presented here ([26]) using broad-band colors compared to observations, and find an agreement of up to $0^{\rm m}.1$ for optical colors. In Fig. 4 the differences among spectra belonging to different libraries are shown for a cool dwarf ($T_{\rm eff}$=4000 K). The upper panel shows the original spectra with full resolution. The BASEL spectrum is only shown for comparison. The huge difference in the slope compared to all synthetic libraries is certainly something to consider; given the fact that the BASEL library is calibrated using broad-band colors of real stars ([25]), the spectral continuum should properly reproduce the actual one used in the literature. The resolving power for the BASEL library is very different from the other libraries and causes a fraction of the apparent discrepancy, however not completely, and it certainly does not explain the different slope of the spectrum at longer wavelengths. The residuals of each library to MARCS are shown slightly smoothed to simplify

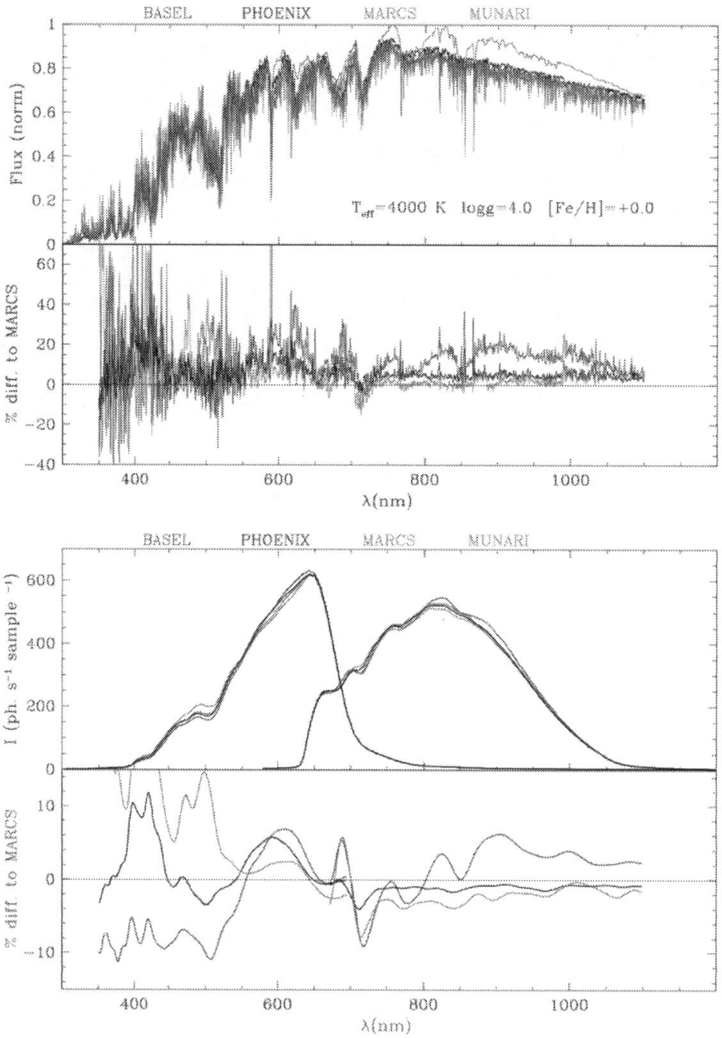

Figure 4. Comparison of spectra from different libraries computed for the reported parameters. The BASEL spectrum is plotted for comparison. In the top panels, the synthetic spectra are plotted and the differences with the MARCS spectrum are shown (smoothed for better visualization, using the same color code). The differences are significant, especially in the blue part of the spectrum. The computation of molecular bands is clearly still a problem. The bottom panels show the GOG simulations of the same spectra and the differences with the MARCS spectrum simulations. Due to the lower resolution most of the differences are washed out, but residuals of up to 10% remain.

the plot. The differences are significant for matching molecular bands at all wavelengths. In the red part of the spectrum the slopes agree, but an offset remains. In the lower panels the same comparison is shown for GOG simulations of Gaia spectra. The huge differences in the blue decrease to 10%, while in the red the libraries are very similar. While the cool stars represent

Journal of Physics: Conference Series **328** (2011) 012006 doi:10.1088/1742-6596/328/1/012006

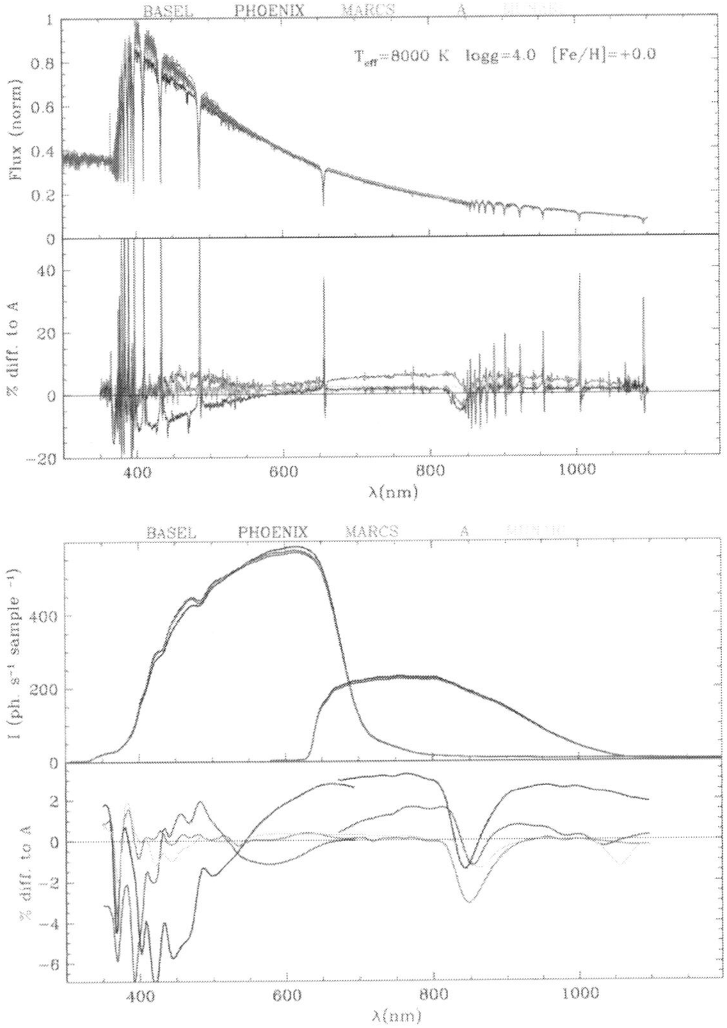

Figure 5. As in Fig. 4, but for a star of larger temperature. Differences are less significant but still present. Note the different slope of the PHOENIX library (top panel).

a problem for the models, the A-type stars are supposed to be well-known. In Fig. 5 the same comparison is shown for $T_{\rm eff}$=8000 K. Except for the low-resolution BASEL spectrum, of course, the lines reproduction is in good agreement among the libraries (residuals are computed with respect to the A stars library). However, the slope of the PHOENIX spectrum is very different. The GOG simulations are performed at constant apparent magnitude (i.e. same area). The strong differences in the blue region do not only translate into differences in the blue part of the simulated spectrum, but also affect the overall simulation. It results in the offset of PHOENIX residuals shown in the bottom panel.

Journal of Physics: Conference Series **328** (2011) 012006 doi:10.1088/1742-6596/328/1/012006

4.1. Impact on algorithms
Given all of this, Gaia will rely on the available models and the algorithms, however complex, and will classify and derive parameters compared to them. The differences among libraries appear to be small at first glance, while in Fig. 2 variations of temperature and gravity appear to have a stronger effect. However, Gaia is supposed to determine $\log g$ with a precision of 0.5 dex for cool stars, i.e *1/4* of the variations shown in the gravity sequence. So, even small differences among the input libraries can have an effect, which must be considered. Tests are now ongoing to determine possible systematic errors due to the mixing of libraries. The DSC algorithm appears to be affected by it, since it assigns to the simulated spectra of the A-stars library a lower rate of misclassification than to the BASEL and MARCS spectra in overlapping $T_{\rm eff}$ regions. Its causes are still under investigation. A first test with GSP-Phot uses a subsample of the BASEL library simulations for training. Next, the algorithm is applied to the remaining BASEL or MARCS library simulations. The conclusion is that a possible problem exists; the dispersion in the determination of APs (compared to the true values) in the BASEL-MARCS case becomes larger by $\sim$30%, and worse for K-type stars. Further tests, not involving BASEL with its very low initial resolution, will explore the issue further and give hints to the providers of the libraries, possibly allowing them to focus their efforts on strongly affected AP regions.

5. Single Stellar Population
Gaia will observe 10^7 unresolved galaxies. Single Stellar Populations (SSPs) are an essential tool to model and interpret the spectral energy distribution of clusters and galaxies. In an SSP, all stars are coeval, share the same chemical composition and the relative proportion among them is set by the Initial Mass Function (IMF). For a given age and metallicity, theoretical stellar models are first used to compute the distribution of stars along the isochrone following an IMF. Next, the appropriate stellar spectrum is assigned to each star and summed up to obtain the integrated light. To simulate a galaxy, SSPs can be combined following a Star Formation History (SFH) and a chemical evolutionary model. For assigning the spectral energy distribution to stars we use the Gaia libraries listed in Table 1. The first part of the table lists the already implemented libraries, while the inclusion of other libraries is ongoing. With the code presented by [37] we compute different SSP datasets, coupling different sets of isochrones (Padova family, [23], [38], [39], and the Teramo family [40]) with the appropriate Gaia libraries according to chemistry (see [41]). Mixing different spectral libraries is always a problem given the above discussed differences. In the literature this is usually a problem of resolution and of coverage. Thanks to the Gaia providers this problem has been overcome. The differences among the different libraries decrease with increasing temperatures. We carefully avoid mixing different libraries at smaller $T_{\rm eff}$, but instead compute different SSP datasets for a given isochrone. It allows us to further compare the libraries in a different way. Extensive tests on the results (Lick indexes calculation and comparison with observed clusters) are ongoing (see [42] for preliminary results). We find for example in Fig. 6 a very good agreement between the Lick indexes computed on our SSPs (Padova isochrones and Gaia libraries dataset) and the observational data of Globular Clusters from [36]. We show the Hβ-Mg2 diagram, where Hβ is a good age indicator for ages above 3 Gyr, while Mg2 is both sensitive to metallicity and α-enhancement.

6. Conclusions
We present a homogeneous set of spectral libraries computed with state-of-the-art codes that are currently used in the Gaia community. We compare them for three cases representative of cool, A-type, and hot stars, and find noticeable differences between them. The impact of these differences on the Gaia algorithms that aim to classify and characterize the observed sources, are non-negligible. We are implementing the different spectral libraries into SSP codes, mixing them in different ways but avoiding a mixture inside critical regions of the HR-diagram. The

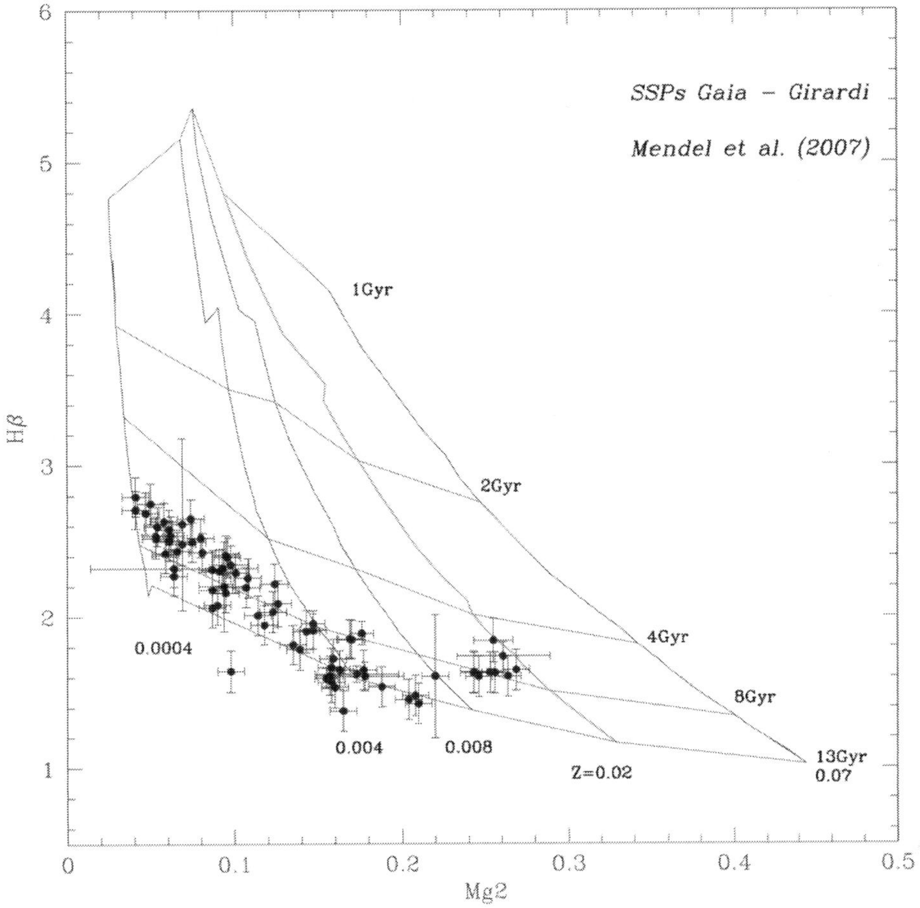

Figure 6. Comparison in the Mg2-Hβ diagram between the indexes computed for the SSP grid (Gaia libraries and Padova isochrones) at different ages and metallicities, and with data of Globular Cluster observations of [36].

synthetic spectral libraries and the SSPs will become available to the community through the Italian Space agency Science Data Center website (ASDC, `www.asdc.asi.it`).

Acknowledgments

The authors would like to thank the referee Dr. U. Heiter (Dept. of Physics and Astronomy, Uppsala Univ., Uppsala, Sweden) for the valuable comments.

References

[1] Perryman M A C, Lindegren L, Kovalevsky J, Hoeg E, Bastian U, Bernacca P L, Crézé M, Donati F, Grenon M, van Leeuwen F, van der Marel H, Mignard F, Murray C A, Le Poole R S, Schrijver H, Turon C, Arenou F, Froeschlé M and Petersen C S 1997 *A&A* **323** L49–L52

[2] Fioc M and Rocca-Volmerange B 1997 *A&A* **326** 950–962

[3] Vallenari A and Sordo R 2011 *MmSAIS* **18** 205–+

Journal of Physics: Conference Series **328** (2011) 012006 doi:10.1088/1742-6596/328/1/012006

[4] Tsalmantza P, Kontizas M, Bailer-Jones C A L, Rocca-Volmerange B, Korakitis R, Kontizas E, Livanou E, Dapergolas A, Bellas-Velidis I, Vallenari A and Fioc M 2007 *A&A* **470** 761–770
[5] Tsalmantza P, Kontizas M, Rocca-Volmerange B, Bailer-Jones C A L, Kontizas E, Bellas-Velidis I, Livanou E, Korakitis R, Dapergolas A, Vallenari A and Fioc M 2009 *A&A* **504** 1071–1084
[6] Munari U 1999 *Baltic Astronomy* **8** 73–96
[7] Steinmetz M, Zwitter T, Siebert A, Watson F G, Freeman K C, Munari U, Campbell R, Williams M, Seabroke G M, Wyse R F G, Parker Q A, Bienaymé O, Roeser S, Gibson B K, Gilmore G, Grebel E K, Helmi A, Navarro J F, Burton D, Cass C J P, Dawe J A, Fiegert K, Hartley M, Russell K S, Saunders W, Enke H, Bailin J, Binney J, Bland-Hawthorn J, Boeche C, Dehnen W, Eisenstein D J, Evans N W, Fiorucci M, Fulbright J P, Gerhard O, Jauregi U, Kelz A, Mijović L, Minchev I, Parmentier G, Peñarrubia J, Quillen A C, Read M A, Ruchti G, Scholz R, Siviero A, Smith M C, Sordo R, Veltz L, Vidrih S, von Berlepsch R, Boyle B J and Schilbach E 2006 *The Astron. J.* **132** 1645–1668
[8] Munari U, Sordo R, Castelli F and Zwitter T 2005 *A&A* **442** 1127–1134
[9] Bouret J, Lanz T, Frémat Y, Martins F, Lefever K, Blomme R, Martayan C, Neiner C, Quinet P and Zorec J 2008 *Revista Mexicana de Astronomia y Astrofisica Conference Series* vol 33 pp 50–50
[10] Kochukhov O and Shulyak D 2008 *Contributions of the Astronomical Observatory Skalnate Pleso* **38** 419–420
[11] Shulyak D, Tsymbal V, Ryabchikova T, Stütz C and Weiss W W 2004 *A&A* **428** 993–1000
[12] Gustafsson B, Edvardsson B, Eriksson K, Jørgensen U G, Nordlund Å and Plez B 2008 *A&A* **486** 951–970
[13] Brott I and Hauschildt P H 2005 *The Three-Dimensional Universe with Gaia* (*ESA Special Publication* vol 576) ed C Turon, K S O'Flaherty, & M A C Perryman pp 565–+
[14] Allard F e a 2011 in preparation
[15] Alvarez R and Plez B 1998 *A&A* **330** 1109–1119
[16] Martayan C, Frémat Y, Blomme R, Jonckheere A, Borges M, de Batz B, Leroy B, Sordo R, Bouret J, Martins F, Zorec J, Neiner C, Nazé Y, Alecian E, Floquet M, Hubert A, Briot D, Miroshnichenko A, Kolka I, Stee P, Lanz T and Meynet G 2008 *SF2A-2008* ed C Charbonnel, F Combes, & R Samadi pp 499–+
[17] Castanheira B G, Kepler S O, Handler G and Koester D 2006 *A&A* **450** 331–337
[18] Hindson L, Napiwotzki R, Heber U and Lemke M 2008 *Hot Subdwarf Stars and Related Objects* (*Astronomical Society of the Pacific Conference Series* vol 392) ed U Heber, C S Jeffery, & R Napiwotzki pp 163–+
[19] Korn A, Edvardsson B and Gustaffson B 2009 priv. comm.
[20] Frémat Y, Zorec J, Hubert A M and Floquet M 2005 *A&A* **440** 305–320
[21] Josselin E 2008 priv. comm.
[22] de Laverny P 2011 priv. comm.
[23] Bertelli G, Bressan A, Chiosi C, Fagotto F and Nasi E 1994 *A&AS* **106** 275–302
[24] Zwitter T, Castelli F and Munari U 2004 *A&A* **417** 1055–1062
[25] Lejeune T, Cuisinier F and Buser R 1997 *A&AS* **125** 229–246
[26] Sordo R, Vallenari A, Bouret J, Brott I, Edvardsson B, Frémat Y, Heber U, Josselin E, Kochukhov O, Korn A, Lanzafame A, Martins F, Schweitzer A, Thévenin F and Zorec J 2009 *MmSAI* **80** 103–+
[27] Luri X, Babusiaux C and Masana E 2005 *The Three-Dimensional Universe with Gaia* (*ESA Special Publication* vol 576) ed C Turon, K S O'Flaherty, & M A C Perryman pp 357–+
[28] Smith K W, Bailer-Jones C A L, Klement R J and Xue X X 2010 *A&A* **522** A88+
[29] Bailer-Jones C A L 2010 *MNRAS* **403** 96–116
[30] Sbordone L, Bonifacio P and Castelli F 2007 *IAU Symposium* (*IAU Symposium* vol 239) ed T Kuroda, H Sugama, R Kanno, & M Okamoto pp 71–73
[31] Kurucz R L 1992 *Revista Mexicana de Astronomia y Astrofisica* **23** 45–+
[32] Martins L P and Coelho P 2007 *MNRAS* **381** 1329–1346
[33] Siviero A, Munari U, Sordo R, Dallaporta S, Marrese P, Zwitter T and Milone E F 2004 *A&A* **417** 1083–1092
[34] Tomasella L, Munari U, Siviero A, Cassisi S, Dallaporta S, Zwitter T and Sordo R 2008 *A&A* **480** 465–473
[35] Tomasella L, Munari U, Cassisi S, Siviero A, Dallaporta S, Sordo R and Zwitter T 2008 *A&A* **483** 263–270
[36] Mendel J T, Proctor R N and Forbes D A 2007 *MNRAS* **379** 1618–1636
[37] Tantalo R 2005 *The Initial Mass Function 50 Years Later* (*Astrophysics and Space Science Library* vol 327) ed E Corbelli, F Palla, & H Zinnecker pp 235–+
[38] Girardi L, Bertelli G, Bressan A, Chiosi C, Groenewegen M, Marigo P, Salasnich B and Weiss A 2003 *MmSAI* **74** 474
[39] Marigo P, Girardi L, Bressan A, Groenewegen M A T, Silva L and Granato G L 2008 *A&A* **482** 883–905
[40] Pietrinferni A, Cassisi S, Salaris M and Castelli F 2006 *ApJ* **642** 797–812
[41] Sordo R e a 2011 in preparation
[42] Sordo R, Vallenari A, Tantalo R, Allard F, Blomme R, Bouret J, Brott I, Fremat Y, Martayan C, Damerdji Y, Edvardsson B, Josselin E, Plez B, Kochukhov O, Kontizas M, Munari U, Saguner T, Zorec J, Schweitzer A and Tsalmantza P 2010 *Astrophys. Space. Sci.* **328** 331–335

Journal of Physics: Conference Series **328** (2011) 012007 doi:10.1088/1742-6596/328/1/012007

The metallicity scale of dwarf and giant stars

Giancarlo Pace

Centro de Astrofsica, Universidade do Porto, Rua das Estrelas, 4050-098 Porto, Portugal

E-mail: gpace@astro.up.pt

Abstract. Differences between metallicity measurements of dwarfs and giants belonging to the same open clusters, as envisaged, would have tremendous implications, among other things, on the planet formation theory and on the stellar populations in the Milky Way. They also have a high importance on their own right, as they may indicate either that there are important processes presently not taken into account in modelling the atmosphere of evolved stars, or that the composition of the stars changes when they leave the main sequence. However, they have not yet been evaluated based on solid and conclusive data. We review literature data, employing different models and line lists, in order to investigate the extent, the origin, and the nature of the aforementioned differences.

1. Introduction

Based on abundance measurements from stars in open clusters at different evolutionary stages, several authors suggested that a difference between dwarfs and giants may arise in the analysis. Some of the authors [1] also state that the differences depend on the particular method used, pointing to systematic errors in the analysis rather than to true abundance differences responsible for this result. This, if confirmed, is not only relevant for its own sake, as it urges for the improvement of stellar atmosphere models, but also has tremendous implications for several fields of astrophysics, since various important studies rely on the comparison of metallicities between dwarf and giants, which, if our hypothesis is confirmed, may lead to spurious results.

2. Implications on the planet formation

Pasquini et al. [2] noticed that the metallicity distribution of planet host giant stars is comparable to that of the overall giant population, while it has long been recognized that planet host dwarfs tend to be much more metal rich than main sequence stars without detected planets. This result was later confirmed [3]. Two classes of scenarios have been invoked to explain the higher metal content of planet host dwarfs: the pollution scenario and the higher probability of metal rich discs forming planets. In the former, planet host dwarfs accrete discs whose composition had been previously modified by the planet formation process, thus accumulating metals only in their convective envelopes. When the stars evolve off the main sequence, the deepening of the convective layer washes out the metal enhancement by mixing enriched and unprocessed material, thus bringing the observed abundance closer to the original one. On the contrary, if planet host stars are intrinsically more metal rich, we expect no accumulation of metals in the external layers, and when the star evolves off the main sequence, the observed chemical abundance is expected to remain unaltered. That is why the result of Pasquini et al. [2] is decisive in favoring the pollution scenario, provided that a direct comparison between abundance measurements in dwarfs and giants is possible and does not lead to spurious results.

3. Implications on the Galactic chemical evolution

Taylor and Croxall [4] noticed that the distribution of metallicity of giant stars in the solar neighborhood is narrower than that of the dwarfs in the same volume of space. Later surveys [3, 5] confirmed that the metallicity distribution of giants lacks the metal-rich tail, for [Fe/H]>0.2 dex, on the contrary to that of dwarfs stars.

Again, we should ask ourselves whether we are facing an important piece of evidence shedding light on the chemical evolution of the Galaxy, or an artifact created by the comparison of different classes of objects.

As a matter of fact, the comparison of different classes of objects already lead to conclusions that had to be revised later on. Early studies claimed that the Galactic bulge is metal-poor. This would have important theoretical consequences, because a lack of metal-poor G-dwarfs in the solar neighborhood is a major discrepancy between observations and predictions of closed-box models, i.e. models that do not include the inflow or outflow of gas. Such a phenomenon, commonly referred to as the G-dwarf problem, would then be absent or less severe in the bulge. However, later on it has become widely accepted that the metallicity distributions of bulge stars and solar neighborhood stars are very similar, and this through studies employing rich samples of giants in both populations [6,7,8], or dwarfs. The study of bulge dwarfs with high-resolution spectroscopy is only possible by exploiting the micro-lensing effect [9].

Another important probe of the Galactic chemical evolution is the metallicity gradient of the Galactic disc, which is best measured through chemical analysis of open-cluster members. Most of the chemical analyses in open clusters are based on giants, especially the older ones, which are typically distant. The BOCCE project was undertaken to study the properties of a sample of mostly old Galactic open clusters [10]. It targeted red clump stars. It is of crucial importance to know how to use the results of this kind of survey, in order to understand whether the chemical composition that we measure in giants truly reflects that of the molecular cloud out of which the open cluster formed. Observations in open clusters are the only tool that allows us to directly investigate the differences between giants and dwarfs. The state of the art and the foreseeable developments about data in open clusters is the topic of the next Section.

4. State of the art of the observations in open clusters

The advent of the 8-meter class telescopes boosted chemical composition analyses of stars in open clusters through high-resolution spectroscopy. Several single studies and coordinated efforts have been dedicated to this effort. To the best of my knowledge, the most updated compilation is given in [11].

In Santos et al. [1] data about 6 clusters with observations of both evolved and unevolved stars were examined. Systematic differences in iron content were found between dwarf and giants, which depended on both the metallicity of the cluster and the list of spectral lines used to determine the chemical abundances. When the list of Sousa et al. [12] was used, the iron abundances in metal rich giants were overestimated with respect to dwarfs, while the opposite occurs for metal-poor giants. The results obtained by employing the line list of Hekker & Melendez [13], on the contrary, lead to an underestimation of the metallicity in giants with respect to dwarfs, of which the degree was proportional to the metallicity itself. Thus, it is important in this type of analysis to analyze the data in a strictly homogeneous way, because one of the effects to detect and correct for is the dependence of the results on the approach adopted.

Carrera & Pancino [11] have recently published results of chemical abundances from high-resolution spectroscopy of evolved stars in 4 open clusters: Berkeley 32, Praesepe, Hyades and NGC 752. For Praesepe and Hyades, abundance measurements from dwarf stars are also available in the literature. These clusters are rich and close, therefore they are well studied down to their main sequence.

As far as the Hyades are concerned, all the studies except one [14], provide [Fe/H] between 0.11 and 0.14 dex, independent on whether they use giants [11] or dwarfs [15, 16, 17, 18]. Therefore, Hyades data do not point to any difference in the metallicity scale between dwarf and giants. However, a homogeneous analysis of stars of both types in this cluster is still not available.

Contrary to the Hyades, iron abundance measurements in *Praesepe* stars show a remarkably large range of values, from solar up to 0.27±0.1 dex in [19]. The latter value is based on 7 dwarfs studied with high-resolution and large signal-to-noise ratio spectra. Carrera & Pancino [11] measure for Praesepe [Fe/H]=0.16 ±0.05 based on 3 giants. The difference between dwarfs in Praesepe measured by Pace et al. and giants measured by Carrera & Pancino, are just what we expected on the basis of the trend in differences found by Santos et al. when using the line list of [12]. We note, however, that such a trend depends on the particular line list used for determining the abundance. As a matter of fact, An et al. [19] also analyzed a sample of main sequence stars in Praesepe using high-quality spectra, and their metallicity is slightly lower than that of Carrera & Pancino. Even lower metallicities are found by other authors who analyzed dwarfs in Praesepe with [Fe/H]=0.04±0.04 dex [21].

Clearly, the dependences of the results on the used method has to be properly understood before drawing conclusions about the differences between dwarfs and giants, and this has to be done by analyzing homogeneously available data, and by testing different models and line lists.

4.1. Abundances of aluminum and sodium

Overabundances of aluminum and sodium in open cluster giants, were found by several authors. In the following clusters overabundances of both elements were found : IC4756, NGC6939, and NGC7142 [22], NGC6475 [23], Collinder261 [24], Berkeley17 [25], Saurer1, and Berkeley29 [26]. In some other clusters, enhancements were found only in sodium: IC4651 [27], M67 [28], NGC7789 [29], NGC1817, NGC1883, NGC2141, and NGC2158 [30]. In some of the aforementioned references silicon enhancements were also measured. These overabundances are either directly attributed to the difference between giants and dwarfs, or to a large abundance ratio: [Na-Al/Fe]>0.1 dex. Heavier neutron-capture elements are believed to remain unchanged during the normal course of stellar evolution, therefore they have been used to check that the molecular clouds out of which open clusters were formed were homogeneous [31]. On the contrary, the overabundances of sodium and aluminum could be the result of deep mixing with layers that underwent the NeNa cycle of hydrogen burning, as suggested, for instance, by [29].

They therefore deserve close attention, but a thorough test must be performed in order to check whether some non-LTE effect stronger in giants than in dwarfs is affecting our analysis.

5. Conclusions

It is of fundamental importance to carry out chemical abundance measurements of both dwarfs and evolved stars for as many open clusters as possible. It would guarantee the accurate knowledge of the chemical gradient in the Galactic disc, and, at the same time, would finally answer the question whether the metallicity scales of giants and dwarfs are different. The Gaia-ESO survey will provide a wealth of precious data.

Acknowledgments

The author is supported by grant SFRH/BPD/39254/2007 and by the project PTDC/CTE-AST/098604/2008, funded by Fundação para a Ciência e a Tecnologia (FCT), Portugal. The author also warmly thanks Dr. E. Pancino (INAF-Osservatorio Astronomico di Bologna), who carefully peer reviewed this paper.

References

[1] Santos N C, Lovis C, Pace G., *et al.* 2009 *A&A* **493** 309
[2] Pasquini L, Döllinger M P, Weiss A, *et al.* 2007 *A&A* **473** 979
[3] Takeda Y, Sato B and Murata D 2008 *PASJ* **60** 781
[4] Taylor B J and Croxall K 2005 *MNRAS* **357** 967
[5] Luck R E and Heiter U 2007 *AJ* **133** 2464
[6] Alves-Brito A, Meléndez J, Asplund M, *et al.* 2010 *A&A* **513** A35
[7] Meléndez J, Asplund M, Alves-Brito A, *et al.* 2008 *A&A* **484** L21
[8] Ryde N, Gustafsson B, Edvardsson B, *et al.* 2010 *A&A* **509** A260000
[9] Bensby T, Asplund M, Johnson J A, *et al.* 2010 *A&A* **521** L57
[10] Bragaglia A and Tosi M 2006 *AJ* **131** 1544
[11] Carrera R and Pancino E 2011 *Preprint arXiv:1107.2242*
[12] Sousa S G, Santos N C, Mayor M, *et al.* 2008 *A&A* **487** 373
[13] Hekker S and Meléndez J 2007 *A&A* **475** 1003
[14] Varenne O and Monier R 1999 *A&A* **351** 247
[15] Boesgaard A M and Friel E D 1990 *ApJ* **351** 467
[16] Paulson D B, Sneden C and Cochran W D 2003 *AJ* **125** 3185
[17] Sestito P, Randich S, Mermilliod J C and Pallavicini R 2003 *A&A* **407** 289
[18] D'Orazi V and Randich S 2009 *A&A* **501** 553
[19] Pace G, Pasquini L and François P 2008 *A&A* **489** 403
[20] An D, Terndrup D M, Pinsonneault M H, *et al.* 2007 *ApJ* **655** 233
[21] Friel E D and Boesgaard A M 1992 *ApJ* **387** 170
[22] Jacobson H R, Friel E D and Pilachowski C A 2007 *AJ* **134** 1216
[23] Villanova S, Carraro G and Saviane I 2009 *A&A* **504** 845
[24] Friel E D, Jacobson H R, Barrett E, *et al.* 2003 *AJ* **126** 2372
[25] Friel E D, Jacobson H R and Pilachowski C A 2005 *AJ* **129** 2725
[26] Carraro G, Bresolin F, Villanova S, *et al.* 2004 *AJ* **128** 1676
[27] Pasquini L, Randich S, Zoccali M, *et al.* 2004 *A&A* **424** 951
[28] Tautvaišiene G, Edvardsson B, Tuominen I and Ilyin I 2000 *A&A* **360** 499
[29] Tautvaišiene G, Edvardsson B, Puzeras E and Ilyin I 2005 *A&A* **431** 933
[30] Jacobson H R, Friel E D and Pilachowski C A 2009 *AJ* **137** 4753
[31] De Silva G M, Sneden C, Paulson D B, *et al.* 2006 *AJ* **131** 455

Journal of Physics: Conference Series **328** (2011) 012008 doi:10.1088/1742-6596/328/1/012008

Abundance analysis of post-AGB stars

Nadya Gorlova

Institute of Astronomy, Celestijnenlaan 200D, 3001, Leuven, Belgium

E-mail: nadya@ster.kuleuven.be

Abstract. Using BD+46 442 as an example, we describe our analysis of the spectra of post-AGB stars (SpTs F-K, luminosity types I-II, [Fe/H]=−2.0...+0.5), obtained with the échelle spectrograph HERMES on the 1.2 m Mercator telescope. We obtain atmospheric parameters and atomic abundances using hydrogen line profiles and equivalent widths (EW) of weak metal lines. Our oscillator strengths and solar abundances for the majority of elements have been consistently adjusted to match the observed solar spectrum. The EW analysis is performed using F. Castelli's modified ATLAS9 photospheric models and C. Sneden's radiative transfer code MOOG. The resulting abundances are compared to those obtained by using the original R. Kurucz's ATLAS9 models and the WIDTH9 code. The outlined procedure can be employed for the spectroscopic analysis of the high-resolution spectra of warm (super-)giant stars in ground-based follow-up programs of Gaia.

1. Introduction

Since 2009 we perform a spectroscopic survey of various types of evolved stars with the échelle spectrograph HERMES on the Flemish 1.2 m Mercator telescope at La Palma [1]. The main goal is to obtain radial velocities for investigating binarity, but the large signal-to-noise ratio (S/N) spectra also allow us to carry out detailed studies of the photospheric parameters, including detailed abundance determinations. BD+46 442 is one of the first single-lined spectroscopic binaries (with an orbital period of 140 days) discovered in this survey [2]. It was selected based on its infra-red excess characteristic of discs around binary post-AGB stars [3]. The disc is believed to have formed when the primary was an AGB star and transferred its outer layers to the companion via Roche overflow or a wind accretion mechanism. The discs are apparently very stable as they are found around a large fraction of the post-AGB population [4]. Furthermore, many of these systems show an interstellar-like abundance pattern in their atmospheres, namely, refractory elements with high condensation temperatures ($T_{\rm cond}$), such as iron, that are deficient compared to the near-solar more volatile elements such as zinc [5], [6]. It can result from the gas-dust fractionation in the disc and the subsequent re-accretion of the depleted gas onto the star [7].

Post-AGB binaries have since long been proposed to be progenitors of symbiotic stars, chemically-peculiar giants and asymmetric planetary and proto-planetary nebulae. While imaging data is abundant thanks to the fast advances in space- and ground-based observations, the information on the orbital parameters and the evolutionary status of these systems is far from complete [8], [9]. Gaia will establish luminosities and provide first-order estimates of the temperature of Galactic post-AGB stars, which is sufficient to put a single star on an evolutionary track. But only through ground-based long-term high-resolution surveys, such as ours, will it become possible to obtain information on binarity, metallicity, depletion, and nucleosynthesis in these stars. This information is essential for placing various groups of peculiar stars in the context of binary evolution.

In this contribution we outline a procedure for the chemical abundance determination of moderately metal-poor high-latitude F-G (super-)giants from high-resolution, large S/N optical spectra that can be employed in Gaia follow-up surveys. First, we briefly describe our spectral material, then the procedures for the determination of the basic photospheric parameters and chemical abundances, emphasizing deviations from local thermodynamical equilibrium (LTE) at low surface gravities. In Section 5 we evaluate the influence of the atmospheric model and the radiative transfer code on the computed abundance values.

2. Observations and data reduction

For the analysis we combined two HERMES exposures of BD+46 442 observed on the same night in August 2009 to obtain a S/N$\sim$130 spectrum. HERMES is a fiber-fed échelle spectrograph with an image slicer and an anti-fringe CCD coating. It provides a continuous coverage over λ =3800–9000 Å and a maximum spectral resolving power $R\sim80\,000$. Two sets of bias, flat, and arc frames are obtained at dawn and at dusk, and the nearest one is used for the calibration of the science target. The typical uncertainty in the wavelength calibration is a few hundredths of an angstrom and is limited by the number of arc lines in a given spectral order. The precision of the radial velocity is $\sim$0.2 km s^{-1}, limited by the instrument drift due to pressure fluctuations during the night. The data reduction is performed using a Python-based pipeline, while the radial velocity is determined via a cross-correlation method with a custom-made line mask (corresponding to a G2-type star in case of BD+46 442). More details on the instrument and software can be found in [10] and online at `http://www.mercator.iac.es/`.

3. Determination of the photospheric parameters and non-LTE effects

According to *Simbad*, BD+46 442 is an F2 III star. It provides an initial range of the effective temperature ($T_{\rm eff}$) from 6200 K to 8000 K. First, we attempted to use the line ratio method of Kovtyukh *et al* [11], [12] to determine $T_{\rm eff}$. We were however unable to identify any suitable pair of metal lines of their calibration method in our spectrum. The reason, as we explain below, is that most metals in BD+46 442 are depleted by a factor of five compared to the Sun. In addition, the lines are rather broad, with FWHM$\sim$25 km s^{-1}, which complicates the EW measurements.

We then turned to the hydrogen lines, which have the advantage over metal lines of being far less sensitive to metallicity and macro-broadening. In Fig. 1 we compare the Hδ line to the synthetic profiles of [13] and the Paschen 14 line to the models of [14]. We find the best agreement for $T_{\rm eff}$ =6000–6500 K and log $g\leq2$. The same result is obtained with Hβ, Hγ, and Pa17 lines. The other hydrogen lines are too noisy or blended in our spectra, while Hα could not be used due to contamination by circumstellar emission. Paschen lines are especially useful for eliminating large surface gravities; to obtain a satisfactory match with the models, however, we have to normalize the observed and the model profiles to pseudo-continuum regions between the lines, as the Paschen continuum flux is too high in the models. Both synthetic libraries [13], [14] are freely online available.

Next we performed an ionization balance analysis using Fe I and Fe II lines. We measured the EWs of metal lines using a line list with oscillator strengths (log(gf)) from V. Kovtyukh (2010, personal comm.), which is a slightly updated version of the list published in [15]. To eliminate blends, we consulted with the interactive spectral atlases in SpectroWeb by A. Lobel [16], [17] online available at `http://spectra.freeshell.org/spectroweb.html`. To convert EWs to abundance values, we used the LTE radiative transfer code MOOG of C. Sneden, available at `http://www.as.utexas.edu/~chris/moog.html`, and the ATLAS9 atmosphere models with the new opacity distribution functions of F. Castelli and R. Kurucz [18].

The typical behavior of iron lines with EW is shown in the left-hand panel of Fig. 2. The trend with EW can be removed by varying the value of the microturbulence velocity (V_{tur}), while the difference between the average iron abundance from Fe I and Fe II lines can be eliminated by varying the log g-value. However, as shown in the right-hand panel, strong Fe I lines require lower V_{tur}- and log g-values compared to Fe II lines. In addition, a plot of the iron abundance value from Fe I and Fe II lines versus the excitation energy of the lower level (Fig. 3), reveals that low-excitation Fe I lines require a larger T_{eff} (>7000 K) incompatible with the Fe II and hydrogen lines. This is a known effect in yellow supergiants [15] and post-AGB stars [19]. It is usually explained by over-ionization in Fe I lines compared to the LTE case. The effect becomes more pronounced at higher temperatures, lower gravities, and lower metallicities (see Bergemann, this Volume). The recipe for the LTE calculations is therefore to utilize ionic lines such as Fe II, and the high-excitation/weak lines of neutral lines. Adopting T_{eff}=6250±250 K, and extrapolating the Fe I abundance values to EW=0, we obtain for BD+46 442 log g=1.5±0.5, V_{tur}=4.0±0.5 km s^{-1}, and [Fe/H]=−0.7±0.2 (Fig. 2), with lower/higher T_{eff} corresponding to lower/higher log g- and [Fe/H]-values.

4. Chemical composition

The abundance values of all 22 elements determined in this work are shown in the upper panel of Fig. 4 where the lines are ordered according the EW-values, and the elements according the T_{cond}. The abundances are given relative to the solar values, or the adjusted photospheric abundances of [20] (see Section 5). We call this 2-D representation an 'abundogram', because it is a good diagnostic of non-LTE effects (such as the dependence of Fe I abundances on the EW-values) and of a possible depletion with T_{cond} (that would be manifested by the color gradient in the vertical direction). Although we do not find strong non-LTE effects in other elements

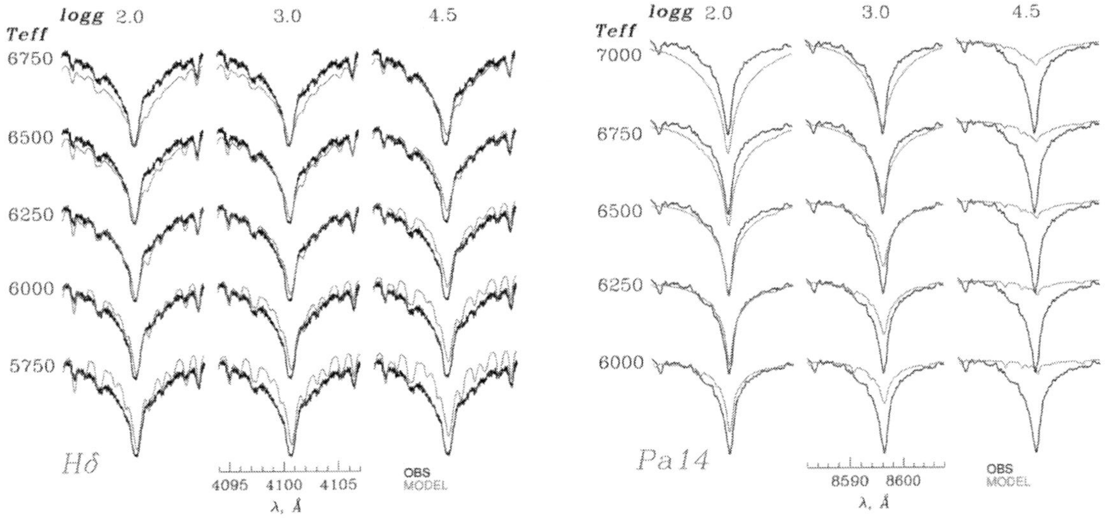

Figure 1. The observed hydrogen line profiles of the post-AGB star BD+46 442 (in black) overlaid with model line profiles (in red) of [13] and [14]. Hδ is shown for R=80 000, while Pa14 for R=20 000. Based on the comparison we estimate the effective temperature of BD+46 442 at 6250±250 K and the surface gravity log g below 3.

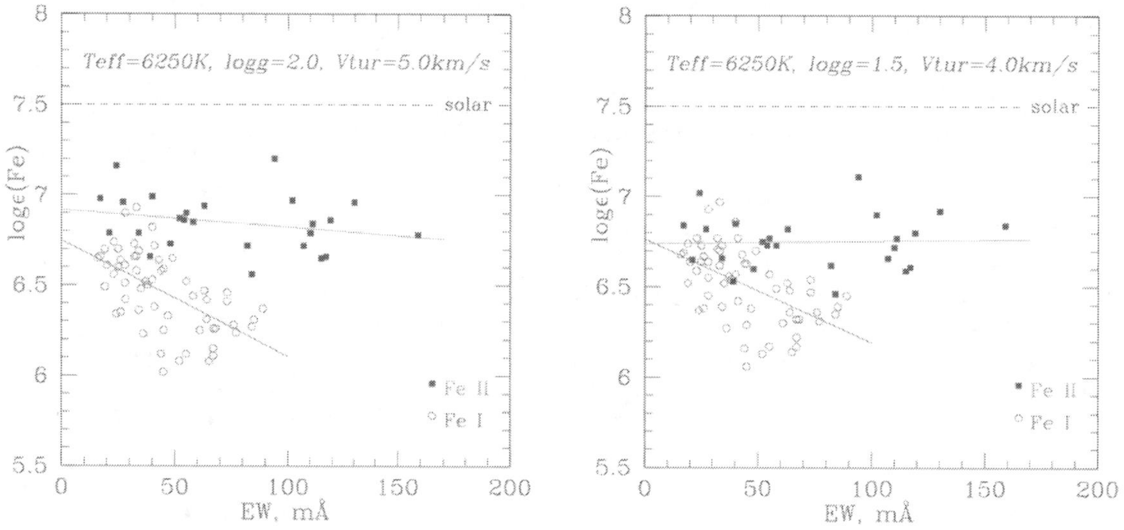

Figure 2. The iron abundance as a function of equivalent width of Fe I (in blue) and Fe II (red) lines separately, computed with MOOG for two sets of photospheric parameters. The right-hand figure represents our best fit model because the Fe II abundances do not show a dependence on the EW (due to the correct microturbulence velocity) and also match the abundances of weak Fe I lines (due to the proper value of the surface gravity). The abundance values from strong Fe I lines deviate because they are affected by non-LTE effects at low surface gravity.

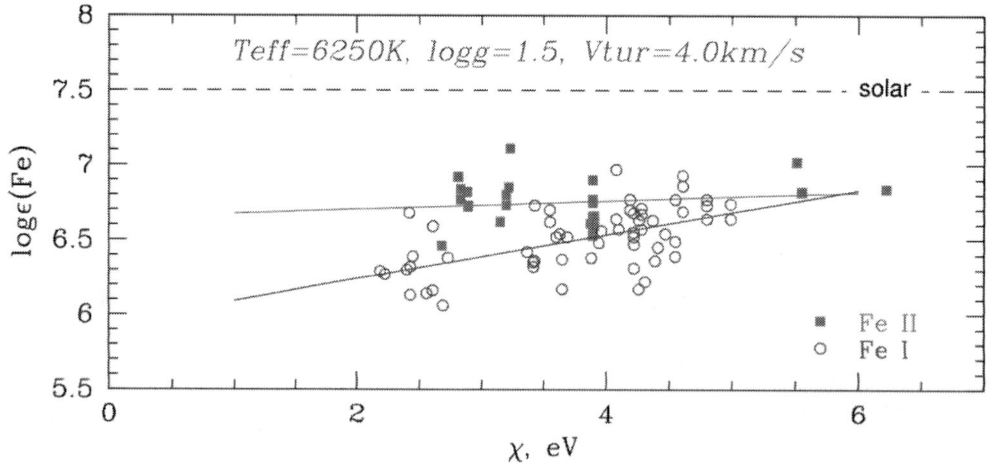

Figure 3. Iron abundance values from Fe I and Fe II lines as a function of the lower excitation potential (χ). This graph is often used to determine the effective temperature. However, Fe I lines show a trend with χ due to non-LTE effects. The Fe II lines, on the other hand, show no trend, confirming the validity of the $T_{\rm eff}$-value obtained from the hydrogen lines.

(except perhaps for the well-known case of N I), they can be masked by the small number of lines. For obtaining an average abundance per ionization stage, we divided all species into three groups: 1) ions that do not show a trend with EW, 2) neutrals for which at least three lines of EW≤50 mÅ are available, and 3) the remaining lines, and averaged them accordingly. The first group provides the most reliable abundance values, while the last group the least reliable.

The averaged abundances are represented in the lower panel of Fig. 4, where the symbol size signals the reliability of the abundance value. The error-bars represent the RMS of the average abundances. The plot reveals that the majority of metals are underabundant relative to the Sun by 0.7 – 1.2 dex. The more volatile elements such as C, S, and Na, on the other hand, show less underabundant values by ~0.3 dex. It could indicate a small depletion with $T_{\rm cond}$, a possible sign of contamination by material from the disc. On the other hand, a more plausible explanation for this pattern is that we measure the original composition of a thick disc star with enhanced abundances of α-process elements (Si, S, Mg, Ti). This is consistent with the low abundance of zinc, a metal with low $T_{\rm cond}$ [7] (although we have only one weak line for this element).

5. Comparison of ATLAS9 & Castelli, WIDTH9 & MOOG

Former studies, including those of Kovtyukh *et al*, employed an earlier version of the ATLAS9 atmosphere models and the LTE WIDTH9 code by R. Kurucz, available at `http://wwwuser.oat.ts.astro.it/castelli/sources/width9.html`. In particular, their log(gf)-values were adapted from the original values by forcing them to match the solar abundance values of [20] using EW-values they measured in the solar spectrum, and utilizing the canonical 5777 K/4.4/1.0 solar model of the earlier ATLAS grids. It may not be possible to re-calculate hundreds of objects that have been studied this way. A more practical approach is to evaluate the differences between their and our set of models and transfer codes, for both BD+46 442 and the Sun.

The result of the comparison for BD+46 442 is shown in Fig. 5. All abundances have been computed with the same line list, including the same oscillator strengths. In the left-hand panel we show the abundance line-by-line difference for the calculations with MOOG using two different atmosphere models: Castelli's (or 'ATLAS9 new', e.g., our adopted model) and the original ATLAS9 model. In the right-hand panel we show the combined difference in case both the atmosphere model and the transfer code differ from ours. The top row shows the difference for iron lines, while the bottom row for several other representative elements. From the comparison we find that the atmosphere model of Castelli yields systematically lower abundances by 0.07–0.11 dex, with the exception of nitrogen with a difference of −0.16 dex. The differences do not depend on the EW-values. On the other hand, using WIDTH9 instead of MOOG results in both negative and positive differences, but of a smaller amount: up to 0.03 dex, and the strong lines are more affected than the weak lines. Overall, the maximum uncertainty caused by the choice of the atmosphere model and transfer code is at most 0.15 dex for EW-values <170 mÅ. We perform a similar analysis for the Sun (the solar EWs are kindly provided by V. Kovtyukh) and obtain differences of the order of a few hundredths of a dex, mostly with negative sign, but which become positive for large EWs. These differences are still smaller in BD+46 442 than both the RMS scatter of the line-to-line abundance of a given element, and the uncertainties due to those in the atmosphere model parameters. The RMS scatter, caused by the uncertainties in the line oscillator strengths and the measured EWs, constitutes 0.1–0.2 dex (Fig. 4), while uncertainties caused by the atmosphere model parameters amount to 0.1–0.3 dex (for $\Delta T_{\rm eff}$=250 K, Δlog g=0.5 dex, see 2).

6. Summary

We have analyzed a S/N~130, R~80,000 spectrum of a post-AGB F giant with a dusty disc; a single-lined spectroscopic binary. The Balmer and Paschen line profiles allow us to estimate the effective temperature of 6250±250 K. The EWs of Fe I and Fe II lines enable us to simultaneously

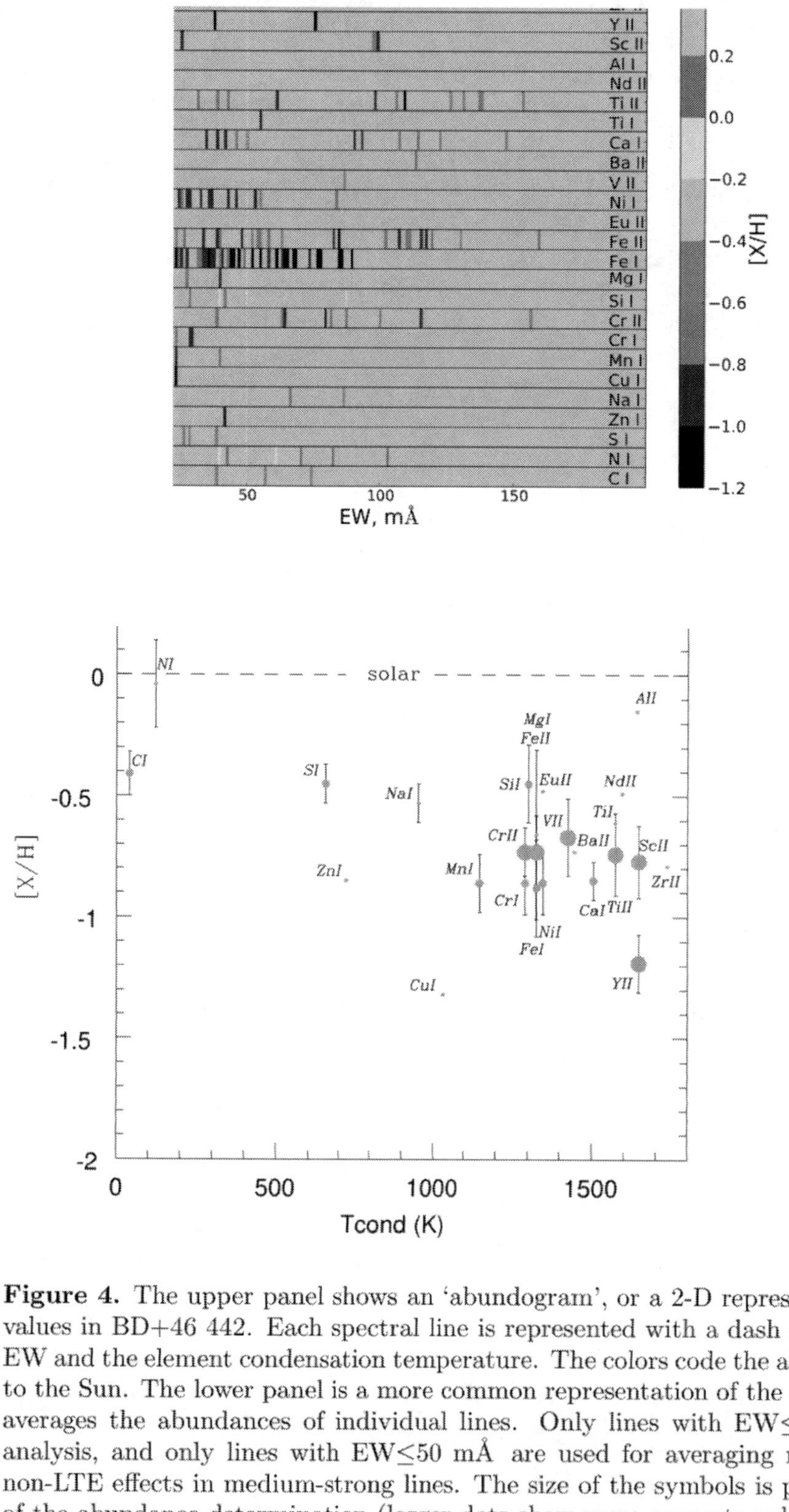

Figure 4. The upper panel shows an 'abundogram', or a 2-D representation of the abundance values in BD+46 442. Each spectral line is represented with a dash symbol at the appropriate EW and the element condensation temperature. The colors code the abundance values compared to the Sun. The lower panel is a more common representation of the chemical composition that averages the abundances of individual lines. Only lines with EW≤170 mÅ are used in our analysis, and only lines with EW≤50 mÅ are used for averaging neutral species because of non-LTE effects in medium-strong lines. The size of the symbols is proportional to the quality of the abundance determination (larger dots show more accurate values). In the plots, BD+46 442 appears to be a slightly metal-poor star ([X/H]∼−0.7) with enhanced abundances of the α-elements and without strong depletion of refractory elements characteristic for some post-AGB stars with discs.

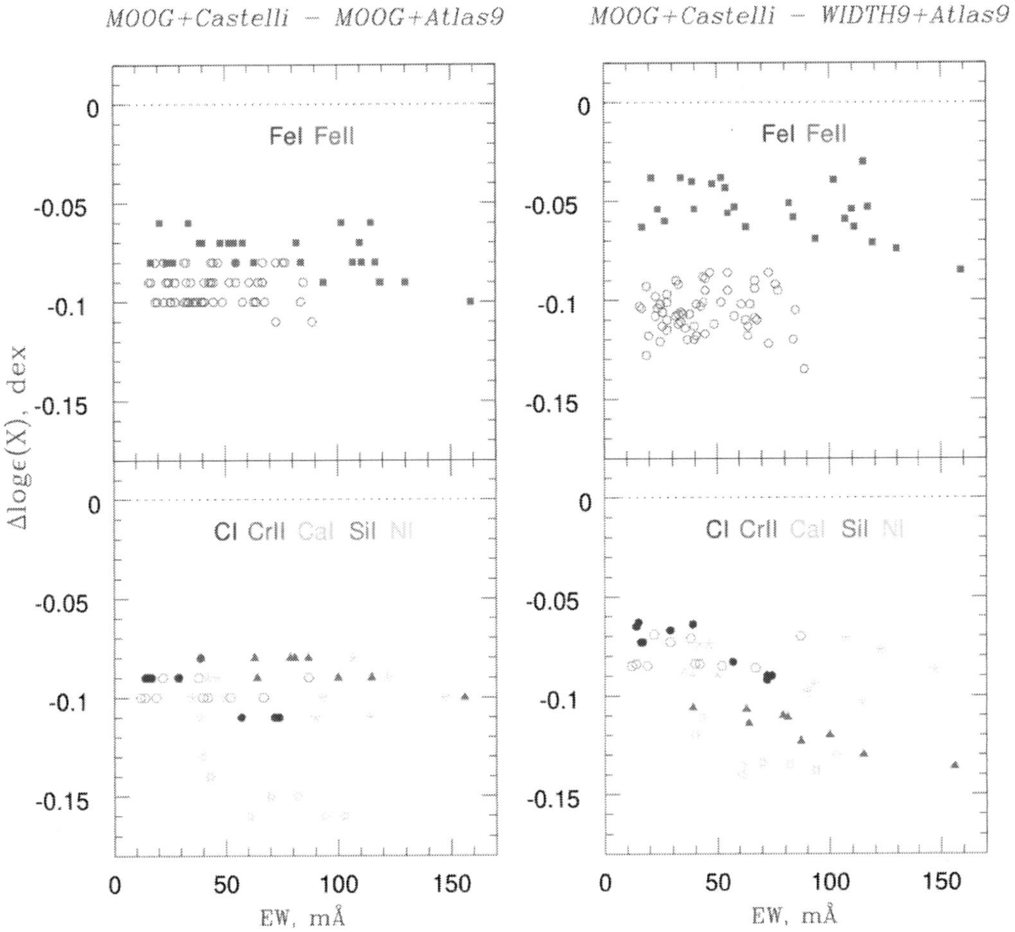

Figure 5. The two panels show the difference in abundances for BD+46 442 by computing abundances using the same line list and the same photospheric parameters but 1) different versions of the atmosphere model (left-hand panel) and 2) different atmosphere models and different transfer codes (right-hand panel). The comparison is shown for the iron lines and for a number of other representative elements. F. Castelli's atmosphere models result in systematically smaller abundances of ~0.1 dex compared to the original ATLAS9 models. The differences can decrease or increase by ~0.03 dex for a given element/ion and EW using WIDTH9 instead of MOOG. These differences are still smaller compared to errors due to uncertainties in the model parameters $T_{\rm eff}$ and log g. Similar effects, but to a smaller degree, are also observed in the Sun.

constrain the surface gravity of log g=1.5±0.5, [Fe/H]=−0.7±0.2 dex, and the microturbulence velocity of $V_{\rm tur}$=4.0±0.5 km s^{-1}. We determine the abundance values of 22 elements in total. For Ti, Cr, and Fe these values are obtained from two ionization stages. Lines with EW≤170 mÅ are used for the ions, while EW<50 mÅ for the neutrals. The dominant source of the abundance errors is the uncertainty in the effective temperature and surface gravity, followed by the uncertainty in the EW measurements and the employed oscillator strengths. Our analysis of Fe I lines confirms an urgent need for non-LTE corrections exceeding 0.2 dex to analyze lines with EW>50 mÅ. We also compare two radiative transfer LTE codes (MOOG and WIDTH9) and two versions of ATLAS9 atmosphere models. Hence, we detect maximum differences of 0.15 dex for N I lines. The errors caused by different models are generally larger than for different transfer codes, but they depend less on the measured EW-values. Despite all these uncertainties, our method (which was originally developed for massive supergiants) is adequate for detecting the metal-poor nature of post-AGB stars and to distinguish it from chemical depletion effects. Considering possible non-LTE effects in these stars, we propose a new 2-D representation of abundance values with EW and e.g., $T_{\rm cond}$ or atomic number. We call this 2-D abundance diagram an *abundogram*, complementing the traditional 1-D diagram of average abundance values. The outlined method can be employed for abundance determinations in low-gravity F-and G-stars with solar and sub-solar metallicity that will be observed in future ground-based surveys in support of the Gaia mission.

Acknowledgments
We thank the members of the HERMES consortium for maintaining the instrument and the excellent reduction pipeline, and all the observers who collect spectra in the Q-mode. The project is sponsored by the Fund for Scientific Research of Flanders (FWO), Belgium; the Research Council of K.U.Leuven, Belgium; the Fonds National Recherche Scientifique (FNRS), Belgium; the Royal Observatory of Belgium; the Observatoire de Geneve, Switzerland; and the Thüringer Landessternwarte Tautenburg, Germany. We also acknowledge support by the KULeuven research council (GOA/2008/04).

References
[1] Van Winckel H, Jorissen A, Gorlova, N, Dermine T, Exter, K, Masseron, T, Østensen, R, Van Eck S and Van de Steene G 2010 *MmSAI* **81** 1022
[2] Gorlova N, Van Winckel H, Jorissen A, Van Eck S, Dermine T, Exter K, Østensen R and Van de Steene G 2011 *Asymmetric Planetary Nebulae V conference*, ed. A. A. Zijlstra *et al* (Manchester: Jodrell Bank Centre for Astrophysics) p 246
[3] Van Winckel H, Lloyd Evans T, Briquet M, De Cat P, Degroote P, De Meester W, De Ridder J, Deroo, P, Desmet M, Drummond R *et al* 2009 *A&A* **505** 1221
[4] van Aarle, E, Van Wincklel H, Lloyd Evans T, Ueta T, Wood P R and Ginsburg A G 2011 *A&A* **530** 90
[5] Giridhar S, Lambert D L, Reddy B E, Gonzalez G and Yong D 2005 *ApJ* **627** 432
[6] Maas T, Van Winckel H and Lloyd Evans T 2005 *A&A* **429** 297
[7] Van Winckel H, Mathis J S and Waelkens C 1992 *Nature* **356** 500
[8] de Marco O 2009 *PASP* **121** 316
[9] Lagadec E, Verhoelst T, Mékarnia D, Suáeez O, Zijlstra A A, Bendjoya P, Szczerba R, Chesneau O, Van Winckel H, Barlow M J *et al* 2011 *MNRAS* **417** 32
[10] Raskin G, Van Winckel H, Hensberge H, Jorissen A, Lehmann H, Waelkens C, Avila G, de Cuyper J-P, Degroote P, Dubosson R *et al* 2011 *A&A* **526** 69
[11] Kovtyukh V V, Soubiran C, Bienaymé O, Mishenina T V and Belik S I 2006 *MNRAS* **371** 879
[12] Kovtyukh V V 2007 *MNRAS* **378** 617
[13] Coelho P, Barbuy B, Meléndez J, Schiavon R P and Castilho B V 2005 *A&A* **443** 735
[14] Munari U and Castelli F 2000 *A&AS* **141** 141
[15] Kovtyukh V V and Andrievsky S M 1999 *A&A* **351** 597
[16] Lobel A 2008 *JPhCS* **130** 2015
[17] Lobel A 2011 *Can. J. Phys.* **89** 395 (*Preprint* `arXiv:astro-ph/1012.5243`)

[18] Castelli F and Kurucz R 2003 *IAU Symp. 210: Modeling of Stellar Atmospheres*, ed N Piskunov, W W Weiss and D F Gray (San Francisco: ASP) p 20
[19] Takeda Y, Taguchi H, Yoshioka K, Hashimoto O, Aikawa T and Kawanomoto S 2007 *PASJ* **59** 1127
[20] Grevesse N, Noels A and Sauval A J 1996 *Proc. of the 6th Ann. October Astrophysics Conf. in College Park* (San Francisco: ASP) **99** p 117

Journal of Physics: Conference Series **328** (2011) 012009 doi:10.1088/1742-6596/328/1/012009

A grid of S stars MARCS model atmospheres

Sophie Van Eck[1], Pieter Neyskens[1], Bertrand Plez[2], Alain Jorissen[1], Bengt Edvardsson[3], Kjell Eriksson[3], Bengt Gustafsson[3], Uffe Gråe Jørgensen[4] and Åke Nordlund[5]

[1] Institut d'Astronomie et d'Astrophysique, Université Libre de Bruxelles, Boulevard du Triomphe, B-1050 Brussels, Belgium

[2] LUPM, Université de Montpellier II, CNRS-UMR 5299, Place Eugène Bataillon CC 72 , F-34095 Montpellier Cédex 05, France

[3] Department of Physics and Astronomy, Uppsala Astronomical Observatory, box 516, S-751 20 Uppsala, Sweden

[4] Niels Bohr Institute, Juliane Maries vej 30, 2100 Copenhagen, Denmark

[5] Centre for Star and Planet Formation, Geological Museum, Øster Voldgade 5, 1350 Copenhagen, Denmark

E-mail: `svaneck@astro.ulb.ac.be`

Abstract.

S stars are cool stars of temperatures similar to those of M giants, but their atmospheres are enriched in carbon and s-process elements because of either extrinsic pollution by a binary companion or intrinsic nucleosynthesis and dredge-up on the thermally-pulsing AGB. Despite numerous attempts to link phenomenological spectral classification criteria to physical parameters ($T_{\rm eff}$, gravity, C/O, [s/Fe], [Fe/H]), the parameter space of S stars is poorly known and this has prevented accurate abundance analysis of S stars until now. Here we present a large grid of S-star model atmospheres. ZrO and TiO band strength indices as well as $VJHKL$ photometry are needed to disentangle the effective temperature, C/O and [s/Fe]. The stellar parameters derived on the basis of low-resolution spectra and photometry are shown to be fairly accurate when compared to high-resolution data of the same stars. The C/O ratio of S stars is found to be between the solar value (0.5) and 0.99, and not 1 as often claimed in the literature. Consistently with stellar evolution expectations, the C/O ratio increases as the effective temperature decreases.

1. Introduction

S stars are giant stars with effective temperatures comparable to those of M giants, but with ZrO bands in addition to TiO bands characterizing M giant spectra. In the past, several attempts to link phenomenological spectral classification criteria to physical parameters ($T_{\rm eff}$, gravity, C/O, [s/Fe], [Fe/H]) [1, 2, 3, 4] only lead to imprecise results, because low-resolution diagnostics are strongly entangled in terms of $T_{\rm eff}$, C/O and [s/Fe] variations. The only in-depth discussion of the thermal structure dates back to the pioneering paper of [5]. Piccirillo already insisted on the strong influence of the C/O ratio on the atmospheric structure and spectra of S stars, in addition to effects due to the s-process elements overabundance. His investigation was however mostly limited to qualitative statements, due to obvious technical limitations. Most subsequent analysis of S stars relied on models designed for M-type stars, not allowing for C/O or [s/Fe] variations.

Journal of Physics: Conference Series **328** (2011) 012009 doi:10.1088/1742-6596/328/1/012009

2. S stars in the GAIA context

S stars are often believed to be transition objects with C/O=1. This is not true, because to form their characteristic ZrO bands, enough free oxygen must be available and the C/O ratio has to be lower than unity. Anticipating results found in Sect. 7, the C/O ratio of S stars in fact spans the complete range 0.5 <C/O< 1. Some rough statistics can then be performed and indicates that S stars are not numerically negligible. Stellar evolution timescales extracted from the STAREVOL code (L. Siess, personal comm.), tagging as S stars all stars with 0.5 <C/O< 1, combined with the initial mass function of [6], show that the number ratio of carbon stars over S stars is around C/S $\approx$ 2.

As an alternative estimate, different galactic catalogs have been examined (GCGCS [7], GCSS2 [8]). Imposing various cuts on the V magnitude in the interval [8..10] leads to the same proportion C/S $\approx$ 2.2, independently of the adopted magnitude cut.

Given this non-negligible proportion, it is important to properly identify and characterize S stars, otherwise important stellar properties, like the third dredge-up luminosity, might be biased, since roughly 50 % of S stars are extrinsic and, as such, do not populate the thermally-pulsing AGB.

3. Model atmosphere grid

A large grid of S-star model atmospheres, superseding the one presented in [9], has been computed. More than 3500 converged MARCS model atmospheres were calculated, covering the following parameter space: $2700 \leq T_{\rm eff}$ (K) ≤ 4000 (steps of 100 K); C/O = 0.5, 0.750, 0.899, 0.925, 0.951, 0.971, 0.991; [s/Fe] = 0., +1., +2. dex ; [Fe/H] = -0.5 and 0. dex ; $\log g$ = 0, 1, 2, 3, 4, 5. A stellar mass of $M = 1$ $M_\odot$ was adopted for all models. Consistently with [10], α-elements were assumed to follow the relation $[\alpha/\mathrm{Fe}] = -0.4 \times [\mathrm{Fe/H}]$. Opacities as complete and accurate as possible were included, including polyatomic molecules and a specific ZrO linelist (described in Plez et al., in preparation). Models were computed through opacity sampling with more than 10^5 wavelength points, local thermodynamic equilibrium, mixing-length theory of convection and spherical symmetry for $\log g \leq 2$.

The model structure has been described elsewhere [11, 12]. C/O has a major influence on the thermal structure, whereas the [s/Fe] ratio has less importance. In summary, the T - $\tau_{\rm Ross}$ relation (governed by the energy balance requirement) reaches higher temperatures at the surface for higher C/O. When C/O increases, $P_{\rm gas}$ at a given T increases, because the important H_2O and TiO opacities decrease, making the atmospheres much more transparent.

4. Comparison with observed photometry

The agreement between synthetic spectra and observational material collected on the large Henize sample of S stars [13] is now examined. From these data, the $(V-K)_0$, $(J-K)_0$ color-color diagram, dereddened according to [14], has been constructed (Fig. 1).

The $(V-K)_0$, $(J-K)_0$ color-color diagram reveals that, for a given $V-K$, the range in effective temperature covered by models of different C/O ratios can be as large as 400 K. Therefore, the application to S stars of the usual M-star temperature scale based on the $V-K$ index (as done in the past when specific S-star models were unavailable) leads to errors on the effective temperature of up to 400 K.

5. Comparison with low-resolution spectroscopic data

Similarly, a set of TiO and ZrO band-strength indices have been computed from the low-resolution spectra, and displayed on Fig. 2. Here again, the indices measured from observations are well reproduced by the indices computed on synthetic spectra.

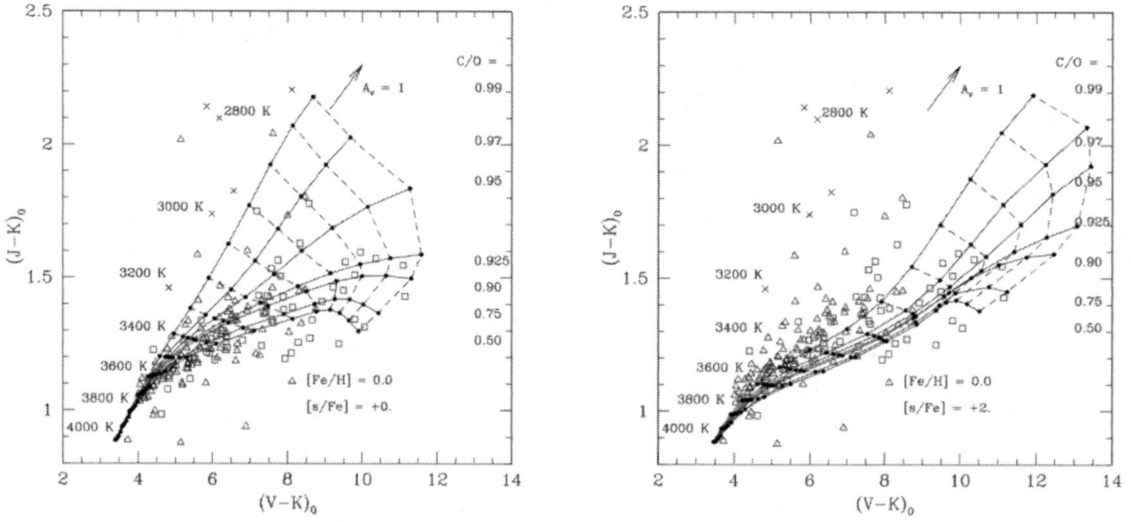

Figure 1. Left-hand panel: Comparison between color indices of observed M (square symbols), S (triangle symbols) and C (cross symbols) stars, and color indices computed on synthetic spectra of S stars for log g=0, [s/Fe] = 0. Right-hand panel: Same for [s/Fe] = 2.

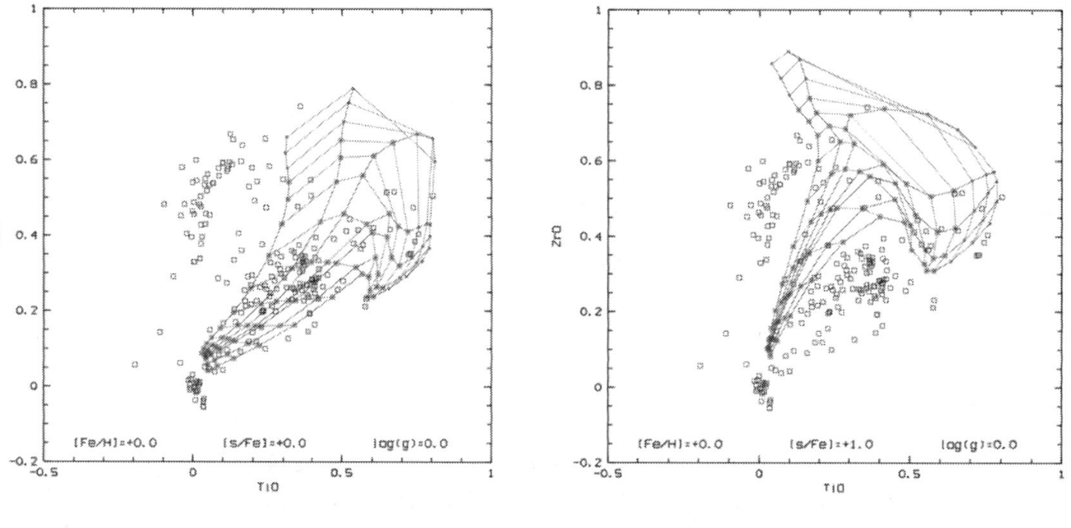

Figure 2. Left-hand panel: ZrO index versus TiO index for [s/Fe]=0. The grid corresponds to solar-metallicity, log g=0. models ranging from $T_{\rm eff}$=4000 K, C/O=0.5 (around coordinates 0.05, 0.05) to $T_{\rm eff}$=2700 K, C/O=0.99 (around 0.3, 0.65). Stars clumping around (TiO,ZrO) = (0,0) are G and K giants. All S stars to the left of the region covered by the grid are SC stars. Right-hand panel: Same for [s/Fe] =1.

6. Comparison with high-resolution spectra

The comparison of photometric colors and spectroscopic indices with the model values makes it possible to estimate $T_{\rm eff}$, C/O and [s/Fe] since: (i) the $(V-K)_0$, $(J-K)_0$ color-color diagram

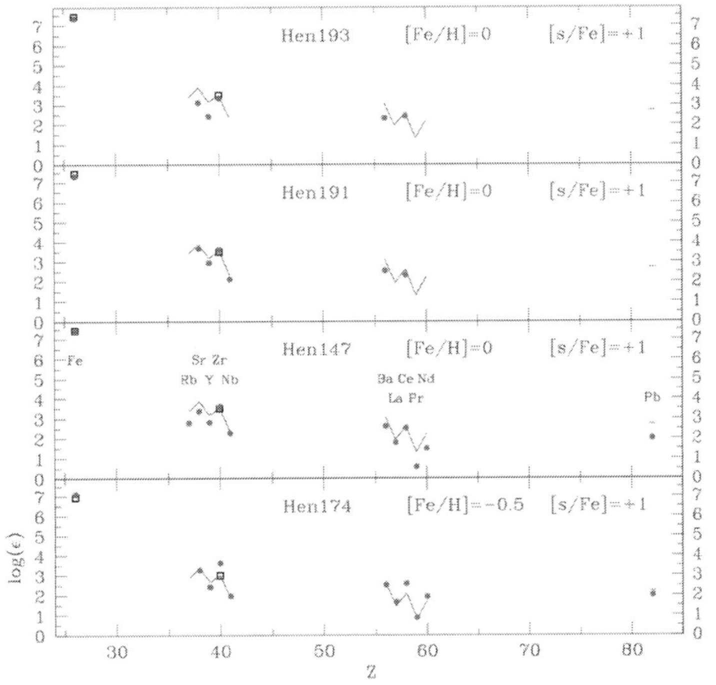

Figure 3. Abundances derived from high-resolution HERMES spectra for 4 Henize S stars (bullet symbols) compared to the Fe and Zr abundances derived from low-resolution spectra (square symbols). The abundances connected by a straight line 'bullet' correspond to a standard s-process scaled to the star metallicity. For the sake of clarity, elements are labelled in the third panel.

disentangles $T_{\rm eff}$ and C/O; (ii) the (TiO, ZrO) diagram disentangles $T_{\rm eff}$ and [s/Fe]. In both cases, there is a good segregation between M and S stars with, however, some degeneracy between C/O and [s/Fe], especially for low $T_{\rm eff}$. The assignment of a 'best model atmosphere' to each observed star is made on the basis of χ-square minimization between the set of photometric and spectroscopic indices computed on the synthetic and observed data. This assignment has been checked on an individual basis to ensure that the derived parameters were meaningful, e.g., were leading to giants with plausible effective temperatures and surface gravities.

For four stars, high-resolution HERMES [15] spectra were available in addition to low-resolution data. The metallicity and zirconium abundance were derived from the HERMES spectra through spectrum synthesis using the best-fitting model atmosphere, and compared to low-resolution estimates. As shown on Fig. 3, the agreement is very satisfactory. The abundances of other s-process elements (e.g. Ba, La) have been also derived through spectral synthesis and are plotted together with their 'scaled-solar' estimates. An agreement is, however, not required since there is no reason for the s-process to be 'scaled-solar'. In particular, the s-process is known to be very sensitive to metallicity: the overall overabundance level, as well as the ratio of heavy-s elements (Ba, La, Ce) to light-s elements (Sr, Y, Zr), are very metallicity-dependent.

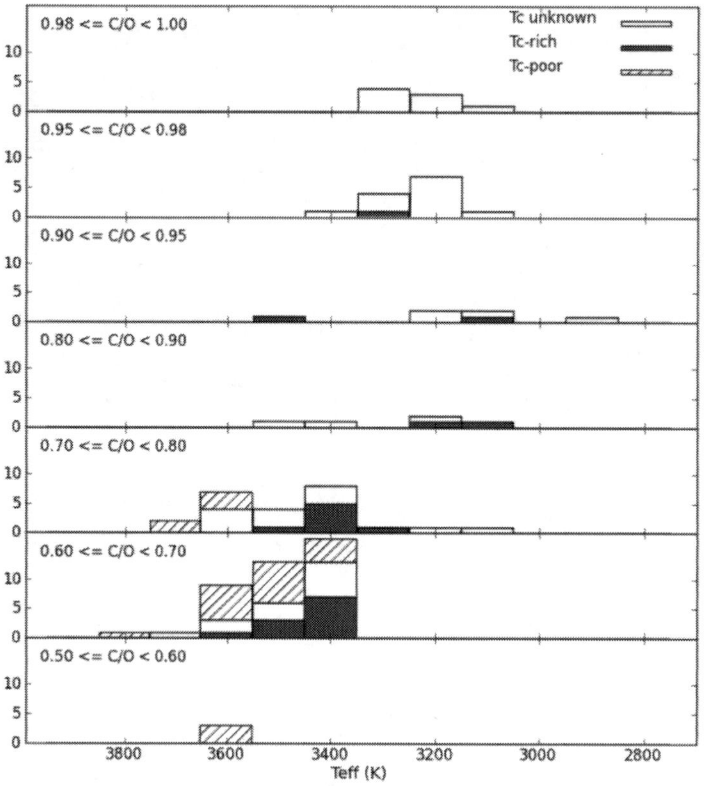

Figure 4. Effective temperature histograms of 100 S stars of the Henize sample, ordered according to C/O ratios. The bins corresponding to the 26 S stars with no technetium are hatched, whereas those corresponding to the 23 Tc-rich S stars are colored. White bins correspond to the 51 stars of unknown technetium content.

7. The effective temperatures and C/O ratios of S stars

Temperature histograms, according to various C/O bins, are presented in Fig. 4. Not all Henize S stars have been analyzed yet, so the subsample of analyzed stars is certainly biased towards hotter and less photometrically variable stars, because the observations are more extensive for those. Hence an effective temperature of 3100 K cannot be considered as a lower temperature limit for S stars, whereas 3900 K probably represents a fair upper limit estimate.

The decrease of effective temperature with increasing C/O ratio, as expected from stellar evolution, is clearly observed. The temperature difference between Tc-poor (polluted binary) S stars and the cooler Tc-rich (genuine thermally-pulsing AGB) S stars is also clearly demonstrated.

This new grid of model atmospheres is an unavoidable prerequisite to reliable spectroscopic chemical analysis of these objects enriched in s-process nucleosynthesis products. It will allow us to pursue on a more quantitative basis the comparison between extrinsic and intrinsic S stars initiated by [16].

Acknowledgments

The authors would like to thank Dr. L. Pompéia (UNIVAP, São José dos Campos - SP, Brazil), for her valuable comments. SVE is FNRS Research Associate and PN benefits from a fellowship 'Boursier FRIA' (Belgium). The HERMES/Mercator project is a collaboration between the K.U. Leuven, the Université libre de Bruxelles and the Royal Observatory of Belgium with contributions from the Observatoire de Genève (Switzerland) and the Thüringer Landessternwarte Tautenburg (Germany). It is funded by the Fund for Scientific Research of Flanders (FWO) under the grant G.0472.04, from the Research Council of K.U. Leuven under grant GST-B4443, from the Fonds National de la Recherche Scientifique (F.R.S.-FNRS) under contracts IISN4.4506.05 and FRFC 2.4533.09, and financial support from Lotto (2004) assigned to the Royal Observatory of Belgium. Support has also been provided by the Actions de recherche concertées (ARC)(Heavy Elements in the Universe: Stellar Evolution, Nucleosynthesis and Abundance Determinations) from the Direction générale de l'Enseignement non obligatoire et de la Recherche scientifique - Direction de la Recherche scientifique - Communauté française de Belgique.

References

[1] Keenan P C 1954 *Ap.J.* **120** 484

[2] Keenan P C and McNeil R C 1976 *An atlas of spectra of the cooler stars: Types G,K,M,S, and C. Part 1: Introduction and tables*

[3] Ake T B 1979 *Ap.J.* **234** 538–547

[4] Keenan P C and Boeshaar P C 1980 *Ap.J.S.S.* **43** 379–391

[5] Piccirillo J 1980 *M.N.R.A.S.* **190** 441–457

[6] Kroupa P and Weidner C 2005 *The Initial Mass Function 50 Years Later* (*Astrophysics and Space Science Library* vol 327) ed E Corbelli, F Palla, & H Zinnecker p 175 (*Preprint* `arXiv:astro-ph/0410461`)

[7] Alksnis A, Balklavs A, Dzervitis U, Eglitis I, Paupers O and Pundure I 2001 Balt. A. **10** *VizieR Online Data Catalog* **3227**

[8] Stephenson C B 1984 *Publications of the Warner & Swasey Observatory* **3** 1

[9] Plez B, Van Eck S, Jorissen A, Edvardsson B, Eriksson K and Gustafsson B 2003 *Modelling of Stellar Atmospheres* (*IAU Symposium* vol 210) ed N Piskunov, W W Weiss, & D F Gray

[10] Gustafsson B, Edvardsson B, Eriksson K, Jørgensen U G, Nordlund Å and Plez B 2008 *A&A* **486** 951–970 (*Preprint* `0805.0554`)

[11] Van Eck S, Neyskens P, Plez B, Jorissen A, Edvardsson B, Eriksson K, Gustafsson B, Jorgensen U G and Nordlund A 2011 *Why Galaxies Care About AGB Stars II: Shining Examples and Common Inhabitants* ed Kerschbaum F (ASP Conference Series 445) 71–76

[12] Neyskens P, Van Eck S, Plez B, Jorissen A, Edvardsson B, Eriksson K, Gustafsson B, Jorgensen U G and Nordlund A 2011 *Why Galaxies Care About AGB Stars II: Shining Examples and Common Inhabitants* ed Kerschbaum F (ASP Conference Series 445) 77-82

[13] Van Eck S, Jorissen A, Udry S, Mayor M, Burki G, Burnet M and Catchpole R 2000 *A&AS* **145** 51–65

[14] Drimmel R, Cabrera-Lavers A and López-Corredoira M 2003 *A&A* **409** 205–215 (*Preprint* `arXiv:astro-ph/0307273`)

[15] Raskin G, van Winckel H, Hensberge H, Jorissen A, Lehmann H, Waelkens C, Avila G, de Cuyper J P, Degroote P, Dubosson R, Dumortier L, Frémat Y, Laux U, Michaud B, Morren J, Perez Padilla J, Pessemier W, Prins S, Smolders K, Van Eck S and Winkler J 2011 *A&A* **526** A69 (*Preprint* `1011.0258`)

[16] Van Eck S and Jorissen A 2000 *A&A* **360** 196–212

Journal of Physics: Conference Series **328** (2011) 012010 — doi:10.1088/1742-6596/328/1/012010

Using seismic targets as benchmarks for spectroscopic analyses of cool stars

Thierry Morel[1], Andrea Miglio[2] and Marica Valentini[1]

[1] Institut d'Astrophysique et de Géophysique, Université de Liège, 4000 Liège, Belgium

[2] School of Physics and Astronomy, University of Birmingham, Birmingham, B15 2TT, UK

Abstract. The frequency of maximum oscillation power measured in dwarfs and giants exhibiting solar-like pulsations provides a precise, and potentially accurate, inference of the stellar surface gravity. An extensive comparison for about 40 well-studied pulsating stars with gravities derived by traditional means (ionization balance, pressure-sensitive spectral features or location with respect to evolutionary tracks) supports the validity of this technique and reveals an overall remarkable agreement with mean differences not exceeding 0.05 dex (although with a dispersion of up to ~0.2 dex). It is argued that interpolation in theoretical isochrones may be the most precise way of estimating the gravity by traditional means in nearby dwarfs. The use of seismic targets as benchmarks in the context of forthcoming large-scale surveys (such as the follow up of the *Gaia* mission) is briefly discussed.

1. Solar-like oscillations as a powerful gravity indicator

It is notoriously difficult to accurately estimate the stellar surface gravity in late-type stars, with systematic differences of the order of 0.2 dex being commonplace depending on the technique used and its exact implementation. This large uncertainty surrounding $\log g$ limits the accuracy with which elemental abundances can be determined. This is especially the case for purely spectroscopic analyses where the determinations of the stellar parameters are intimately coupled. In that case, the use of a model atmosphere with an inappropriate gravity adversely impacts on the estimation of the other parameters (i.e., effective temperature and microturbulence) and, ultimately, chemical abundances.

However, the properties of the p-mode pulsations exhibited by cool stars on the main sequence and during the red-giant phase can be used to derive values that are precise to a level rivaling that obtained for eclipsing binaries. In this study, we consider the frequency of maximum oscillation power, $\nu_{\rm max}$, as a surface gravity indicator (see, e.g., Kallinger et al. 2010a for definition and further details on how this quantity can be derived). As first suggested by Brown et al. (1991), $\nu_{\rm max}$ is expected to scale as the acoustic cut-off frequency:

$$\frac{\nu_{\rm max}}{\nu_{\rm max,\odot}} = \left(\frac{M}{{\rm M}_\odot}\right)\left(\frac{R}{{\rm R}_\odot}\right)^{-2}\left(\frac{T_{\rm eff}}{{\rm T}_{\rm eff,\odot}}\right)^{-1/2}. \quad (1)$$

This leads to:

$$\log g = \log g_\odot + \log\left(\frac{\nu_{\rm max}}{\nu_{\rm max,\odot}}\right) + \frac{1}{2}\log\left(\frac{T_{\rm eff}}{{\rm T}_{\rm eff,\odot}}\right). \quad (2)$$

Journal of Physics: Conference Series **328** (2011) 012010 doi:10.1088/1742-6596/328/1/012010

This relation is largely insensitive to the $T_{\rm eff}$ assumed ($\Delta T_{\rm eff}$ = 100 K leads to $\Delta \log g \sim 0.004$ dex only for Sun-like stars). On the other hand, $\nu_{\rm max}$ can usually be measured with an error below 5% from high-quality time series (e.g., Kallinger et al. 2010a). It follows that $\log g$ can be precise to better than 0.03 dex. If confirmed in terms of accuracy, this would be far better than what can be achieved by other means in single stars. Indeed, seismic gravities are beginning to be adopted in spectroscopic analyses as an alternative to values derived from traditional methods in order to narrow down the uncertainties in the other fundamental stellar parameters and chemical abundances (e.g., Batalha et al. 2011).

The high accuracy of the gravities obtained from asteroseismology is supported by a comparison with values obtained using completely independent techniques (e.g., as shown in the case of a few binaries by Bruntt et al. 2010). However, the validity of the scalings relating the stellar parameters (mass, radius) and the seismic observables has yet to be thoroughly investigated for stars occupying different parts of the H-R diagram and having various properties in terms of metallicity and activity level, for instance. This work is an effort towards this goal (see also Miglio 2011) and also aims at drawing attention to the usefulness of seismic targets for validation purposes in the context of large-scale stellar surveys.

2. A sample of well-studied stars with a precise seismic gravity

About 40 bright, well-studied solar-like and red-giant stars have an accurate estimate of the frequency of maximum power, either from ground-based radial-velocity monitoring or from ultra-precise photometric observations from space (Table 1). These have been used, along with mean literature $T_{\rm eff}$ values (see Section 3) and assuming $\nu_{\rm max,\odot}$ = 3100 μHz, to compute the seismic gravities (the exact choice of $\nu_{\rm max,\odot}$, which has an uncertainty of ~50 μHz, has a negligible impact on our results). The temperatures adopted are marginally higher than those derived from angular diameter and bolometric flux measurements (Bruntt et al. 2010): $<\Delta T_{\rm eff}>=+44\pm56$ K (1σ, 10 stars). Adopting these values would lead to negligible differences in the seismic $\log g$ (well below 0.01 dex). The uncertainty in $\nu_{\rm max}$, which is the main source of error, is often not quoted in the original literature source or its estimation relies on widely different criteria and assumptions. It is therefore impossible to properly account for the star-to-star differences in the data quality and provide a homogeneous set of uncertainties. Adopting various procedures for the determination of $\nu_{\rm max}$ and taking into account the different signal realizations arising from the stochastic nature of the oscillations, Hekker et al. (2011) inferred an uncertainty in the range 1–10% for stars observed by the *Kepler* mission. Based on the type of data collected for the stars in our sample, we estimate a typical uncertainty of 5%. This translates into an error in the seismic gravities of ~0.03 dex only. These figures are supported by a comparison with the values for the three stars in binaries with dynamical masses and interferometric radii (Bruntt et al. 2010): the gravities agree to within 0.02 dex.

3. The classical gravity diagnostics used in late-type stars put to the test

As the stars in Table 1 are amongst the brightest in the sky and are even sometimes regarded as standards (e.g., α Boo or Procyon A), a large number of independent determinations from classical techniques can be found in the literature. This offers an opportunity to empirically assess the reliability of the most popular gravity diagnostics used in cool stars: ionization balance of a given chemical species (usually iron), fitting the wings of strong, pressure-sensitive metal lines or interpolation in theoretical isochrones. These three approaches suffer to different extents from drawbacks. First, non-LTE effects can bias the values obtained from ionization equilibrium, the problem becoming more acute for stars with extended atmospheres and/or metal poor (e.g., Allende Prieto et al. 1999). Another caveat is the neglect of granulation in 1-D model atmospheres. On the other hand, values obtained from fitting the wings of pressure-sensitive lines are generally affected by quite large uncertainties related, for instance, to

Table 1. Values of the frequency of maximum power, $\nu_{\rm max}$, for the stars in our sample. The typical uncertainty is 5% (see text). The original references for the seismic data are given. When not explicitly quoted in these papers, the $\nu_{\rm max}$ values were taken from Bruntt et al. (2010), Kallinger et al. (2010a) or Mosser et al. (2010). The value for 18 Sco was computed from the original data by A.-M. Broomhall (personal comm.).

	Name	$\nu_{\rm max}$ [μHz]	Ref.		Name	$\nu_{\rm max}$ [μHz]	Ref.
	Dwarfs			HD 165341	70 Oph A	4500	17
HD 2151	β Hyi	1000	1	HD 170987		930	18
HD 10700	τ Cet	4490	2	HD 175726		2000	19
HD 17051	ι Hor	2700	3	HD 181420		1500	20
HD 20010	α For	1100	1	HD 181906		1912	21
HD 23249	δ Eri	700	4	HD 190248	δ Pav	2300	1
HD 49385		1013	5	HD 203608	γ Pav	2600	22
HD 49933		1657	6	HD 210302	τ Psa	1950	9
HD 52265		2090	7				
HD 61421	Procyon A	1000	8		**Subgiants and giants**		
HD 63077	171 Pup	2050	9	HD 71878	β Vol	51	23
HD 102870	β Vir	1400	10	HD 100407	ξ Hya	92.3	24
HD 121370	η Boo	750	11	HD 124897	α Boo	3.47	25
HD 128620	α Cen A	2400	1	HD 146791	ϵ Oph	53.5	26
HD 128621	α Cen B	4100	1	HD 153210	κ Oph	35	23
HD 139211	HR 5803	2800	12	HD 161096	β Oph	46	27
HD 142860	γ Ser	1600	1	HD 163588	ξ Dra	36	23
HD 146233	18 Sco	3170	13	HD 168723	η Ser	125	28
HD 150680	ζ Her A	700	14	HD 188512	β Aql	410	1
HD 160691	μ Ara	2000	15	HD 211998	ν Ind	313	29
HD 161797	μ Her	1200	16		M67 S1305	208.9	27

Key to references: [1] Kjeldsen et al. (2008); [2] Texeira et al. (2009); [3] Vauclair et al. (2008); [4] Bouchy & Carrier (2003); [5] Deheuvels et al. (2010); [6] Kallinger et al. (2010b); [7] Ballot et al. (2011); [8] Arentoft et al. (2008); [9] Bruntt et al. (2010); [10] Carrier et al. (2005a); [11] Carrier et al. (2005b); [12] Carrier et al. (2008); [13] Bazot et al. (2011); [14] Martić et al. (2001); [15] Bouchy et al. (2005); [16] Bonanno et al. (2008); [17] Carrier & Eggenberger (2006); [18] Mathur et al. (2010); [19] Mosser et al. (2009); [20] Barban et al. (2009); [21] García et al. (2009); [22] Mosser et al. (2008); [23] Stello et al. (2009); [24] Frandsen et al. (2002); [25] Tarrant et al. (2007); [26] Kallinger et al. (2008); [27] Kallinger et al. (2010a); [28] Barban et al. (2004); [29] Bedding et al. (2006).

difficulties in continuum placement (e.g., Bruntt et al. 2010). Finally, although for very nearby stars parallaxes and reddening are not a major concern, values estimated from interpolation in theoretical isochrones are strongly model dependent and may suffer from degeneracy problems (as a result, the applicability of this method is limited for stars on the red-giant branch).

The $T_{\rm eff}$, [Fe/H] and $\log g$ literature values for the stars in Table 1 were primarily extracted from the PASTEL catalogue (Soubiran et al. 2010), but were supplemented by data from several missing sources. Only studies published after 1990 were considered, as older ones may be based on poor-quality data or inadequate model atmospheres. Each original reference was inspected to evaluate the method used for the $\log g$ determination. In some instances, a single value was quoted in PASTEL whereas estimates based on different techniques were reported in the original paper (e.g., Santos et al. 2004, 2005 where the gravities estimated from isochrone fitting are missing). These values were added. Finally, duplicate entries from the same authors were omitted; only the value in the most recent paper was used. This roughly totals to 360 individual measurements from 80 independent literature sources. The results are presented in Table 2.

Table 2. Mean effective temperature, iron content and mean surface gravities from the four different methods for the stars in Table 1. The error bars are the quadratic sum of the standard deviation of the individual measurements and the typical uncertainty in the parameter determination (80 K for $T_{\rm eff}$, 0.1 dex for [Fe/H], 0.1 dex for the ionization and isochrone gravities, and 0.15 dex for the strong-line gravities). The numbers in brackets are the number of measurements. The typical uncertainty in the seismic $\log g$ is 0.03 dex (see text).

				$\log g$			
	Name	$T_{\rm eff}$ [K]	[Fe/H]	seismology	ionization	wings	isochrone
			Dwarfs				
HD 2151	β Hyi	5829±107 (10)	−0.09±0.12 (11)	3.95	4.02±0.18 (6)	3.76±0.15 (1)	3.98±0.10 (8)
HD 10700	τ Cet	5334±103 (14)	−0.53±0.11 (14)	4.58	4.48±0.21 (12)	4.45±0.15 (1)	4.51±0.14 (5)
HD 17051	ι Hor	6136±120 (9)	0.15±0.12 (8)	4.39	4.48±0.15 (6)	4.40±0.15 (1)	4.40±0.13 (5)
HD 20010	α For	6154±141 (7)	−0.26±0.12 (7)	4.00	4.07±0.25 (4)	3.79±0.15 (1)	3.97±0.11 (4)
HD 23249	δ Eri	5060±111 (11)	0.12±0.13 (11)	3.76	3.86±0.18 (5)	3.95±0.27 (2)	3.82±0.21 (5)
HD 49385		6131±94 (2)	0.09±0.10 (1)	3.97	4.00±0.10 (1)	4.03±0.15 (1)	4.08±0.10 (1)
HD 49933		6580±120 (7)	−0.44±0.10 (5)	4.20	4.24±0.19 (4)	4.00±0.15 (1)	4.23±0.14 (3)
HD 52265		6097±92 (12)	0.19±0.11 (11)	4.28	4.31±0.16 (9)		4.29±0.11 (6)
HD 61421	Procyon A	6590±131 (13)	−0.03±0.11 (15)	3.98	4.06±0.32 (9)	3.92±0.20 (2)	4.01±0.11 (8)
HD 63077	171 Pup	5783±135 (8)	−0.86±0.14 (6)	4.26	4.16±0.19 (3)	4.00±0.15 (1)	4.22±0.15 (5)
HD 102870	β Vir	6131±107 (11)	0.13±0.11 (11)	4.11	4.11±0.16 (7)	3.97±0.15 (1)	4.13±0.11 (7)
HD 121370	η Boo	6059±143 (9)	0.23±0.11 (9)	3.83	3.83±0.29 (7)	3.90±0.15 (1)	3.80±0.11 (4)
HD 128620	α Cen A	5745±138 (14)	0.21±0.13 (14)	4.33	4.21±0.21 (9)	4.32±0.15 (1)	4.31±0.11 (6)
HD 128621	α Cen B	5191±126 (9)	0.24±0.11 (9)	4.54	4.46±0.12 (5)	4.52±0.15 (1)	4.54±0.11 (5)
HD 139211		6296±161 (3)	−0.15±0.18 (2)	4.41	4.05±0.10 (1)	4.10±0.15 (1)	4.20±0.15 (2)
HD 142860	γ Ser	6253±108 (10)	−0.19±0.12 (10)	4.17	4.05±0.17 (6)	4.02±0.15 (1)	4.20±0.12 (6)
HD 146233	18 Sco	5783±92 (13)	0.03±0.11 (13)	4.45	4.40±0.13 (10)		4.43±0.11 (7)
HD 150680	ζ Her A	5762±110 (6)	0.01±0.12 (6)	3.79	3.85±0.18 (3)		3.71±0.11 (5)
HD 160691	μ Ara	5732±104 (12)	0.26±0.11 (11)	4.25	4.20±0.20 (6)	4.07±0.15 (1)	4.23±0.11 (9)
HD 161797	μ Her	5532±105 (7)	0.23±0.13 (7)	4.02	3.98±0.10 (3)		3.94±0.17 (5)
HD 165341	70 Oph A	5221±135 (7)	0.00±0.15 (7)	4.58	4.38±0.19 (5)	4.56±0.15 (1)	4.52±0.11 (3)
HD 170987		6540±80 (1)	−0.15±0.10 (1)	3.94		4.35±0.15 (1)	
HD 175726		6031±88 (3)	−0.07±0.10 (2)	4.26	4.53±0.10 (1)		4.38±0.10 (2)
HD 181420		6671±151 (2)	0.00±0.10 (1)	4.15	4.26±0.10 (1)		4.23±0.10 (1)
HD 181906		6607±80 (1)		4.26			4.24±0.10 (1)
HD 190248	δ Pav	5558±129 (9)	0.30±0.16 (9)	4.30	4.23±0.15 (5)	4.32±0.15 (1)	4.32±0.12 (6)
HD 203608	γ Pav	6065±109 (11)	−0.73±0.13 (10)	4.37	4.22±0.35 (4)	4.15±0.15 (1)	4.33±0.12 (7)
HD 210302	τ Psa	6295±96 (3)	0.05±0.11 (2)	4.26	4.09±0.10 (1)	4.11±0.15 (1)	4.25±0.12 (2)
			Subgiants and giants				
HD 71878	β Vol	4736±246 (2)	−0.01±0.10 (1)	2.61	3.00±0.10 (1)		2.42±0.10 (1)
HD 100407	ξ Hya	5002±106 (5)	0.11±0.15 (4)	2.88	2.86±0.17 (2)	2.88±0.15 (1)	2.69±0.23 (3)
HD 124897	α Boo	4292±97 (17)	−0.58±0.11 (17)	1.42	1.61±0.25 (11)		1.84±0.29 (7)
HD 146791	ϵ Oph	4921±98 (8)	−0.09±0.13 (7)	2.64	2.82±0.20 (4)		2.73±0.23 (5)
HD 153210	κ Oph	4559±116 (4)	0.06±0.13 (3)	2.44	2.50±0.30 (2)		2.47±0.24 (2)
HD 161096	β Oph	4580±112 (6)	0.14±0.13 (5)	2.56	2.67±0.26 (3)		2.38±0.24 (3)
HD 163588	ξ Dra	4464±123 (3)	−0.05±0.11 (2)	2.45	2.40±0.10 (1)		2.46±0.24 (2)
HD 168723	η Ser	4927±89 (10)	−0.18±0.14 (9)	3.01	3.06±0.15 (7)	2.95±0.15 (1)	3.09±0.13 (6)
HD 188512	β Aql	5100±93 (8)	−0.20±0.12 (8)	3.53	3.58±0.14 (3)	3.69±0.15 (1)	3.55±0.11 (5)
HD 211998	ν Ind	5244±101 (7)	−1.54±0.14 (6)	3.42	3.31±0.18 (3)	3.70±0.15 (1)	3.40±0.11 (4)
	M67 S1305	4940±80 (1)	−0.08±0.10 (1)	3.23	3.20±0.10 (1)		

The comparison between the seismic $\log g$ values and those obtained through traditional techniques is shown in Fig. 1. Overall, there is a remarkably good agreement with systematic differences not exceeding 0.04 dex. The significant 1-σ dispersion of up to 0.19 dex with respect to the reference seismic values may have been expected considering the heterogeneous nature of the data and the diversity of analyses performed. By averaging results from a large number of

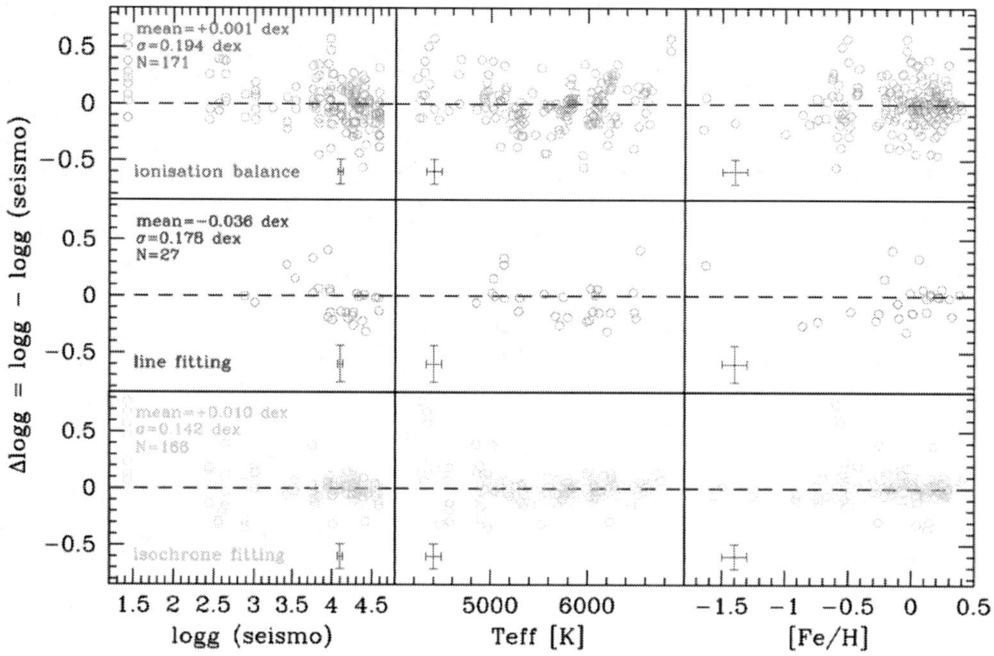

Figure 1. Difference between the seismic $\log g$ values and those obtained through ionization balance of iron (*top panels*), fitting of the wings of pressure-sensitive lines (*middle panels*) and isochrone fitting (*bottom panels*), as a function of the seismic gravities, effective temperature and metallicity. Representative error bars are shown.

independent studies (as is the case here for the ionization and isochrone gravities, but *not* for the strong-line ones), one can hope that the systematic errors partly cancel out and that the mean offset with respect to the seismic values provides a better appraisal of the true accuracy of the method. It should be kept in mind that the systematic differences which may exist between the various studies (discussed in Morel & Miglio 2011) might not be completely related to the method used, but instead to other assumptions in the modeling (e.g., $T_{\rm eff}$ scale). As a matter of fact, the $\log g$ values from the literature were associated to different temperatures than those adopted here. It can readily be seen that the scatter is lower for the gravities estimated from isochrone fitting. The same conclusion holds when considering for each star the average of the measurements obtained using a given method (Fig. 2), especially when one excludes the evolved objects ($\log g < 3.2$) for which the determination through the position of the star with respect to evolutionary tracks is ill defined. In that case, the difference scatter is a mere ~15%: $<\Delta \log g>=-0.006\pm0.065$ dex (1σ, 29 stars). Although this method is generally the most precise, it must be stressed that the mean difference with respect to the seismic gravities is less than 0.05 dex irrespective of the technique used.

There is no evidence for trends as a function of $\log g$, $T_{\rm eff}$ or [Fe/H]. An underestimation of $\log g$ through ionization balance may be expected for very metal-poor stars ([Fe/H] < -1) because of non-LTE effects (Allende Prieto et al. 1999). We only have one such star in our sample (ν Ind), but the ionization gravity does not appear discrepant. The $\log g$ values are systematically underestimated in the dwarfs by up to 0.3 dex when fitting the wings of strong metal lines. However, the bulk of the data comes from a single source (Bruntt et al. 2010) and large line-to-line differences are observed (a weighted mean has been used here). Some of these results have recently been criticized by Fossati et al. (2011).

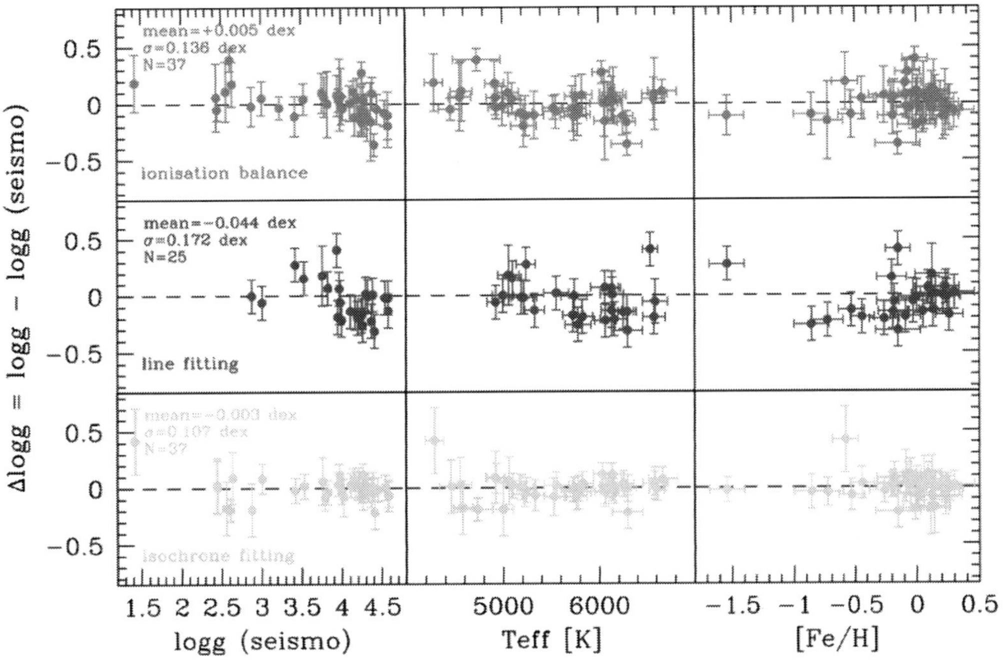

Figure 2. As Fig. 1, but with the data averaged on a star-to-star basis.

4. The case of the red-giant *CoRoT* targets

The *CoRoT* satellite has detected solar-like pulsations in hundreds of red giants in the so-called exoplanet fields (e.g., De Ridder et al. 2009). These stars are relatively faint ($11 < V < 16$) and a determination of their fundamental parameters from spectroscopic analyses is usually not available. However, a spectral analysis of some of these stars has recently been presented by Gazzano et al. (2010) based on GIRAFFE spectra and the automated software MATISSE. To assess the reliability of these results, we have determined the seismic gravities using the $\nu_{\rm max}$ values from Mosser et al. (2011) and temperatures computed from dereddened 2MASS (J–K_s) colors (Alonso et al. 1999). As can be seen in Fig. 3, the gravities appear largely overestimated with respect to the seismic ones, especially for stars in the Galactic centre direction (the LRc01 field) with the poorest signal-to-noise ratio (in the range 10–55). An independent analysis using the software developed by Valentini & Munari (2010) is underway. The benefit of a manual analysis of high-quality FEROS or HARPS spectra can clearly be seen in Fig. 3 for red giants in the seismology fields (Morel et al. 2011). The agreement is much better in that case.

5. Seismic targets as benchmark stars in the *Gaia* era

An accurate determination of the physical parameters of the sources detected by the forthcoming *Gaia* satellite is required for optimising the scientific return of the mission. These parameters ($T_{\rm eff}$, $\log g$, metallicity) will be derived within the coordination unit CU8 using data acquired with both the BP/RP (down to $V{\sim}19$) and RVS (down to $V{\sim}13$) photometric and spectroscopic onboard instruments (for the brightest sources, this will also be achieved through ambitious ground-based observing campaigns, most notably the so-called 'Gaia-ESO Public survey').

The algorithms developed for this purpose by CU8 (GSP-phot and GSP-spec) will be calibrated using a set of well-studied benchmark stars with an accurate determination of the fundamental parameters in the literature. The preliminary list of candidates includes many of

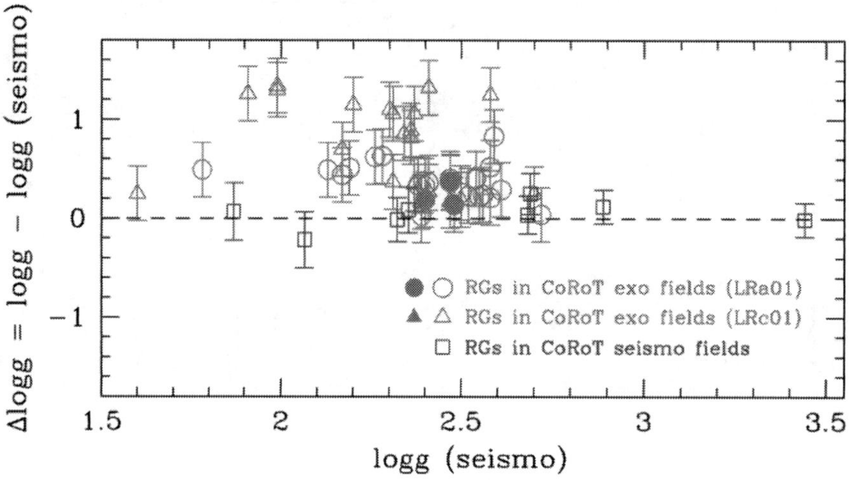

Figure 3. Difference for the red-giant *CoRoT* targets between the seismic $\log g$ values and those derived from spectral synthesis of GIRAFFE spectra with MATISSE (Gazzano et al. 2010; *red*) and from ionization balance of iron using high-resolution spectra (Morel et al. 2011; *blue*). The stars in the exoplanet fields belonging to the red clump (as diagnosed by the $l = 1$ pulsation modes; see Mosser et al. 2011) are indicated with filled symbols.

the solar-like pulsators in Table 1. The seismic gravities can hence constitute a valuable piece of information in this context. Of particular interest in this respect are the stars with an accurate $T_{\rm eff}$ and $\log g$ estimate from interferometric and seismic observations, respectively.

6. Conclusions

The good agreement between the gravities inferred from asteroseismology and from classical methods supports the applicability of the scaling law linking $\log g$ and $\nu_{\rm max}$ for stars spanning a relatively wide range in temperature and evolutionary status (its validity in the low-metallicity and low-gravity regimes cannot, however, be meaningfully investigated here owing to the limited number of objects). In turn, this also suggests that the global seismic properties can be used to retrieve the stellar mass and radius to high accuracy (see, e.g., Miglio 2011). Although the scope of this comparison is limited in some cases by the fundamental difficulties plaguing the classical techniques (e.g., isochrone fitting in giants), seismic gravities may therefore provide a promising alternative in the case of the faint *CoRoT* (see Section 4) or *Kepler* (see, e.g., Bruntt et al. 2011) targets whose spectroscopic gravities may be attached by large uncertainties.

A comparison with data for detached eclipsing binaries has already illustrated the power of isochrone fitting as gravity indicator (Allende Prieto & Lambert 1999), and our study indeed identifies it as being the most precise classical method for nearby dwarfs.

Finally, several large-scale spectroscopic surveys are being planned or are presently conducted. The pipelines developed for that purpose should be able to recover the parameters determined through completely different and, as much as possible, model-independent methods for a set of training stars before embarking on the automatic analysis of large samples of potentially faint objects. Asteroseismic targets may be of interest in this context.

Acknowledgments

T. M. acknowledges financial support from Belspo for contract PRODEX Gaia-DPAC. We wish to thank an anonymous referee for valuable comments, K. Furhmann for providing details on his results and A.-M. Broomhall for recomputing the $\nu_{\max}$ value of 18 Sco.

References

Allende Prieto C and Lambert D L 1999 *A&A* **352** 555
Allende Prieto C, García López R J, Lambert D L and Gustafsson B 1999 *ApJ* **527** 879
Alonso A, Arribas S and Martínez-Roger C 1999 *A&AS* **140** 261
Arentoft T, Kjeldsen H, Bedding T R, et al. 2008 *ApJ* **687** 1180
Ballot J, Gizon L, Samadi R, et al. 2011 *A&A* **530** A97
Barban C, De Ridder J, Mazumdar A, Carrier F, Eggenberger P, De Ruyter S, Vanautgaerden J, Bouchy F and Aerts C 2004 *ESASP* **559** 113
Barban C, Deheuvels S, Baudin F, et al. 2009 *A&A* **506** 51
Batalha N M, Borucki W J, Bryson S T, et al. 2011 *ApJ* **729** 27
Bazot M, Ireland M J, Huber D, et al. 2011 *A&A* **526** L4
Bedding T R, Butler R P, Carrier F, et al. 2006 *ApJ* **647** 558
Bonanno A, Benatti S, Claudi R, Desidera S, Gratton R, Leccia S and Paternò L 2008 *ApJ* **676** 1248
Bouchy F and Carrier F 2003 *Ap&SS* **284** 21
Bouchy F, Bazot M, Santos N C, Vauclair S and Sosnowska D 2005 *A&A* **440** 609
Brown T M, Gilliland R L, Noyes R W and Ramsey L W 1991 *ApJ* **368** 599
Bruntt H, Bedding T R, Quirion P-O, Lo Curto G, Carrier F, Smalley B, Dall T H, Arentoft T, Bazot M and Butler R P 2010 *MNRAS* **405** 1907
Bruntt H, Frandsen S and Thygesen A O 2011, *A&A* **528** A121
Carrier F, Eggenberger P, D'Alessandro A and Weber L 2005a *New Astronomy* **10** 315
Carrier F, Eggenberger P and Bouchy F 2005b *A&A* **434** 1085
Carrier F and Eggenberger P 2006 *A&A* **450** 695
Carrier F, Eggenberger P and Leyder J-C 2008 *J. Phys.: Conf. Series* **118** 012047
Deheuvels S, Bruntt H, Michel E, et al. 2010 *A&A* **515** A87
De Ridder J, Barban C, Baudin F, et al. 2009 *Nature* **459** 398
Fossati L, Ryabchikova T, Shulyak D V, Haswell C A, Elmasli A, Pandey C P, Barnes T G and Zwintz K 2011, *MNRAS*, in press
Frandsen S, Carrier F, Aerts C, et al. 2002 *A&A* **394** L5
García R A, Régulo C, Samadi R, et al. 2009 *A&A* **506** 41
Gazzano J-C, de Laverny P, Deleuil M, Recio-Blanco A, Bouchy F, Moutou C, Bijaoui A, Ordenovic C, Gandolfi D and Loeillet B 2010 *A&A* **523** A91
Hekker S, Elsworth Y, De Ridder J, et al. 2011 *A&A* **525** A131
Kallinger T, Guenther D B, Matthews J M, Weiss W W, Huber D, Kuschnig R, Moffat A F J, Rucinski S M and Sasselov D 2008 *A&A* **478** 497
Kallinger T, Weiss W W, Barban C, et al. 2010a *A&A* **509** A77
Kallinger T, Gruberbauer M, Guenther D B, Fossati L and Weiss W W 2010b *A&A* **510** A106
Kjeldsen H, Bedding T R, Arentoft T, Butler R P, Dall T H, Karoff C, Kiss L L, Tinney C G and Chaplin W J 2008 *ApJ* **682** 1370
Martić M, Lebrun J C, Schmitt J, Appourchaux T and Bertaux J L 2001 *ESASP* **464** 431
Mathur S, García R A, Catala C, et al. 2010 *A&A* **518** A53
Miglio A 2011 *in Red giants as probes of the structure and evolution of the Milky Way Preprint* arXiv:1108.4555
Morel T, et al. 2011, in preparation
Morel T and Miglio A 2011, *MNRAS*, in press
Mosser B, Deheuvels S, Michel E, Thévenin F, Dupret M A, Samadi R, Barban C and Goupil M J 2008 *A&A* **488** 635
Mosser B, Michel E, Appourchaux T, et al. 2009 *A&A* **506** 33
Mosser B, Belkacem K, Goupil M J, et al. 2010 *A&A* **517** A22
Mosser B, Barban C, Montalbán J, et al. 2011, *A&A* **532** A86
Santos N C, Israelian G and Mayor M 2004 *A&A* **415** 1153
Santos N C, Israelian G, Mayor M, Bento J P, Almeida P C, Sousa S G and Ecuvillon A 2005 *A&A* **437** 1127
Soubiran C, Le Campion J-F, Cayrel de Strobel G and Caillo A 2010 *A&A* **515** A111
Stello D, Chaplin W J, Basu S, Elsworth Y and Bedding T R 2009 *MNRAS* **400** L80
Tarrant N J, Chaplin W J, Elsworth Y, Spreckley S A and Stevens I R 2007 *MNRAS* **382** L48
Teixeira T C, Kjeldsen H, Bedding T R, et al. 2009 *A&A* **494** 237
Valentini M and Munari U 2010, *A&A* **522** A79
Vauclair S, Laymand M, Bouchy F, Vauclair G, Hui Bon Hoa A, Charpinet S and Bazot M 2008 *A&A* **482** L5

Journal of Physics: Conference Series **328** (2011) 012011 doi:10.1088/1742-6596/328/1/012011

Spectroscopic properties of stars with debris discs

J Maldonado[1], A Mora[2], B Montesinos[3], E Villaver[1] and C Eiroa[1]

[1] Universidad Autónoma de Madrid, Dpto. Física Teórica, Módulo 15, Facultad de Ciencias, Campus de Cantoblanco, E-28049 Madrid, Spain
[2] ESA-ESAC Gaia SOC. P.O. Box 78, E-28691 Villanueva de la Cañada, Madrid, Spain
[3] Centro de Astrobíologia (INTA-CSIC), LAEFF Campus, European Space Astronomy Center (ESAC), P.O. Box 78, E-28691 Villanueva de la Cañada, Madrid, Spain

E-mail: `jesus.maldonado@uam.es`

Abstract. The question of the origin and evolution of planetary systems is of fundamental importance for astrophysics. Dusty debris discs are signatures of planetary systems and, therefore, constitute valuable tools to provide new light in our understanding of how planetary systems form and evolve. We present the first results of a spectroscopic programme of a sample of stars with debris discs. High-resolution echelle spectra are used to determine metallicities and abundances. Properties of stars with debris discs, are compared with those of stars hosting planets, as well as 'normal' stars.

1. Introduction

Understanding the properties of planet-host stars is of the utmost importance to constrain theoretical models of planetary systems formation and dynamical evolution. At the time of writing more than 570 extrasolar planets have been discovered[1], and the number is expected to increase in the immediate future thanks to the development of both ground-based and space deployed new facilities (e.g., ELTs, SPICA, EUCLID, ECHO). In particular, the ESA-Gaia mission is expected to detect around 8000 giant planets orbiting solar-type stars located at distances closer than 200 pc [5]. For more distant stars, the unprecedented microarcsecond-precision astrometric measurements of Gaia will significantly improve our knowledge of orbital parameters and mass distributions.

Dusty debris discs are generated by collisions of large solid bodies orbiting main-sequence stars [3]. They are direct signatures of planetesimal systems and, therefore, constitute valuable tools to understand the formation and evolution of planets. Current statistics show that around 15 % of the solar-type (spectral types FGK) stars are attended by debris discs. Typical disc characteristics are grain blackbody temperatures of ∼50 K to 100 K, radii between 10 AU and 100 AU, and fractional dust luminosities $L_{\rm dust}/L_{\star}\sim 10^{-5}$ and larger [20].

Whether stars with planets show (or not) special properties when compared to 'normal' stars has been the subject of an intensive study, and the topic is still developing. It is well known for instance, that giant planets detected by radial-velocity techniques are more likely to be found around stars with high metallicity content (e.g., [21]). When considering different chemical species individually, the results are inconclusive, for example, the finding of a higher depletion of lithium in planet host stars is the subject of an ongoing debate (e.g., [4, 10, 18]).

[1] http://exoplanet.eu/

There are relatively few works dealing with the properties of stars harboring debris discs [7, 11]. In this contribution, we present the first results of an observational programme aimed to characterize the properties of stars with debris discs and compare them to stars known to host planets.

2. Observational data

High-resolution échelle spectra of stars with/without debris discs were observed with the FOCES spectrograph at the 2.2 m telescope of the German-Spanish Observatory in Calar Alto (CAHA, Almería, Spain); the SARG spectrograph at the the 3.56 m Telescopio Nazionale Galileo (TNG), and the FIES instrument at the 2.56 m Nordic Optical Telescope (NOT), both on La Palma (Canary Islands, Spain). Additional spectra from the public library 'S^4N' [1] and FEROS spectra from the ESO/ST-ECF Science Achieve Facility[2] were also used. Part of these data, including the observing runs and data reduction procedures are detailed in [14]. All these spectra have high-resolution $\lambda/\Delta\lambda\sim57\,000$ and large signal-to-noise ratio ($\sim$100), covering a wide spectral range.

3. Spectroscopic parameters

Basic stellar parameters $T_{\rm eff}$, log g, microturbulence velocity ($\xi_{\rm t}$), and [Fe/H] are computed by using the code TGV developed by [19]. The code implements the iron excitation and ionization conditions for a set of 65 Fe I and 13 Fe II isolated lines: *i)* $T_{\rm eff}$ is adjusted until no dependence is found between the derived Fe I abundances and the excitation potential of the lines; *ii)* $\xi_{\rm t}$ is changed until no dependence is found between the derived Fe I abundances and the equivalent width of the lines; *iii)* the ionization condition requires that log g should be changed until the averaged abundances obtained from Fe I and Fe II lines are the same.

Starting from an initial guess of the stellar parameters, the code uses an iterative algorithm to find the set of stellar parameters which simultaneously minimizes the standard deviation of the obtained Fe I lines (conditions *i)* and *ii)*), and the difference between the averaged Fe I and Fe II abundances (condition *iii)*).

ATLAS9 models [13] are used and abundances are computed under LTE assumptions. The list of lines was specifically chosen for solar-like stars (spectral types F5/K2-K3). An example is given in Fig. 1 where the obtained abundances for a solar-type spectrum are plotted versus the excitation potential (left-hand panel) and the reduced equivalent widths of the lines (right-hand panel).

The normalized metallicity distribution of the metallicity of the stars with debris discs is plotted against the distribution of those stars without discs in Fig. 2. Both distributions look quite similar, suggesting there is no dependence of the debris disc phenomena on the stellar metallicity. This behavior is different from those stars which host gas-giant planets (e.g., [8]), although stars hosting less massive planets have recently found not to be metal rich [9].

4. Other chemical abundances

Abundances of several elements (Na, Mg, Al, Si, Ca, Ti I, Ti II, Cr, Mn, Co, Ni, and Zn) are computed by using the WIDTH9 program written by R. L. Kurucz [13], updated to work under Linux by [6]. Atmosphere models are computed for each star with the ATLAS9 code [13], updated to work under the Linux platform by [17, 16].

An exhaustive search was performed in order to identify narrow, non-blended lines for each of the aforementioned ions (e.g., [15]) although we kept the line's parameters (i.e., excitation potential, log(gf)) as given in WIDTH9 which are mostly from the Kurucz line lists. Final abundance values are expressed relative to the solar values from [2].

[2] http::/archive.eso.org/cms/

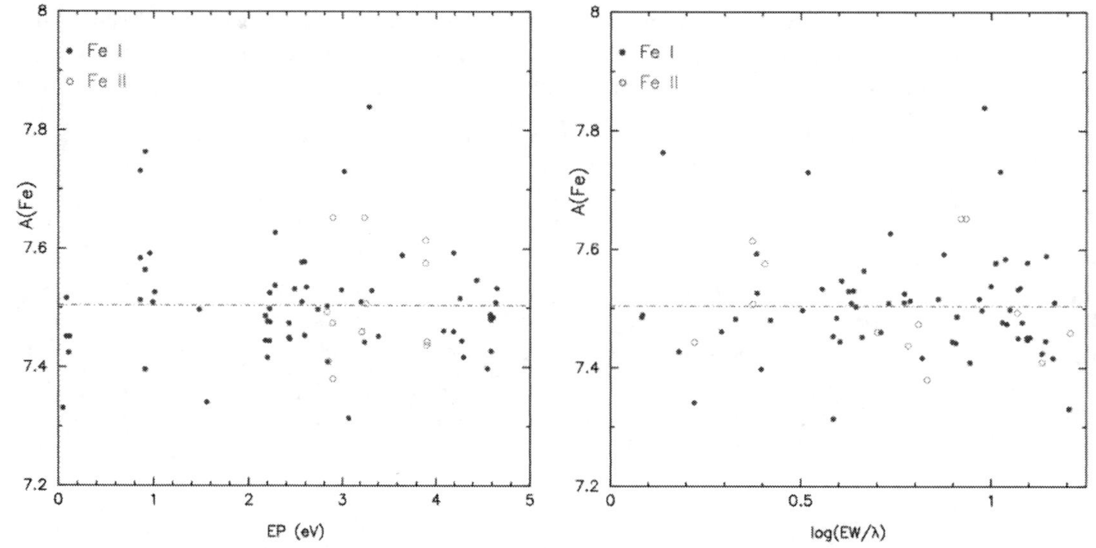

Figure 1. Solar iron abundances obtained for each individual Fe I (blue circles) and Fe II (open red circles) line versus the lower excitation potential of the lines (left-hand panel), and versus the reduced equivalent width of the lines (right-hand panel). The horizontal dashed-dotted lines represent the mean iron abundance.

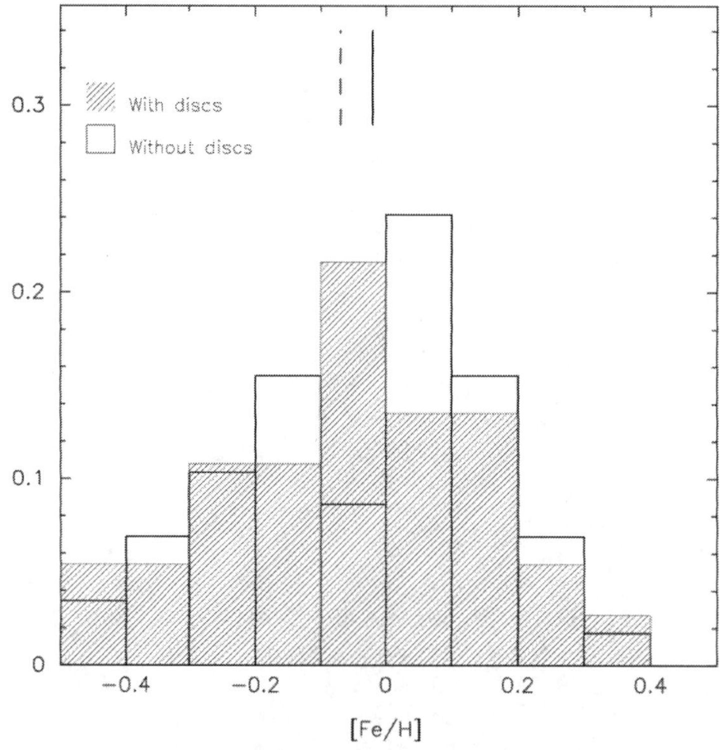

Figure 2. Normalized metallicity distribution of the stars without debris discs (empty histogram) and the stars with debris discs (grey histogram shaded at 45 degrees). Median values of the distributions are shown with vertical lines.

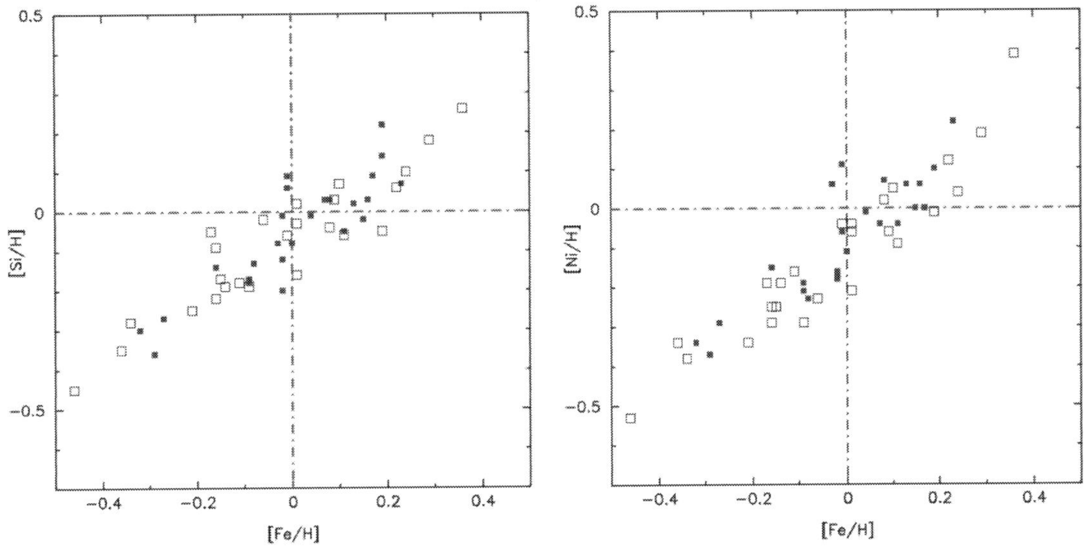

Figure 3. Abundance trend [Si/H] vs. [Fe/H] (left-hand panel) and [Ni/H] vs. [Fe/H] (right-hand panel) for stars with debris discs (blue filled squares) and stars without discs (open squares).

As an example, the abundance trends [Si/H] vs. [Fe/H] and [Ni/H] vs. [Fe/H] are plotted in Fig. 3. Again, we do not find any special distinction between stars with and without debris discs in any of the analyzed ions.

5. Discussion

The well-known planet-metallicity correlation for stars hosting giant radial-velocity planets, fits well into the framework of core-accretion models (e.g., [12]). The formation of planetary cores in this context is favored in discs with high-metal content. Here we find a lack of correlation between the presence of debris discs and high-metallicity. Although it can be explained within the current theoretical understanding of planet formation, its full implications will be discussed at length in a forthcoming paper.

Acknowledgments
This work was supported by the Spanish Ministerio de Ciencia e Innovación (MICINN), Plan Nacional de Astronomía y Astrofísica, under grant *AYA2008-01727*. J. M. acknowledges support from the Universidad Autónoma de Madrid (Department of Theoretical Physics).

References

[1] Allende Prieto C, Barklem P S, Lambert D L and Cunha K 2004 *Astron. Astroph.* **420** 183
[2] Asplund M, Grevesse, N, Sauval, A J and Scott P 2009 *Annual Review of Astronomy and Astrophysics* **47** 481
[3] Backman D E and Paresce F 1993 *Protostars and Planets III* 1253
[4] Baumann P, Ramírez I, Meléndez J, Asplund M, and Lind K 2010 *Astron. Astroph.* **519** A87
[5] Casertano S, Lattanzi M G, Sozzetti A, Spagna A, Jancart S, Morbidelli R, Pannunzio R, Pourbaix D and Queloz D 2008 *Astron. Astroph.* **482** 699
[6] Castelli F 2005 *Memorie della Societa Astronomica Italiana Supplementi* **8** 44
[7] Chavero C, Gómez M and de La Reza R 2006 *Revista Mexicana de Astronomia y Astrofisica Conference Series* **26** 167
[8] Fischer D A and Valenti J 2005 *Astrophys. J.* **622** 1102
[9] Ghezzi L, Cunha K, Smith V V, de Araújo F X, Schuler S C, and de la Reza R 2010 *Astrophys. J.* **720** 1290
[10] González R E and Padilla N D 2010 *Monthly Notices of the Royal Astronomical Society* **407** 1449

[11] Greaves J S, Fischer D A and Wyatt M C 2006 *Monthly Notices of the Royal Astronomical Society* **366** 283
[12] Hubickyj O, Bodenheimer P and Lissauer J J 2005 *Icarus* **179** 415
[13] Kurucz R 1993 *ATLAS9 Stellar Atmosphere Programs and 2 km/s grid Kurucz CD-ROM No 13 Cambridge, Mass. 1993* **13**
[14] Maldonado J, Martínez-Arnáiz R M, Eiroa C, Montes D and Montesinos B 2010 *Astron. Astroph.* **521** A12
[15] Neves V, Santos N C, Sousa S G, Correia A C M and Israelian G 2009 *Astron. Astroph.* **497** 563
[16] Sbordone L 2005 *Memorie della Societa Astronomica Italiana Supplementi* **8** 61
[17] Sbordone L, Bonifacio P, Castelli F and Kurucz R L 2004 *Memorie della Societa Astronomica Italiana Supplementi* **5** 93
[18] Sousa S G, Fernandes J, Israelian G and Santos N C 2010 *Astron. Astroph.* **512** L5
[19] Takeda Y, Ohkubo M and Sadakane K 2002 *Publications of the Astronomical Society of Japan* **54** 451
[20] Trilling D E, Bryden G, Beichman C A, Rieke G H, Su K Y L, Stansberry J A, Blaylock M, Stapelfeldt K R, Beeman J W and Haller E E 2008 *Astrophys. J.* **674** 1086
[21] Udry S and Santos N C 2007 *Annual Review of Astronomy and Astrophysics* **45** 397

GREAT-ESF Workshop: Stellar Atmospheres in the Gaia Era IOP Publishing
Journal of Physics: Conference Series **328** (2011) 012012 doi:10.1088/1742-6596/328/1/012012

3-D hydrodynamical model atmospheres: a tool to correct radial velocities and parallaxes for Gaia

A Chiavassa[1], L Bigot[2], F Thévenin[2], R Collet[3], G Jasniewicz[4], Z Magic[3] and M Asplund[3]

[1] Institut d'Astronomie et d'Astrophysique, Université Libre de Bruxelles, CP. 226, Boulevard du Triomphe, B-1050 Bruxelles, Belgium
[2] Université de Nice Sophia-antipolis, CNRS Laboratoire Cassiopée, Observatoire de la Côte d'Azur, B.P. 4229, 06304 Nice Cedex 4, France
[3] Max-Planck-Institut für Astrophysik, Karl-Schwarzschild-Str. 1, Postfach 1317, D-85741 Garching b. München, Germany
[4] LUPM, Laboratoire Univers et Particules, Université de Montpellier II, CNRS, Place Eugéne Bataillon, 34095 Montpellier Cedex 05, France

E-mail: achiavas@ulb.ac.be

Abstract. Convection plays an essential role in the emerging intensity for many stars that will be observed by Gaia. Convective-related surface structures affect the shape, shift, and asymmetry of absorption lines, the photocentric and photometric variability causing bias in Gaia measurements. Regarding the importance of Gaia mission and its goals, it is mandatory to have the best models of the observed stars. 3-D time-dependent hydrodynamical simulations of surface convection are crucial to model the photosphere of late type stars in a very realistic way. These simulations are an important tool to correct the radial velocities and to better estimate the parallaxes and photometric variability.

1. Introduction

The main goal of the Gaia mission [1, 2] is to determine high-precision photocentric parameters (i.e., positions, parallaxes, and proper motions) for one billion objects with apparent magnitudes in the range $5.6 \leq V \leq 20$ and kinematic velocities of about 100 millions of stars with a precision of $\sim$1 km s^{-1} up to $V \leq 13$. These data along with multi-band and multi-epoch photometric data will allow to reconstruct the formation history, structure, and evolution of the Galaxy. Convection plays a crucial role in the formation of spectral lines and deeply influences the shape, shift, and asymmetries of lines in late type stars which will represent most of the objects that will be observed by Gaia. In addition to this, granulation-related variability that is considered as "noise" must be quantified in order to better characterize any resulting systematic error on the parallax and photometric determinations.

Realistic modelling of stellar atmospheres is therefore crucial for a better interpretation of future Gaia data and, in this context, three-dimensional radiative-hydrodynamical models are needed for a quantitative correction of the radial velocities (few hundreds km s^{-1}) for all the stars observed and, in evolved stars, for the determination of the photocenter positions.

Journal of Physics: Conference Series **328** (2011) 012012 doi:10.1088/1742-6596/328/1/012012

2. Surface convection simulations and radiative transfer calculations

We adopted here time-dependent, three-dimensional (3-D), radiative-hydrodynamical (RHD) surface convection simulations of different stars across the Hertzsprung-Russell diagram (Table 1). The simulations have been carried out using:

- The BOX-IN-A-STAR setup with STAGGER-CODE[1]. These models cover only a small section of the surface layers atop the deep convection zone, and the numerical box includes a number of convective cells proportional to the surface gravity. They have constant gravity, the lateral boundaries are periodic, and the radiation transport module relies on a Feautrier scheme applied to long characteristics.
- The STAR-IN-A-BOX setup with CO^5BOLD code [3, 4, 5]. These simulations cover the whole convective envelope of the star and have been used to model red supergiant (RSG) stars [3, 5] and Asymptotic Giant Branch stars [4] so far. The computational domain is an equidistant cubic grid in all directions, and the same open boundary condition is employed for all sides of the computational box.

The transition between the box-in-a-star and star-in-a-box occurs around $\log g \sim 1$, when the influence of sphericity and the ratio between granule size versus stellar diameter become important; the star-in-a-box global models are then needed, but they are highly computer-time demanding and difficult to run so that there are only very few models available so far.

We used the 3-D pure-LTE radiative transfer code OPTIM3D [6] to compute spectra and intensity maps from the snapshots of the RHD simulations listed in Table 1. The code takes into account the Doppler shifts due to convective motions. The radiative transfer equation is solved monochromatically using extinction coefficients pre-tabulated with the same chemical compositions as the RHD simulations and using the same extensive atomic and molecular opacity data as the latest generation of MARCS models [7]. We assumed a zero micro-turbulence since the velocity fields inherent in 3-D models are expected to self-consistently and adequately account for non-thermal Doppler broadening of spectral lines.

3. Kinematic radial velocities and correction of convective shifts

A convenient and usual way to estimate the kinematic radial velocity of a star is commonly made from a measurement of its spectroscopic radial velocity. Gaia is provided with a dedicated radial velocity spectrometer (RVS, [11]) with a resolving power of (at best) 11 500 centered on the Ca II triplet in the spectral band from 8480 Å to 8750 Å. It is well known that photospheric absorption lines in many stars are blueshifted as a result of convective movements in the stellar atmosphere. In fact, hot, bright, and rising (i.e., blueshifted) convective elements contribute more photons than cool dark shrinking gas, and as a consequence, the absorption lines appear blueshifted [12, 13]. This effect is one of the systematic errors of spectroscopic measurements with RVS [14, 15] and will be taken into account by the Gaia analysis pipeline using 3-D RHD simulations [16] which account for turbulent movements and therefore do not require the traditional micro- and macro-turbulence velocities used in hydrostatic models.

We computed synthetic spectra in RVS domain using OPTIM3D for all the models reported in Table 1. The spectra were computed along rays of four $\mu-$angles $[0.88, 0.65, 0.55, 0.34]$ and four $\phi-$angles $[0°, 90°, 180°, 270°]$, after which we performed a disk integration and a temporal average over all selected snapshots. It is important that the total time covered by the simulations is such that there are no trends in the lineshift if a subset of snapshots is used for the calculations. This has been verified by [17] for the simulation with $T_{\rm eff}$=4630 K and $\log g$=1.6

[1] www.astro.ku.dk/~aake/papers/95.ps.gz

Journal of Physics: Conference Series **328** (2011) 012012 doi:10.1088/1742-6596/328/1/012012

Table 1. 3-D hydrodynamical model atmospheres used in this work. The symbols refer to Figs. 5, 6, 7, and 8.

Stellar type	$T_{\rm eff}$ [K]	$\log g$	[Fe/H]	Mass [$M_\odot$]	$R_\star$ [$R_\odot$]	$\Delta\lambda_{\rm FeI}$ [km s^{-1}]	$\Delta\lambda_{\rm CaII}$ [km s^{-1}]	Ref.	Symbol
			Local simulations with STAGGER-CODE						
K giant	4700	2.2	0.0	...	...	−0.36	+0.29	[8]	●
K giant	4720	2.2	−1.0	...	...	−0.45	+0.23	[8]	▲
K giant	5035	2.2	−2.0	...	...	−0.58	+0.25	[8]	▼
K giant	5130	2.2	−3.0	...	...	−0.28	+0.31	[8]	◆
K giant	4630	1.6	−3.0	...	...	−0.22	+1.55	[9]	■
F star	6500	4.0	0.0	...	...	−0.75	+3.4	[10]	★
			Global simulations with CO^{5}BOLD						
RSG	3430	−0.35	0.0	12	846	+0.75	−1.89	[5]	●
RSG	3660	0.02	0.0	6	386	+2.80	−7.95	[5]	★

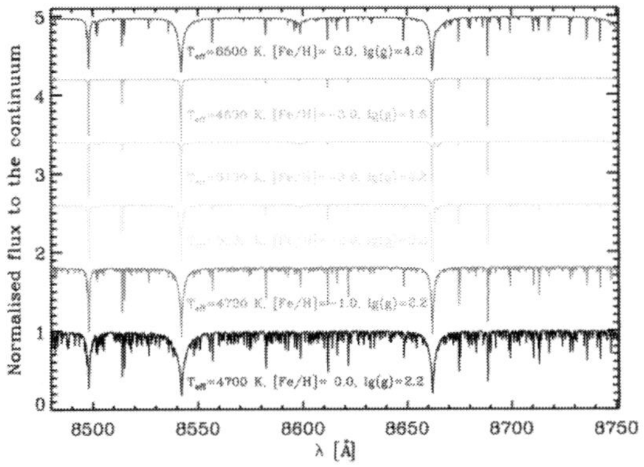

Figure 1. Spectra in the Gaia-RVS domain for the local simulations of Table 1. An offset has been added to the colored curves.

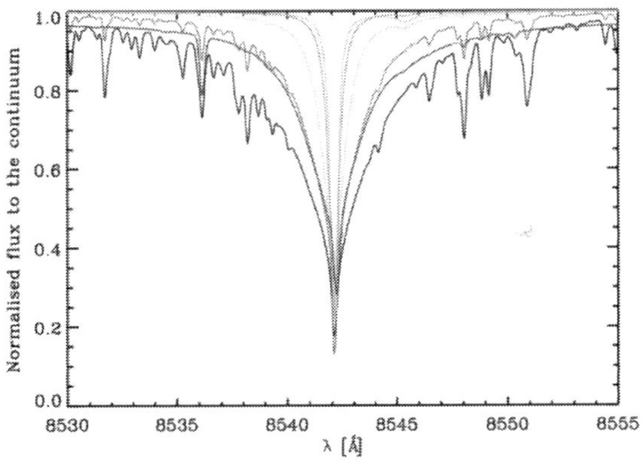

Figure 2. Enlargements of Fig. 1 around a Ca II line.

in Table 1. Ramirez et al. [17] found that the sequence was sufficiently long to ensure that the results are statistically significant and the comparison with observations is also very good. However, for the other K giant models we have not yet done any comparisons with observations. Figures 1 and 2 show the results for the local simulations, and Fig. 3 and 4 the results for the global ones. It is noticeable how the Ca II triplet (λ =8498.02, 8542.09, 8662.14 Å) is well visible in all the stars, even at low metallicity.

Following the work of [18], we determined the position, the depth and the width of Ca II triplet and of 20 Fe I lines selected to be non-blended lines and to cover the spectral domain of the RVS [19]. For this preliminary work, we fitted the absorption lines with a Gaussian function. Figure 5 shows that Fe I lines are mostly blueshifted (except for few cases with low-excitation

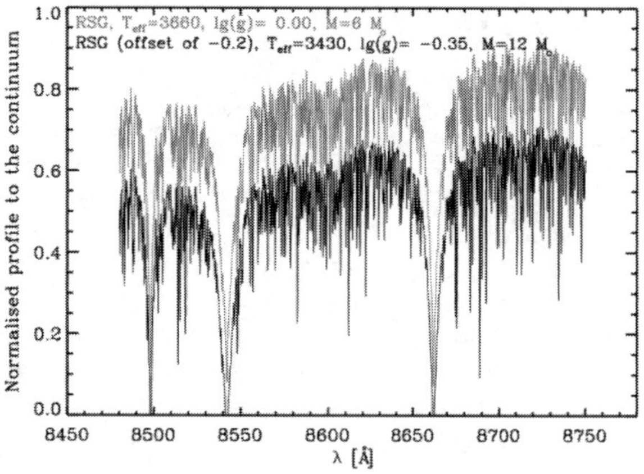

Figure 3. Spectra in the Gaia-RVS domain for the global simulations of Table 1.

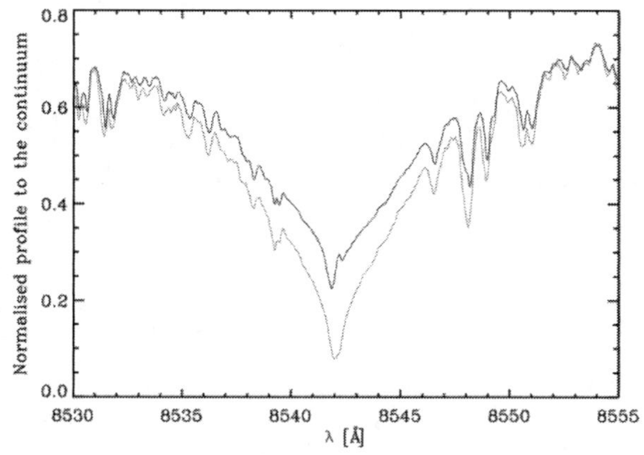

Figure 4. Enlargements of Fig. 3 around a Ca II line.

potentials). The convective lineshift (Table 1, $\Delta\lambda_{\rm FeI}$) appears stronger for high-excitation lines and ranges, on average, from -0.22 km s^{-1} (K giant with [Fe/H]=-3.0) to -0.75 km s^{-1} (F star). The amplitude of the shift increases from K-type giants to the F-type star as a consequence of the more vigorous convective movements. Moreover, the Ca II triplet lines are redshifted because these lines are formed in the upper part of the photosphere where the granulation pattern is reversed. The lineshift is particularly strong in the case of the F-type star (Table 1, $\Delta\lambda_{\rm CaII}$).

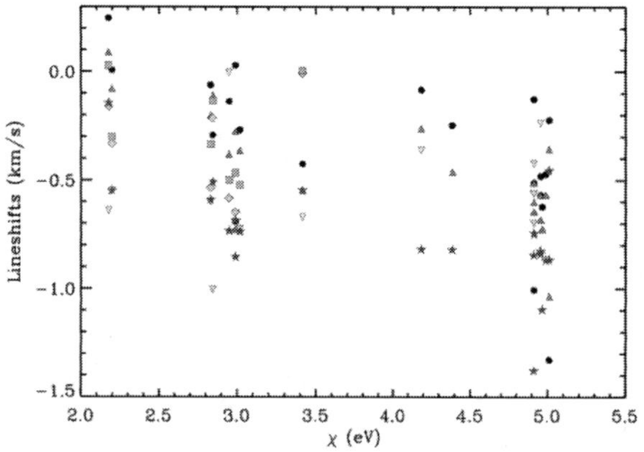

Figure 5. Convective shifts of 20 Fe I from [19] as a function of excitation potentials for local simulations of Table 1. For metal-poor stars, only visible lines in the spectra are reported.

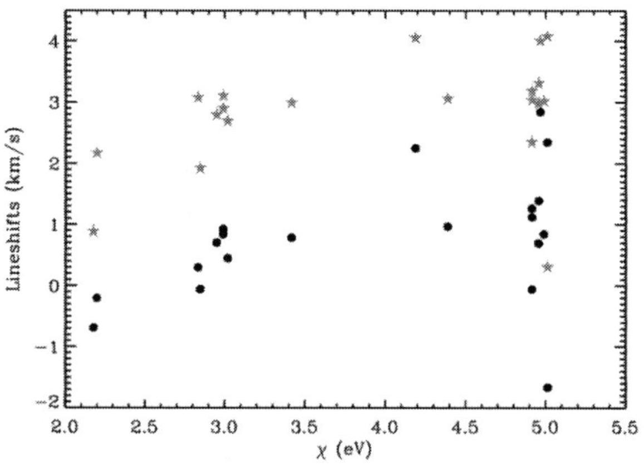

Figure 6. Convective shifts of 20 Fe I from [19] as a function of excitation potentials for global simulations of Table 1.

The lineshifts are not the same for all the lines in the RVS spectra. The precise amount of shift depends on the strength of the absorption line (and hence on the stellar metallicity), since different lines are formed at different atmospheric depths and thus experience different granulation contrasts and convective velocities. In fact, Fig. 7 shows that there is a correlation between lineshifts and line depth indicating that weaker lines (with high excitation potential) are in general more shifted than deep lines (low excitation potential). This picture is also supported by Fig. 5. In addition to this, Fig. 8 shows that there is no apparent connection with the line

Journal of Physics: Conference Series **328** (2011) 012012 doi:10.1088/1742-6596/328/1/012012

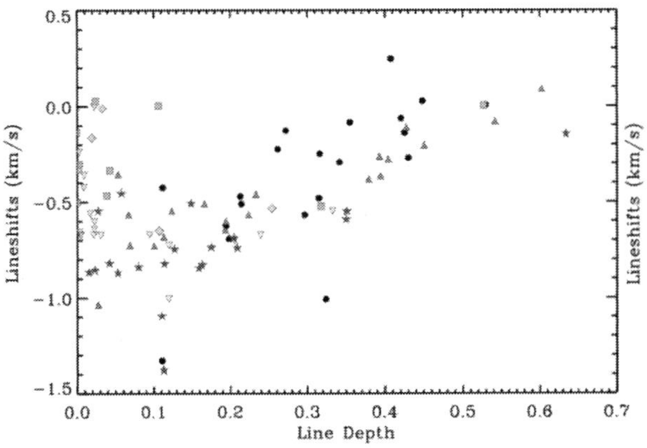

Figure 7. Same as in Fig. 5 but as a function of line depth.

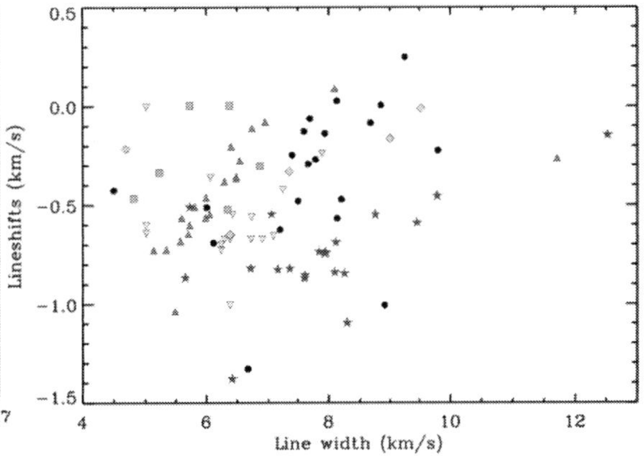

Figure 8. Same as in Fig. 5 but as a function of line width.

width as found by [20] for the Sun.

RSGs show a completely different behavior (Fig. 6) with strong redshifted Fe I up to 2.80 $\mathrm{km\,s^{-1}}$ and blueshifted Ca II up to 8 $\mathrm{km\,s^{-1}}$ (Table 1). The reason is likely connected with the reversed-C shape of line bisectors observed by [21] in the prototypical RSG Betelgeuse and caused by a peculiar and vigorous convection, but further investigations are needed and still in progress with RHD simulations.

In conclusion, velocity shifts of F- and K-giant stars are of the order of RVS accuracy ($\sim$1 $\mathrm{km\,s^{-1}}$), and they are even larger for RSG stars.

4. Photometric and photocentric variabilities and impact on parallaxes

Massive evolved stars such as RSG stars give rise to large granules comparable to the stellar radius in the H and K bands, and an irregular pattern in the optical region [22]. These surface inhomogeneities vary with time and may strongly affect the photometric and astrometric measurements of Gaia.

Figure 9 displays the variability of a RSG during a period comparable with the Gaia mission in the blue and red photometric bands of Gaia. The photometric system of Gaia will be used to characterize the stellar effective temperature, surface gravity and metallicity [23]. The simulations show fluctuations of up to $0^{\mathrm{m}}.28$ in the blue and up to $0^{\mathrm{m}}.15$ in the red bands over 5 years. The photometric fluctuations of RGSs are not negligible and the uncertainties in [Fe/H], T_{eff} and $\log g$ for these stars with $G < 15$ should be revised upwards due to their convective movements [24].

The simulated surface of RSGs (Fig. 10) displays high-contrast structures with spots up to 50 times brighter than the dark ones with strong changes over some weeks. This aspect is connected with the underlying granulation pattern, but also with dynamical effects such as shocks and waves which dominate at optical depths smaller than 1 [22]. The convective-related surface structures in the Gaia G band strongly affect the position of the photocenter and cause temporal fluctuations (left-hand panel of Fig. 11). The photocenter excursion is large, since it goes from 0.005 AU to 0.3 AU over 5 years of simulation and it is on average 0.132 AU (i.e., $\sim$ 3% of the stellar radius). This value corresponds to a photocenter excursion of $\sim$10 μas for

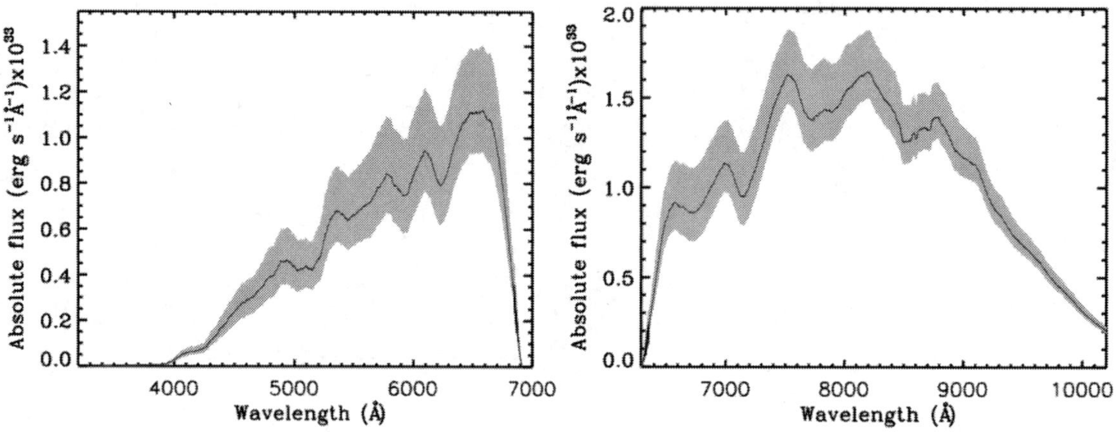

Figure 9. Spectral fluctuations in the blue (*left-hand panel*) and red (*right-hand panel*) Gaia photometric band [25] for a RSG simulation. The black curve is the average flux over ∼5 years covered by the simulation, while the shades areas mark the range of maximum and minimum fluctuations. The spectra have been smoothed to the Gaia photometric resolution of $R \sim 50$ [23]. Figures are adopted from [24].

a star at 1 kpc, which is comparable to the Gaia accuracy of ∼10 μas for stars brighter than G=10.

Chiavassa et al. [24] calculated the Gaia parallax using the Gaia Object Generator v7.0, GOG[2] [26] for RSGs with surface brightness asymmetries from RHD simulations ($\varpi_{\rm spot}$) and without (ϖ). Figure 11 (right-hand panel) shows that there is a systematic error of a few percent. The systematic error may be up to 15 times the formal error σ_ϖ [24]. Although this errorbar should be used to revise σ_ϖ, there is little hope for correcting the Gaia parallaxes of RSGs for this parallax error without knowing the run of the photocentric shift of each star under considerarion. Thus, it might be of interest to monitor the photocentric deviations of a number of selected RSGs during the Gaia mission using interferometry and/or spectroscopy to obtain valuable information about the photocentric position and to correct the resulting parallaxes using RHD predictions.

5. Conclusions and perspectives

3-D hydrodynamical simulations of surface convection are a very important tool to account for convective shift corrections on the kinematic radial velocities, parallaxes and photometry.

With the increase of compute power, it is currently possible to compute 3-D model grids rather easily on reasonable timescales. In this framework, the 3-D local model grid computed with STAGGER-CODE (see the contribution of R. Collet in this Volume) will soon be available, together with the synthetic spectra in RVS domain (before Gaia launch date, mid 2013). The grid will include about 130 models with $T_{\rm eff}$ ranging from 4000 K to 6500 K, $\log g$ from 1.5 to 5 and [Fe/H]=0.0, −1.0, −2.0, −3.0. These models will not be sufficient to represent all the stars observed by Gaia and thus, following [16], we plan to apply 3-D convective corrections to spectroscopic radial velocities obtained with less sophisticated models and templates.

[2] http://gaia-gog.cnes.fr

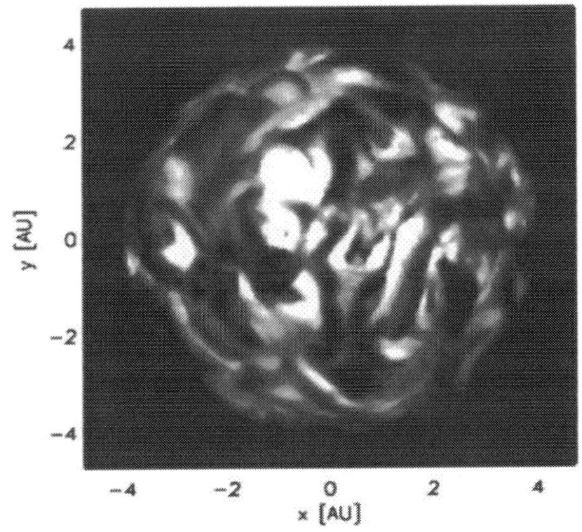

Figure 10. Synthetic map of intensity (the range is [0 – 230000] erg s^{-1} cm^{-2} $Å^{-1}$) in the Gaia G band [25]. Figure adopted from [24].

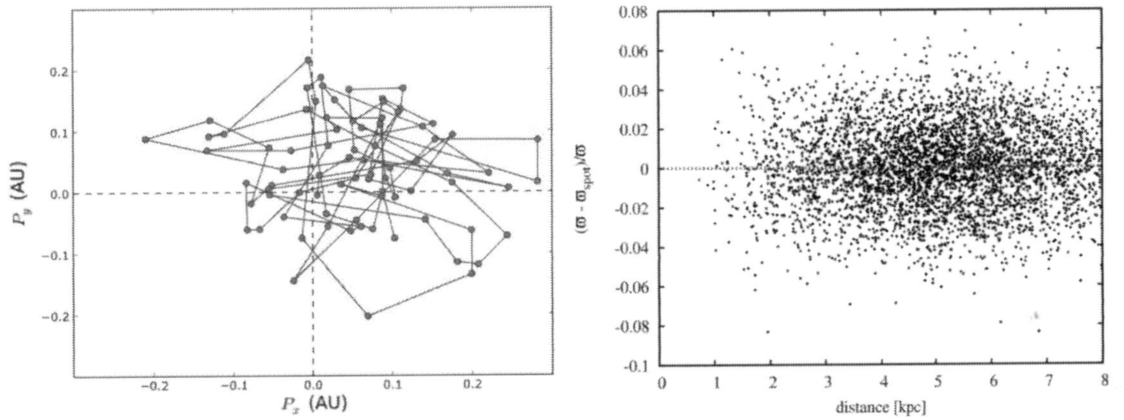

Figure 11. *Left-hand panel:* photocenter displacement as a function of time for a RSG simulation. Each point is a snapshot 23 days apart for a total of 5 years of simulation (comparable to the duration of the Gaia mission). *Right-hand panel:* relative difference between the parallaxes computed with and without the photocentric motion of the left-hand panel, as a function of the distance. Figures adopted from [24].

Acknowledgments

The authors thank the Rechenzentrum Garching (RZG) for providing the computational resources necessary for this work. A.C. is supported in part by an *Action de recherche concertée* (ARC) grant from the *Direction générale de l'Enseignement non obligatoire et de la Recherche scientifique - Direction de la Recherche scientifique - Communauté française de Belgique.* A.C. is supported by the F.R.S.-FNRS FRFC grant 2.4513.11.

References

[1] Perryman M A C, de Boer K S, Gilmore G, Høg E, Lattanzi M G, Lindegren L, Luri X, Mignard F, Pace O and de Zeeuw P T 2001 *A&A* **369** 339–363 (*Preprint* `arXiv:astro-ph/0101235`)

[2] Lindegren L, Babusiaux C, Bailer-Jones C, Bastian U, Brown A G A, Cropper M, Høg E, Jordi C, Katz D, van Leeuwen F, Luri X, Mignard F, de Bruijne J H J and Prusti T 2008 *A Giant Step: from Milli-*

to Micro-arcsecond Astrometry (IAU Symp. 248) (*IAU Symposium* vol 248) ed W J Jin, I Platais, & M A C Perryman (Cambridge: Cambridge University Press) pp 217–223
[3] Freytag B, Steffen M and Dorch B 2002 *Astronomische Nachrichten* **323** 213–219
[4] Freytag B and Höfner S 2008 *A&A* **483** 571–583
[5] Chiavassa A, Freytag B, Masseron T and Plez B *A&A, in press*
[6] Chiavassa A, Plez B, Josselin E and Freytag B 2009 *A&A* **506** 1351–1365 (*Preprint* 0907.1860)
[7] Gustafsson B, Edvardsson B, Eriksson K, Jørgensen U G, Nordlund Å and Plez B 2008 *A&A* **486** 951–970 (*Preprint* arXiv:0805.0554)
[8] Collet R, Asplund M and Trampedach R 2007 *A&A* **469** 687–706 (*Preprint* arXiv:astro-ph/0703652)
[9] Collet R, Nordlund Å, Asplund M, Hayek W and Trampedach R 2009 *Memorie della Societa Astronomica Italiana* **80** 719–+ (*Preprint* 0909.0690)
[10] Collet R et al *personal communication*
[11] Katz D, Munari U, Cropper M, Zwitter T, Thévenin F, David M, Viala Y, Crifo F, Gomboc A, Royer F, Arenou F, Marrese P, Sordo R, Wilkinson M, Vallenari A, Turon C, Helmi A, Bono G, Perryman M, Gómez A, Tomasella L, Boschi F, Morin D, Haywood M, Soubiran C, Castelli F, Bijaoui A, Bertelli G, Prsa A, Mignot S, Sellier A, Baylac M O, Lebreton Y, Jauregi U, Siviero A, Bingham R, Chemla F, Coker J, Dibbens T, Hancock B, Holland A, Horville D, Huet J M, Laporte P, Melse T, Sayède F, Stevenson T J, Vola P, Walton D and Winter B 2004 *MNRAS* **354** 1223–1238 (*Preprint* arXiv:astro-ph/0409709)
[12] Dravins D 1982 *ARA&A* **20** 61–89
[13] Allende Prieto C, Lambert D L, Tull R G and MacQueen P J 2002 *ApJ* **566** L93–L96 (*Preprint* arXiv:astro-ph/0201355)
[14] Lindegren L and Dravins D 2003 *A&A* **401** 1185–1201 (*Preprint* arXiv:astro-ph/0302522)
[15] Allende Prieto C, Koesterke L, Ramírez I, Ludwig H G and Asplund M 2009 *Memorie della societá astronomica italiana* **80** 622–+ (*Preprint* 0909.0470)
[16] Jasniewicz G, Crifo F, Soubiran C, Hestroffer D, Siebert A, Veltz L, Bigot L, Chemin L, David P, Guerrier A, Katz D, Ludwig H G, Richard P, Royer F, Sartoretti P and Udry S 2011 *EAS Publications Series* (*EAS Publications Series* vol 45) pp 195–200
[17] Ramírez I, Collet R, Lambert D L, Allende Prieto C and Asplund M 2010 *ApJL* **725** L223–L227 (*Preprint* 1011.4077)
[18] Bigot L and Thévenin F 2008 *SF2A-2008* ed C Charbonnel, F Combes, & R Samadi pp 3–+
[19] Bigot L and Thévenin F 2006 *MNRAS* **372** 609–614
[20] Asplund M, Nordlund Å, Trampedach R, Allende Prieto C and Stein R F 2000 *A&A* **359** 729–742 (*Preprint* arXiv:astro-ph/0005320)
[21] Gray D F 2008 *AJ* **135** 1450–1458
[22] Chiavassa A, Haubois X, Young J S, Plez B, Josselin E, Perrin G and Freytag B 2010 *A&A* **515** A12+ (*Preprint* 1003.1407)
[23] Thévenin F 2008 *Physica Scripta Volume T* **133** 014010–+
[24] Chiavassa A, Pasquato E, Jorissen A, Sacuto S, Babusiaux C, Freytag B, Ludwig H G, Cruzalèbes P, Rabbia Y, Spang A and Chesneau O 2011 *A&A* **528** A120+ (*Preprint* 1012.5234)
[25] Jordi C, Gebran M, Carrasco J M, de Bruijne J, Voss H, Fabricius C, Knude J, Vallenari A, Kohley R and Mora A 2010 *A&A* **523** A48+ (*Preprint* 1008.0815)
[26] Isasi Y, Figueras F, Luri X and Robin A C 2010 *Highlights of Spanish Astrophysics V* ed J M Diego, L J Goicoechea, J I González-Serrano, & J Gorgas pp 415–+

Journal of Physics: Conference Series **328** (2011) 012013 doi:10.1088/1742-6596/328/1/012013

Comparison between MAFAGS-OS spectra and Kurucz-ODF spectra

Jiannan Zhang, Ali Luo, Yihan Song and Fang Zuo

National Astronomical Observatory, Chinese Academy of Sciences, Beijing, China

E-mail: jnzhang@bao.ac.cn

Abstract. Grids of theoretical stellar spectra are fundamental for estimating basic stellar parameters from photometric and spectroscopic data observed in large sky surveys such as SDSS, LAMOST, Gaia, etc. Do the different atmosphere models influence the parameters estimation? We compute the Lick indices and uvby color indices using the MAFAGS-OS grid of model atmospheres and fluxes provided by F. Grupp (personal comm.) and the Kurucz grids [1]. A spectrum comparison reveals the behavior of spectra from the MAFAGS and Kurucz grids.

We find that using the (b-y) index, consistent effective temperatures can be determined from both the Kurucz and MAFAGS grids of theoretical spectra. The m1 index, together with color index, can be used to determine the metallicity of F- and G-type stars, but the measurements of the Kurucz and MAFAGS grids show systematic discrepancies for cool stars. The c1 indices computed with both grids show small discrepancies for $T_{\rm eff} < 6000$ K, while for $T_{\rm eff} > 6000$ K, the c1 indices agree well.

The Lick indices of the Kurucz grid and the MAFAGS grid tend to be in agreement for warm stars with temperatures above 5000 K, while for cool stars with temperatures ranging from 4000 K to 5000 K, the difference of Lick indices for both models is apparently large.

We also compare the MAFAGS spectrum and Kurucz spectrum of the same temperature, surface gravity, and metallicity using a correlation coefficient for the complete spectrum. For warm stars, the MAFAGS and Kurucz spectra are almost the same, while for cool stars below 5000 K, there are some discrepancies between the MAFAGS and Kurucz spectra that induce internal discrepancies in the parameters determination.

1. Introduction

Many large sky surveys focus on unraveling the structure, formation history, kinematics, and evolution of the Milky Way. For sky surveys such as SDSS, Gaia, LAMOST, etc., accurate and reliable stellar parameters automatically derived from a huge amount of low-resolution spectra are important for accomplishing the science goals. Many methods have been developed to solve this problem. The methods based on spectroscopy usually require a set of spectra for which the parameters have been accurately calibrated by constructing a grid of template spectra or training sets. There are two ways to construct a grid of template spectra. One is based on libraries of empirically observed spectra from ELODIE, MILES, etc. The other uses theoretical stellar spectra computed from stellar atmosphere models, such as the Kurucz models [2,3], the MARCS models [4], and the PHOENIX models [5], etc. Compared to observed spectral grids, theoretical stellar grids are more complete and uniform, having a large density of parameters coverage. While theoretical spectral grids have their shortcomings (although various stellar atmosphere models are widely used in stellar evolution and flux analyses, etc.) their development is still in progress. Atmosphere models that are based on a series of simplified assumptions yield theoretical spectra different from observed spectra. Furthermore, different treatments of the basic assumptions for the calculation of the models make them different from each other. Does the difference between stellar atmosphere models influence the measurement of stellar parameters using low-resolution theoretical spectra as templates? In case theoretical spectral grids are employed for parameter estimations with low-resolution spectra, the systematic errors that result from differences in the theoretical spectral grids should be studied.

In this paper we select the MAFAGS and Kurucz grids of theoretical spectra to investigate if there are differences in the measurement of parameters from low-resolution spectra. The method compares photometric color indices, absorption line indices, and a complete spectrum matching is performed to determine the temperature, surface gravity, and metallicity from both grids. In this paper we do not consider different α-enhancements in the models.

2. MAFAGS-OS and ATLAS9-newODF atmosphere models

We use the MAFAGS-OS atmosphere models of the spectral grid provided by Grupp [6,7,8], and the ATLAS9-newODF atmosphere models of [1] (online available from `http://wwwuser.oat.ts.astro.it/castelli`) and SPECTRUM [10] to construct the grid of theoretical spectra. As was shown by [6,7] and [11], the MAFAGS models are almost identical to the commonly used Kurucz models. Both model grids are valid for A-, F- and G-type stars. They produce almost identical spectra for the same stellar parameters. Besides a number of special treatments for the calculation of convection and the wavelength grid sampling, the main difference between both models is the opacity calculation. MAFAGS-OS is an opacity sampling (OS) code, while ATLAS9-newODF is based on opacity distribution functions (ODF). [6] summarized the difference between OS and ODF models and illustrated the theoretical flux distribution in OS and ODF models. Convection in the MAFAGS models is treated following [12], using a convective efficiency parameter of $\alpha_{cm} = 0.82$, whereas Kurucz (ATLAS9-newODF) models use the mixing-length approach (ML) with a mixing-length to the pressure scale height ratio of $l/H_{\rm p}$=1.25 [13]. Both theoretical spectral grids provide models with the same wavelength coverage of SDSS and LAMOST. They range from 300 nm to 1000 nm, and are decreased from high-resolution spectra to a resolution of R=2000 with a 0.1 nm wavelength sampling step. α-enhancement is not considered for both model grids because the effects are small for low-resolution spectra of R=2000. The microturbulence velocity is held constant at 2.0 km s^{-1} in both grids. The MAFAGS-OS grid comprises of 9841 spectra with $T_{\rm eff}$ ranging from 4600 K to 15000 K in steps of 200 K for cool stars, while 500 K for warm stars. The log g parameter ranges from 0.0 dex to 5.0 dex in steps of 0.2 dex, and the metallicity [Fe/H] ranges from 0.0 dex to −4.8 dex in steps of 0.3 dex. The Kurucz grid comprises of 3929 spectra with $T_{\rm eff}$, ranging from 3500 K to 50 000 K in steps of 250 K for cool stars, and 1000 K for warm stars. The log g parameter ranges from 0.0 dex to 5.0 dex in steps of 0.5 dex, and the metallicity [Fe/H] ranges from 0.5 dex to −4.0 dex in steps of 0.5 dex.

3. Comparison method

With multi-fibre sky surveys employed in succession, various methods have been developed to automatically determine parameters from low-resolution spectra. The most popular methods generally consist of three types; parameter calibrations based on absorption line indices, photometric color indices, and template matching or statistical learning methods that use spectroscopy. To check if different theoretical spectral grids induce systematic discrepancies in the parameter measurements we compare low-resolution grids of MAFAGS and Kurucz from three points of view; theoretical colors, theoretical line indices, and a full spectrum comparison.

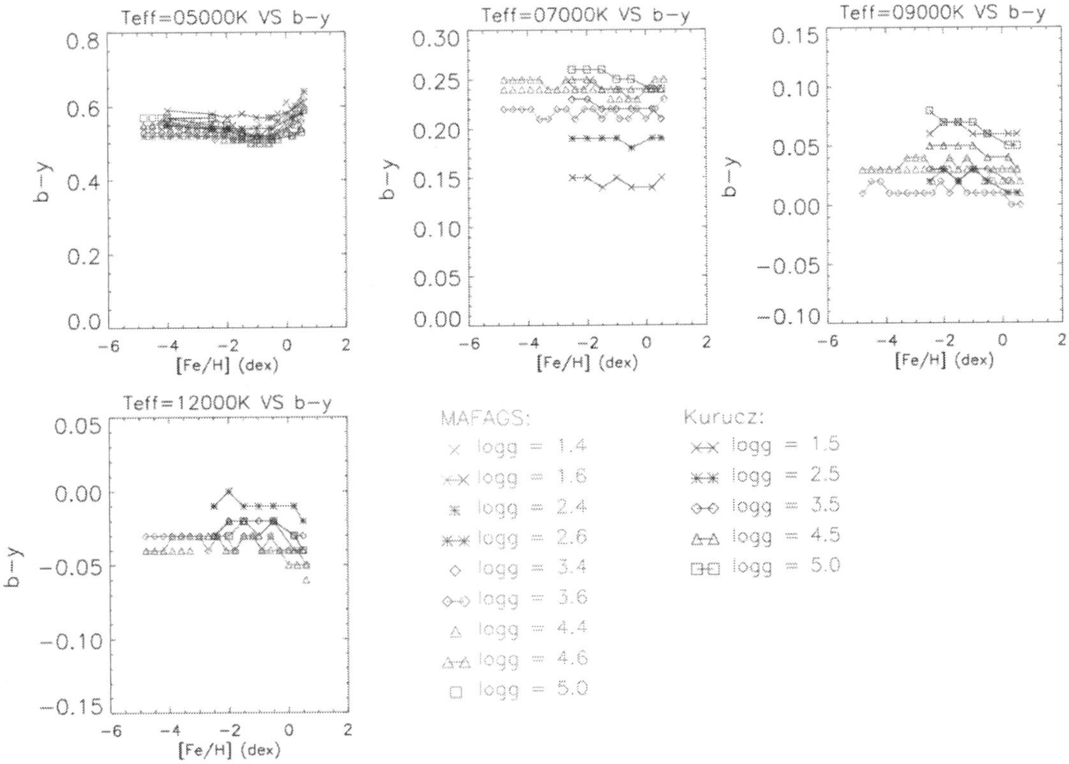

Figure 1. Comparison of the (b−y)-index for the Kurucz and MAFAGS grids. Each plot shows (b−y) of both spectral grids in a (b−y) - metallicity diagram for $T_{\rm eff}$=5000 K, 7000 K, 9000 K, and 12 000 K. In each plot of constant $T_{\rm eff}$ blue symbols mark the Kurucz grid, while red symbols mark the MAFAGS grid. Different symbols show different surface gravity values.

3.1. Comparison with uvby color indices

The uvby color indices c1 = (v−b)−(b−y) and m1 = (u−v)−(v−b) are verified by many researchers for being sensitive to temperature, surface gravity, and metallicity. Several empirical calibration formulae are offered in [14,15], [16], [17], and others, for converting (b−y) to $T_{\rm eff}$. [18] and [19] calibrated [Fe/H] with the m1-index for dwarf stars, while [20, 21] determined [Fe/H] with the m1-index for giant stars. The c1-index can be used to determine the surface gravity of A- and F-type stars [22].

We synthesize the uvby photometric indices of both theoretical spectral grids using the theoretical spectrum of Vega as zero-point. Figure 1 shows the (b−y) of both spectral grids in a (b−y)-metallicity diagram for $T_{\rm eff}$= 5000 K, 7000 K, 9000 K, and 12 000 K. In each plot, the blue points mark the (b−y) vs. metallicity for the Kurucz grid, while the red points are the (b−y) vs. metallicity for the MAFAGS grid. The (b−y) indices are consistent in $T_{\rm eff}$ for stars below 9000 K showing a small influence of gravity in the first three plots of Fig. 1. For stars with $T_{\rm eff}$ above 9000 K (b−y) is almost invariant. The gravity parameter is little influenced by (b−y). For gravity increasing from 1.5 dex to 5.0 dex, the (b−y) decreases by at most 0.1, shown for $T_{\rm eff}$=7000 K. Although the gravities in both grids do not coincide, it is clear that the $\triangle$(b−y)s result from the differences between the MAFAGS and Kurucz grid. $\triangle$(b−y) is smaller than what can be attributed to the different gravity step of the MAFAGS grid (0.2 dex) and the Kurucz grid (0.5 dex). The (b−y) indices of the Kurucz and MAFAGS grids are in good

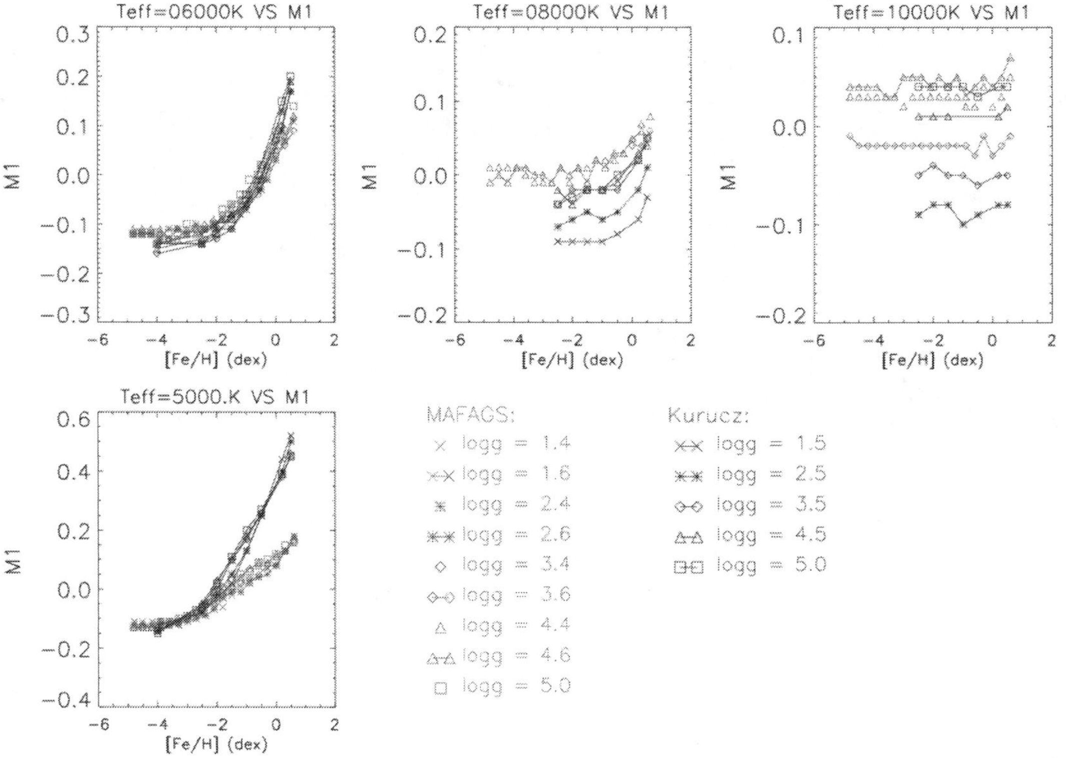

Figure 2. Comparison of the m1-index between the Kurucz and MAFAGS grids. In each plot of constant $T_{\rm eff}$ red symbols mark the MAFAGS grid, while blue symbols mark the Kurucz grid. Others symbols as in Fig. 1.

agreement. Considering the accuracy of the observed photometric data, which shifts (b−y) by ~1–5 %, e.g., most of the galactic reddening value in the map of [23] is small ($E(b-y)_{uvby} <$ 0.04), and its influence is smaller than the influence of the surface gravity parameter. Both the Kurucz and MAFAGS grids of low-resolution spectra can therefore utilize the (b−y) index for consistently determining $T_{\rm eff}$.

Figure 2 shows the m1-index of both spectral grids in a m1 - metallicity diagram with $T_{\rm eff}$=5000 K, 6000 K, 8000 K, and 10 000 K. In the plots for $T_{\rm eff}$>6000 K, the m1-index of the Kurucz grid agrees well with the MAFAGS grid in the 3-D parameter space of $T_{\rm eff}$, surface gravity, and metallicity. In the plot with $T_{\rm eff}$=5000 K, the m1-index for the Kurucz grid and MAFAGS grids are discrepant for metallicity values exceeding −3.0 dex. The m1-index is sensitive to the metallicity. The sensitivity is stronger for smaller $T_{\rm eff}$. The m1-index combined with the (b−y) color index can be used to determine the metallicity of F- and G-type stars. However, there are systematic discrepancies between the Kurucz and MAFAGS grids for cool stars of $T_{\rm eff}$<6000 K. If $T_{\rm eff}$ is exactly 5000 K, for a m1-index between 0.05 and 0.2, the difference in metallicity determined from the MAFAGS grid and the Kurucz grid is as large as 2.0 dex. Therefore, considering the error in the $T_{\rm eff}$ determination, the different m1-indices for stars with $T_{\rm eff}$<6000 K obtained from the MAFAGS and Kurucz grids yield large differences in the metallicity that are clearly discrepant.

The c1-index is a good indicator of the surface gravity in A- and F-type stars. Figure 3 compares c1 to metallicity for the Kurucz and MAFAGS grids with $T_{\rm eff}$=5000 K, 6000 K, 8000 K, and 10 000 K. The plots for $T_{\rm eff}$>6000 K illustrate that the c1 indices for both theoretical

Journal of Physics: Conference Series **328** (2011) 012013 doi:10.1088/1742-6596/328/1/012013

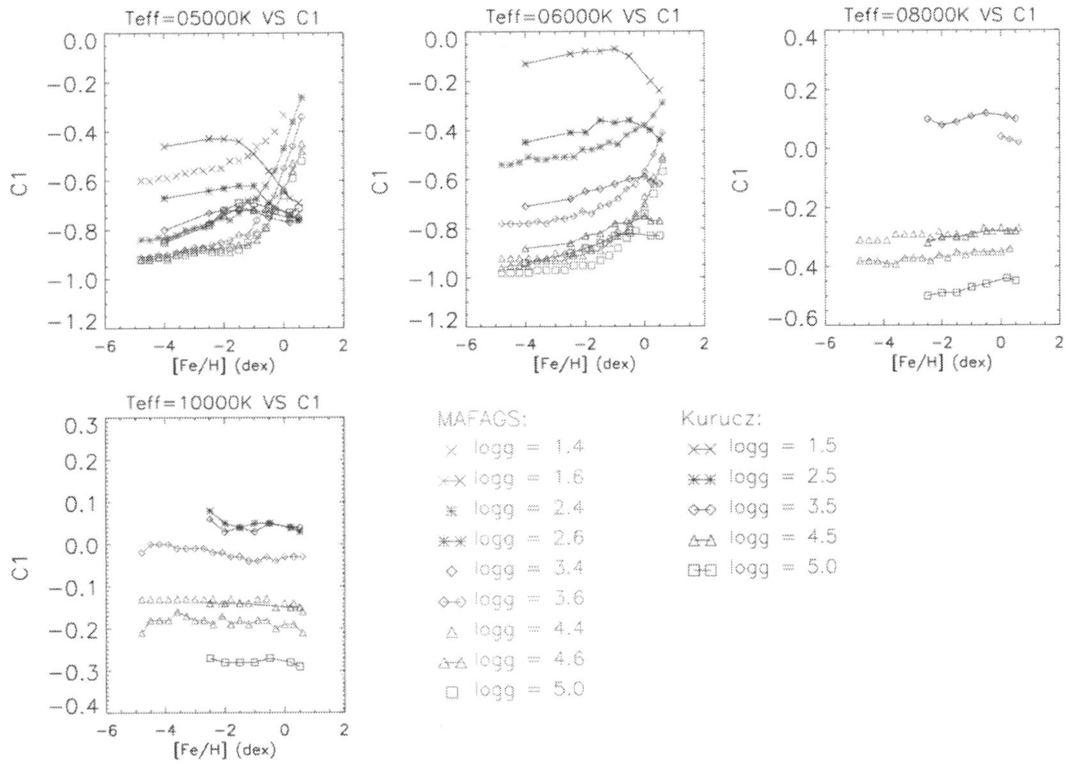

Figure 3. Comparison of the c1-index between the Kurucz and MAFAGS grids. In each plot of constant $T_{\rm eff}$ red symbols mark the MAFAGS grid, while blue symbols mark the Kurucz grid. Others symbols as in Fig. 1.

spectral grids are in a agreement and are sensitive to the surface gravity. However, for stars of $T_{\rm eff}>10\,000$ K the sensitivity to the gravity parameter decreases with decreasing $T_{\rm eff}$. In the second plot with $T_{\rm eff}=6000$ K, the c1 indices of both grids agree for metal-poor stars. In case the metallicity exceeds -1.0 dex, the c1-index of the MAFAGS grid increases with increasing metallicity, whereas the c1-index of the Kurucz grid decreases with increasing metallicity. The differences in the c1-index for the MAFAGS and Kurucz models are maximum for the maximum metallicity value of $+0.5$ dex. These differences yield a maximum difference of $\sim$1.1 dex for the gravity parameter determined from both model grids.

3.2. Comparison with Lick line indices

The Lick line indices [24,25] are widely used for temperature, surface gravity, and metallicity measurements. They are well-suited for automatic determinations of basic stellar parameters with low-resolution spectra in a large sample of data. We calculate 25 Lick line indices following the standard Lick definition, and compare differences in Lick indices between both grids. In this paper we select three indices to investigate the differences. The indices are Fe4668, H_beta, and Mg_b. The nine contour plots in Fig. 4 illustrate the three Lick indices for Kurucz, MAFAGS, and the differences between them. The contour colors code the strength of the Lick indices of the lines, and the amount of difference between each pair of lines.

In Fig. 4 we show the difference of the line indices between both grids. There are many possible causes for these differences. They may for example result from the model grids, from the spectrum calculations, or from the computation of the indices. For practical applications,

Journal of Physics: Conference Series **328** (2011) 012013 doi:10.1088/1742-6596/328/1/012013

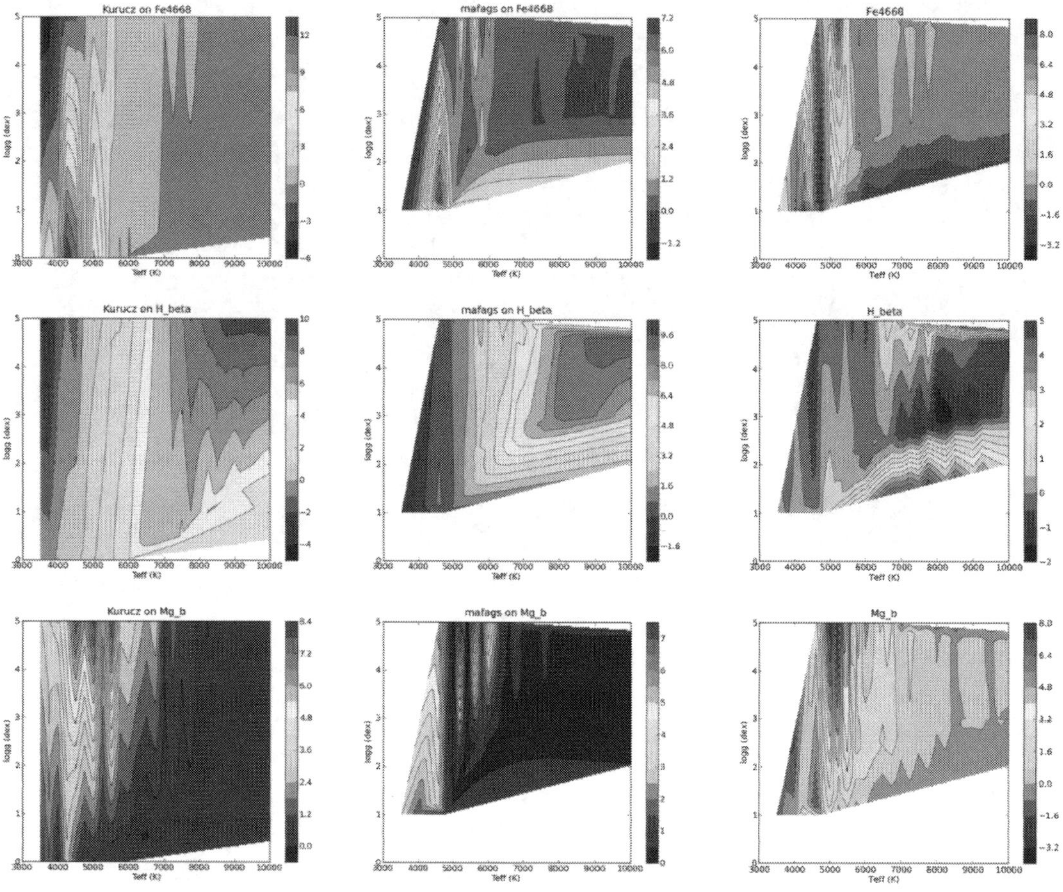

Figure 4. The nine contour plots show three Lick indices of the Kurucz and MAFAGS models, Fe4668, H_beta, and Mg_b in the $T_{\rm eff}$ - log g diagram. The differences between the lines are also shown. From top to the bottom: Fe4668, H_beta, and Mg_b. From left to right: Lick indices of Kurucz, MAFAGS, and the difference between the lines of both models.

the line index calculations with real spectra can yield large errors because of noise in the local continuum flux. The errors in the Lick indices are however sufficiently small for low-resolution spectra and are useful for successfully determining stellar parameters in large surveys such as SDSS. For SDSS/SEGUE, the accuracy of $T_{\rm eff}$ is ~200 K. The accuracy of the gravity is 0.2 dex, and also 0.2 dex for the metallicity. We find that in most contour regions the differences between two line indices stay rather small. A more precise accuracy criterion should be determined for various spectral resolutions. For the current experiment we roughly determine a criterion at $T_{\rm eff}$=5000 K.

3.3. Comparison of the complete spectrum

The complete spectrum or a selected wavelength region can be utilized to obtain the stellar parameters with low-resolution spectra using statistical methods. This type of approach includes template matching, ANN (Artificial Neural Network), and regression methods. We currently employ a template matching method based on a correlation coefficient for comparing the complete spectra of both theoretical model grids. We select 46 pairs of spectra from both grids

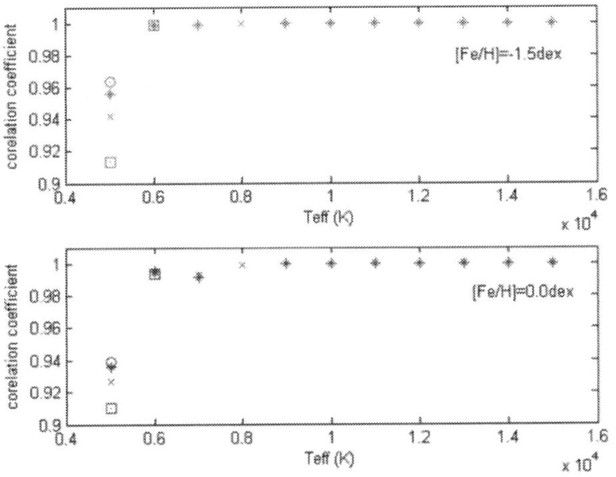

Figure 5. Correlation coefficients computed for 46 pairs of MAFAGS and Kurucz spectra. Upper panel: correlation coefficient with $T_{\rm eff}$ for spectra of [Fe/H]=−1.5 dex. Lower panel: correlation coefficient with $T_{\rm eff}$ for spectra of [Fe/H]=0.0 dex.
○ log g=2.0 dex; $*$ log g=3.0 dex; × log g=4.0 dex; □ log g=5.0 dex.

with the same $T_{\rm eff}$, surface gravity, and [Fe/H]. The 46 pairs of spectra range from low to high metallicity, from small to large $T_{\rm eff}$, for surface gravities of dwarfs to giants. The comparison of MAFAGS and Kurucz spectra with the same parameters is performed by template matching based on a correlation coefficient.

Figure 5 shows the correlation coefficients of the 46 pairs of spectra from MAFAGS and Kurucz. The upper panel shows the correlation coefficient with $T_{\rm eff}$ for spectra of [Fe/H]=−1.5 dex, while the lower panel shows spectra of [Fe/H]=0.0 dex. The different symbols mark the surface gravity. For spectra above 8000 K, the correlation coefficients of the MAFAGS and Kurucz spectra are above 0.99, while for spectra with $T_{\rm eff}$ between 6000 K and 8000 K, the correlation coefficients are in the range of 0.98–0.99. For the spectra with $T_{\rm eff}$=5000 K, the correlation coefficients are in the range of 0.90–0.97. Hence, for warm stars, the MAFAGS and Kurucz spectra are almost identical, while for cool stars below 5000 K, there are some differences.

4. Summary and conclusions

We compare the low-resolution spectral grids of MAFAGS and Kurucz to test if differences between these theoretical model grids influence our measurements of stellar parameters. We employ photometric color indices, absorption line indices, and the complete spectrum with a statistical learning method, for selecting uvby color indices, Lick indices, and template matching of the complete spectrum to investigate the differences between both model grids. We summarize the results as follows:

- both low-resolution Kurucz and MAFAGS grids can employ the (b−y)-index for determining $T_{\rm eff}$.
- the m1-index can be used together with a color index to determine [Fe/H] in F- and G-type stars, although our measurements with both grids reveal systematic discrepancies for cool stars.
- the c1 indices for both grids yield small discrepancies for $T_{\rm eff}$<6000 K. For $T_{\rm eff}$>6000 K the c1 indices of both grids agree well.
- the Lick indices of the Kurucz and MAFAGS grids tend to be in agreement for warm stars of $T_{\rm eff}$>5000 K, while for cooler stars of $T_{\rm eff}$<5000 K the differences in Lick indices for both models are apparently large.
- for warm stars of $T_{\rm eff}$ > 5000 K the MAFAGS and Kurucz spectra are almost the same, while for cooler stars below 5000 K, the MAFAGS and Kurucz spectra are clearly discrepant.

Acknowledgments
We thank Dr. M. Bergemann (Max Planck Institute for Astrophysics, Garching, Germany) for peer reviewing the paper, her insightful and critical comments helped to improve the content and clarity of the paper. The authors also thank Dr. Frank Grupp for providing the MAFAGS grid and helpful suggestions. This work is supported by the National Natural Science Foundation of China (NSFC) under grant No. 10973021 and the Young Researcher Grant of the National Astronomical Observatories, Chinese Academy of Sciences.

5. References

[1] Castelli F and Kurucz R L 2003 *IAU Symposium: Modelling of Stellar Atmospheres* ed N E Piskunov, W W Weiss, and D F Gray **210**
[2] Kurucz R L 1990 *Stellar Atmospheres: Beyond Classical Models* (NATO Asi Ser) ed L Crivellari et al. 441
[3] Kurucz R L 2004 *Mem. S. A. It.* **75** 1
[4] Gustafsson B, Edvardsson B, Eriksson K and Jørgensen U G *et al.* 2008 *A&A* **486** 951
[5] Allard F and Hauschildt P H 1995 *ApJ* **445** 433
[6] Grupp F 2004 *A&A* a **420** 289
[7] Grupp F 2004 *A&A* b **426** 309
[8] Grupp F 2009 *personal comm.*
[9] Castelli F and Kurucz R L 1994 *A&A* **281** 817
[10] Gray R O and Corbally C J 1994 *AJ* **107** 742
[11] Fuhrmann K, Axer M and Gehren T 1993 *A&A* a **271**
[12] Canuto V M and Mazzitelli I 1991*ApJ* **370** 295
[13] Castelli F, Gratton G R and Kurucz R L 1997 *A&A* a **318** 841
[14] Alonso A, Arribas S and Martinez-Roger C 1996 *A&A* **313** 873
[15] Alonso A, Arribas S and Martinez-Roger C 1999 *A&AS* **140** 261
[16] Clem J L, VandenBerg D A, Grundahl F and Bell R A 2004 *AJ* **127** 1227
[17] Casagrande L, Ramírez I, Meléndez J, Bessell M and Asplund M 2010 *A&A* A **512** 54
[18] Gustafsson B and Nissen P E 1972 *A&A* **19** 261
[19] Schuster W J and Nissen P E 1989 *A&A* **221** 65
[20] Calamida A, Bono G Stetson P B, *et al.* 2007 *ApJ* **670** 400
[21] Calamida A, Bono G Stetson P B, *et al.* 2009 *ApJ* **706** 1277
[22] Breger M 1974 *ApJ* **192** 75
[23] Tobin W 1985 *A&A* **142** 189
[24] Worthey G, Faber S M, Gonzalez J J and Burstein D 1994 *ApJS* **94** 687
[25] Worthey G and Ottaviani D L 1997 *ApJS* **111** 377

GREAT-ESF Workshop: Stellar Atmospheres in the Gaia Era | IOP Publishing
Journal of Physics: Conference Series **328** (2011) 012014 | doi:10.1088/1742-6596/328/1/012014

Accurate abundance determinations in S stars

P Neyskens[1], S Van Eck[1], B Plez[2], S Goriely[1], L Siess[1] and A Jorissen[1]

[1]Institut d'Astronomie et d'Astrophysique, Université Libre de Bruxelles,CP 226, Boulevard du Triomphe, B-1050 Bruxelles, Belgium
[2]LUPM, Université Montpellier-II, CNRS-UMR 5024,Place Eugène Bataillon, F-34095 Montpellier, France

E-mail: pieter.neyskens@ulb.ac.be

Abstract. S-type stars are thought to be the first objects, during their evolution on the asymptotic giant branch (AGB), to experience s-process nucleosynthesis and third dredge-ups, and therefore to exhibit s-process signatures in their atmospheres. Until present, the modeling of these processes is subject to large uncertainties. Precise abundance determinations in S stars are of extreme importance for constraining e.g., the depth and the formation of the ^{13}C pocket. In this paper a large grid of MARCS model atmospheres for S stars is used to derive precise abundances of key s-process elements and iron. A first estimation of the atmospheric parameters is obtained using a set of well-chosen photometric and spectroscopic indices for selecting the *best model atmosphere* of each S star. Abundances are derived from spectral line synthesis, using the selected model atmosphere. Special interest is paid to technetium, an element without stable isotopes. Its detection in stars is considered as the best possible signature that the star effectively populates the thermally-pulsing AGB (TP-AGB) phase of evolution. The derived Tc/Zr abundances are compared, as a function of the derived [Zr/Fe] overabundances, with AGB stellar model predictions. The computed [Zr/Fe] overabundances are in good agreement with the AGB stellar evolution model predictions, while the Tc/Zr abundances are slightly over-predicted. This discrepancy can help to set stronger constraints on nucleosynthesis and mixing mechanisms in AGB stars.

1. Introduction: S stars as transition objects

The last nuclear burning stage for stars with an initial mass between ~ 0.8 M$_\odot$ and 8 M$_\odot$ occurs during the asymptotic giant branch (AGB) phase. AGB stars are known to be evolved, cool ($T_{\rm eff} < 4000$ K) and luminous (luminosity $\sim 10^4$ L$_\odot$) giant stars. The rather complex interior of AGB stars is the site of rich nucleosynthesis and mixing processes. The synthesized elements, mainly neutron-rich s-process elements and carbon, are transported to the stellar surface by the so-called third dredge-up (3DUP). The repeated occurrence of thermal pulses and 3DUPs leads to considerable changes in the chemical composition of the stellar atmosphere during its evolution on the AGB. Therefore, AGB stars can be classified into three main groups: M, S, and C stars [1]. Spectra of oxygen-rich M stars are characterized by strong TiO absorption bands, while spectra of carbon-rich C stars are dominated by C_2, CN, and CH absorption bands. Stars showing absorption bands of ZrO (in addition to TiO absorption bands) are classified as S. S stars have a carbon-over-oxygen (C/O) ratio between those of M (~ 0.5) and C stars (> 1) [2], hence they serve as interesting study objects to constrain the mixing and nucleosynthesis processes in AGB stars.

Journal of Physics: Conference Series **328** (2011) 012014 doi:10.1088/1742-6596/328/1/012014

2. Deriving abundances of S stars

Nucleosynthesis calculations in AGB stars still suffer from many uncertainties. For example, neutrons for the s-process can be released by two reactions: $^{13}C(\alpha,n)^{16}O$ and $^{22}Ne(\alpha,n)^{25}Mg$. The former reaction is activated at temperatures around 10^8 K, while the later at larger temperatures around 3×10^8 K. Details about the dominant s-process neutron source are still lacking, however observations and models hint at $^{13}C(\alpha,n)^{16}O$ being the dominant neutron source in low-mass TP-AGB stars (≤ 3 $M_\odot$) [3, 4, 5]. ^{22}Ne is the dominant neutron source in more massive TP-AGB stars since the temperature during the thermal pulses is sufficiently large to activate $^{22}Ne(\alpha,n)^{25}Mg$ [6]. Clues about the dominant neutron source can be obtained by deriving abundances of s-process elements (e.g., Rb, Sr) in S stars. Since ^{22}Ne produces a larger neutron flux density than ^{13}C, branching points at ^{85}Kr and ^{86}Rb can be passed to produce Rb. However, the large Rb/Zr ratios derived in massive (6–7 $M_\odot$) Large Magellanic Cloud AGB stars [7] indicate, if real, severe problems for the s-process in massive AGB stars. Even if ^{22}Ne is the dominant neutron source, it is not possible to synthesize these large amounts of Rb without also overproducing Zr. In addition, it may demonstrate the currently incomplete understanding of the analysis of the atmospheres of massive AGB stars.

The efficiency of the ^{13}C neutron source depends on the amount of protons mixed down from the H-rich envelope into the ^{12}C-rich layers. The mixed-down protons are captured by ^{12}C nuclei to trigger the chain $^{12}C(p,\gamma)^{13}N(\beta^+)^{13}C$. Large uncertainties exist on the depth and the formation the ^{13}C pocket. Different extra mixing mechanisms have been proposed: convective overshooting, rotation, gravity waves, and thermohaline mixing [8, 9, 10, 11]. Although these mixing models are able to reproduce qualitatively the observed global distribution of s-elements, they are not able to reproduce quantitatively the observed overabundance of s-elements and ^{12}C.

Special attention is paid to Tc (Z=43), an element without stable isotopes. ^{99}Tc, the only isotope produced by the s-process in AGB stars, has a half-life of 2.1 $\times 10^5$ years. It is of the order of the time spent by low- and intermediate-mass stars on the TP-AGB. The increasing surface enrichment of Tc is reduced by its radioactive decay and gives rise to an almost time-independent predicted surface abundance of Tc during the ascent of the TP-AGB, while the predicted abundances of all other s-elements increase progressively [4, 12].

Consequently, it is of utmost importance to derive precise abundances of key s-elements in S stars. However, it is a difficult task because of the influence of non-solar C/O ratios on the atmospheric thermal structure [13] and due to strong TiO and ZrO molecular blending that prevent the use of equivalent widths.

Constraints on the efficiency of the 3DUP could also be set by fitting carbon-star luminosity functions (CSLF) with predictions of synthetic AGB models, using a constant free parameter for the efficiency of the 3DUP. However, it is beyond the scope of this paper.

In this paper we present derived Tc/Zr abundances, as a function of derived [Zr/Fe] overabundances, for seven intrinsic (Tc-rich) S stars belonging to the Henize sample of S stars [14]. The results are compared to the predictions of the stellar evolution code STAREVOL [15].

Constraining the initial mass of our stars is a hard task. When available, abundances of Rb and Sr can be used because temperatures that activate the ^{22}Ne neutron source are only reached in more massive AGB stars. However, mass determinations of extrinsic (Tc-poor) S stars are easier since these stars must be members of a binary star system. From an inspection of the orbital parameters of extrinsic S stars, [16] found an average mass of 1.6 $M_\odot$ with the assumption that the star's companion is a white dwarf of 0.6 $M_\odot$. However, extrinsic S stars are not populating the TP-AGB.

3. Observations

Seven Tc-rich S stars were selected from the sample of 205 Henize S stars. The sample contains both *intrinsic* and *extrinsic* S stars with $R \leq 10.5$ and $\delta \leq -25°$. Optical low-resolution Boller &

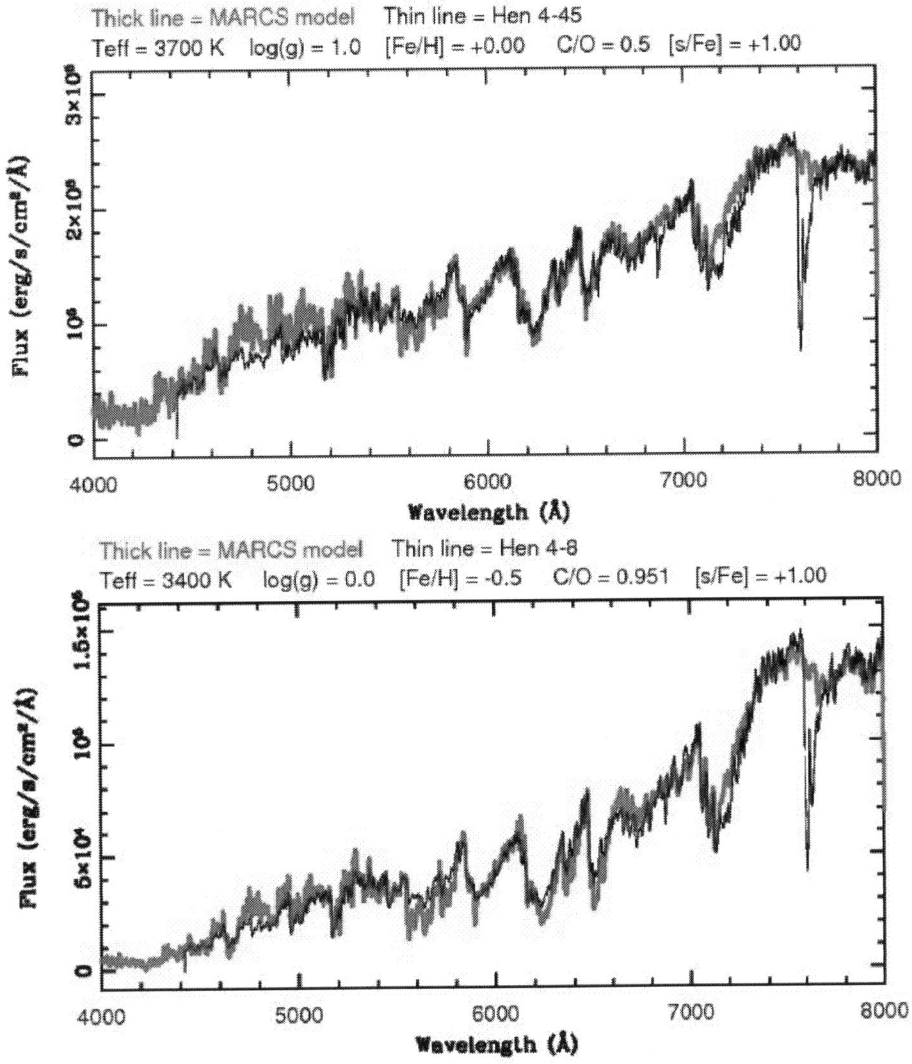

Figure 1. A comparison between the observed low-resolution optical spectrum (thin drawn black line) and the selected synthetic spectrum (boldly drawn red line) for Hen 4-45 (top panel) and Hen 4-8 (bottom panel). The atmospheric parameters of the synthetic spectra are listed in the top labels of each panel. The 7600 Å feature is from telluric O_2.

Chivens spectra ($\Delta\lambda$ =3 Å, 4400 Å – 8200 Å) were observed with the 1.52 m ESO telescope for a large fraction of the Henize S stars. High-resolution Coudé Echelle spectra ($\Delta\lambda \approx$ 0.08 Å, 4220 Å – 4280 Å) were also observed with the 1.4 m CAT telescope at ESO. The presence or absence of Tc could be checked in a consistent way since the high-resolution spectra are covering two Tc I lines: Tc I 4238.191 Å and Tc I 4262.270 Å. Optical Geneva photometry and infrared SAAO *JHKL* photometry (Catchpole, personal comm.) are also available [18, 19]. The photometric and spectroscopic data are dereddened using [20] and [21].

Journal of Physics: Conference Series **328** (2011) 012014 doi:10.1088/1742-6596/328/1/012014

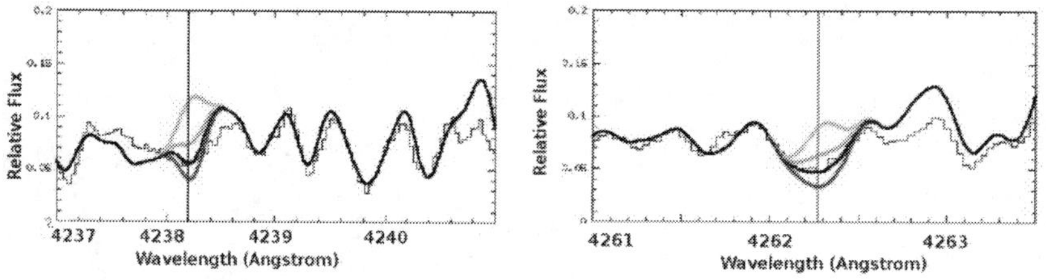

Figure 2. Tc I 4238.191 Å (left-hand panel) and Tc I 4262.270 Å (right-hand panel) for four different Tc abundances (without Tc, log ε(Tc) = 0.0, log ε(Tc) = 0.3 and log ε(Tc) = 0.6) compared to the high-resolution observation (thin drawn line) of Hen 4-162.

4. The grid of S-star MARCS atmosphere models

Dedicated model atmospheres for S stars were computed with the latest version of the MARCS code for late-type stars [22]. The constructed grid contains 3522 models and covers the following parameter space:

- 2700 K $\leq T_{\rm eff} \leq$ 4000 K (with steps of 100 K)
- 0.0 $\leq \log g \leq$ 5.0 (with steps of 1.0)
- [Fe/H] = 0.0 and −0.5
- C/O = 0.501, 0.751, 0.899, 0.925, 0.951, 0.971, and 0.991
- [s/Fe] = 0., 1., 2.
- M = 1 $M_\odot$
- [α/Fe] = −0.4 × [Fe/H]

The one-dimensional S-star model atmospheres are computed for different assumptions such as homogeneous stationary layers, hydrostatic and local thermodynamic equilibrium, and energy conservation by radiative and convective fluxes. Opacity sampling is applied and different continuous and line opacities are taken into account. Synthetic S stars spectra were computed using the radiative transfer code TURBOSPECTRUM [23].

5. Selecting the best atmosphere model

A first estimation of the atmospheric parameters is based on a χ^2-minimization between observed and synthetic colors (UBV & $JHKL$), and between observed with synthetic band/line-strength indices (TiO, ZrO, Na), using a so-called *best-model-finding tool*. As demonstrated by [13], the effective temperature, the C/O ratio, and the s-process enrichment are disentangled by comparing (1) observed $(V-K)_0$, $(J-K)_0$ colors with synthetic ones, and (2) observed TiO and ZrO indices with synthetic ones. Figure 1 shows the agreement, for two different S stars, between the observed low-resolution Boller & Chivens spectrum and the synthetic spectrum derived from the selected *best model atmosphere*. The agreement between the selected synthetic spectra and observed spectra is in most cases very good. However, it is sometimes difficult to choose one out of first few 'best-fit' synthetic spectra. Therefore, the error of the atmospheric parameters is computed from the range covering the first five best-fit models (i.e., the five models with the lowest χ^2 in the atmospheric parameter space). The errors are typically 100 K for $T_{\rm eff}$, 0.2 (for C/O $<$ 0.9) or 0.02 (for C/O $>$ 0.9) for C/O, and 0.5 dex for [s/Fe]. Additional uncertainties (stellar variability, errors on reddening) are inevitably still present and reliably estimating them is delicate.

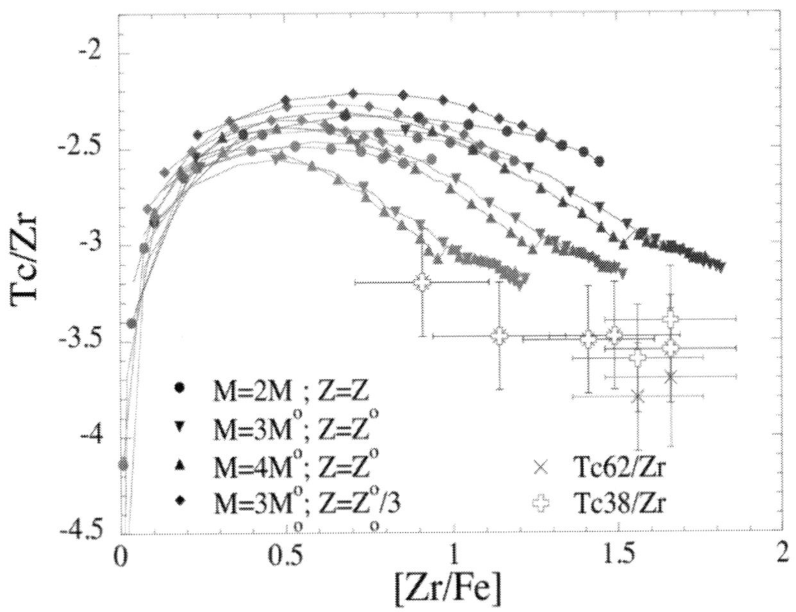

Figure 3. Comparison between observed and predicted Tc/Zr ratios as a function of the [Zr/Fe] overabundances. Four model stars with different masses and metallicities are considered (M = 2, 3, 4 $M_\odot$ at solar metallicity and M = 3 $M_\odot$ for $Z=Z_\odot/3$). For each star, the envelope s-process enrichment is calculated all along the AGB phase (the symbols correspond to the abundance ratio immediately after a 3DUP event) assuming 3 different values for the extent of the partial mixing zone λ_{pm}: for each (M,Z) model (i.e., given symbols), the leftmost curve corresponds to λ_{pm}=0.05, the rightmost to λ_{pm}=0.20, and the middle one to λ_{pm}=0.10. Observations are shown with two cross symbol types. Tc62 and Tc38 concern the Tc abundances derived from the Tc I 4262.270 Å and Tc I 4238.191 Å lines, respectively.

6. Deriving abundances of intrinsic S stars

Synthetic high-resolution spectra, based on the models computed with the derived atmospheric parameters, are next compared to high-resolution observed spectra. It allows us to fine-tune the individual chemical abundances for obtaining a good match between the observed and synthetic spectra. This way abundances of Tc, Zr, and Fe were computed. The agreement between an observed spectrum with two Tc I lines, at 4238.191 Å and 4262.270 Å, respectively, and four synthetic spectra with different Tc abundances is shown in Fig. 2.

7. Results and discussion: observed abundances vs. model predictions

Figure 3 shows the calculated Tc/Zr abundances as a function of the [Zr/Fe] overabundances. The observations are compared with s-process calculations performed on 4 AGB models computed with the stellar evolution code STAREVOL [15]. The s-process model we consider parametrizes the partial mixing of protons into the ^{12}C-rich region at the time of a 3DUP, as described in detail in [4]. The λ_{pm} parameter corresponds to the ratio of the mass of the partial mixing zone and the mass of the thermal pulse at its maximum extension. It is varied between 5% and 20% of the extent of the thermal pulse. Each symbol traces the abundance ratio immediately after a 3DUP event.

The s-process enrichment in the stellar envelope reaches values compatible with the observations, i.e., about 1 to 2 dex enrichment for [Zr/Fe]. As is shown in Fig. 3, the Tc/Zr ratio is systematically over-predicted. Different explanations for the discrepancy can be given, in particular an increased partial decay of Tc, either during longer inter-pulse phases (e.g., in lower-mass stars), or during hotter thermal pulses.

Another problem concerns the C/O ratio. As is shown in Fig. 12 of [4], the models are already carbon-rich after four or five 3DUPs. However, the strength of the observed TiO and ZrO absorption bands of the 7 target stars clearly indicates a C/O ratio below unity. It would imply, as shown in Fig. 3, that the maximum observed [Zr/Fe] abundance has to be around 0.2 for a $M=2M_{\odot}$ model of solar metallicity, which is clearly not the case. This disagreement between the observed and predicted C and s-process abundances is well-known (see [4]); S stars however, with intermediate C/O ratio, allow us to investigate it with greater accuracy.

8. Perspectives

The present data will be supplemented with new data and further analyzed in a forthcoming paper. New observations, obtained with the high-resolution HERMES spectrograph [24] mounted on the Mercator telescope (La Palma), can provide additional constraints on the s-process in AGB stars. These spectra are covering a large wavelength region in the red part of the optical domain (between 7000 Å and 9000 Å) where molecular blending is less dominant than in the blue part. Abundances of C, O, and s-elements will be derived. The derivation of S stars atmospheric parameters and individual abundances, and the comparison with stellar evolution AGB models, will be performed for a larger sample of S stars.

Acknowledgments

P.N. is *Boursier F.R.I.A.*, Belgium. S.V.E., S.G. and L.S. are F.R.S.-F.N.R.S. research associates. This research has been supported by the *Communité Française de Belgique - Actions de Recherche Concertées.*

References

[1] Keenan P C 1954 *ApJ* **120** 484
[2] Iben I Jr and Renzini A 1983 *ARA&A* **21** 271
[3] Smith V V and Lambert D L 1986 *ApJ* **311** 843
[4] Goriely S and Mowlavi N 2000 *A&A* **362** 599
[5] Goriely S and Siess L 2004 *A&A* **421** 25
[6] Masseron T, Johnson J A, Plez B, Van Eck S, Primas F, Goriely S and Jorissen A 2010, *A&A* **509** 93
[7] García-Hernández D A, Manchado A, Lambert D L, Plez B, García-Lario P, D'Antona F, Lugaro M, Karakas A I and van Raai M 2009 *ApJ* **705** L31
[8] Freytag B, Ludwig H and Steffen M 1996 *A&A* **313** 497
[9] Denissenkov P A and Tout C A 2003 *MNRAS* **340** 722
[10] Siess L, Goriely S and Langer N 2004 *A&A* **415** 1089
[11] Eggleton P P, Dearborn D S P and Lattanzio J C 2006 *Science* **314** 1580
[12] Van Eck S, Jorissen A, Goriely S and Plez B 2001 *Nucl. Phys. A* **688** 45
[13] Van Eck S, Neyskens P, Plez B, Jorissen A, Edvardsson B, Eriksson K, Gustafsson B, Jørgensen U G and Nordlund Å 2011 *Proc. Why Galaxies Care About AGB STARS II: Shining Examples And Common Inhabitants (Vienna)* **445** (San Francisco : Sheridan books) p 71
[14] Henize K G 1960 *AJ* **65** 491
[15] Siess L 2007 *A&A* **476** 893
[16] Jorissen A, Van Eck S, Mayor M and Udry S 1998 *A&A* **332** 877
[17] Van Eck S, Jorissen A, Udry S, Mayor M and Pernier B 1998 *A&A* **329** 971
[18] Van Eck S and Jorissen A 1999 *A&A* **345** 127
– 2000 *A&A* **360** 196
[19] Van Eck S, Jorissen A, Udry S, Mayor M, Burki G, Burnet M and Catchpole R 2000 *A&A* **145** 51

[20] Cardelli J A, Clayton G C and Mathis J S 1989 *ApJ* **345** 245
[21] Drimmel R, Cabrera-Lavers A and López-Corredoira M 2003 *A&A* **409** 205
[22] Gustafsson B, Edvardsson B, Eriksson K, Jørgensen U G, Nordlund Å and Plez B 2008 *A&A* **486** 951
[23] Alvarez R and Plez B 1998 *A&A* **330** 1109
[24] Raskin G, Van Winckel H, Hensberge H, Jorissen A, Lehmann H, Waelkens C, Avila G, de Cuyper J P, Degroote P, Dubosson R, *et al* 2011 *A&A* **526** A69

Journal of Physics: Conference Series **328** (2011) 012015 doi:10.1088/1742-6596/328/1/012015

Testing common classical LTE and NLTE model atmosphere and line-formation codes for quantitative spectroscopy of early-type stars

Norbert Przybilla[1], Maria-Fernanda Nieva[1,2] and Keith Butler[3]

[1] Dr. Karl Remeis-Sternwarte & ECAP, Sternwartstr. 7, D-96049 Bamberg, Germany
[2] Max-Planck-Institut für Astrophysik, Karl-Schwarzschild-Str. 1, D-85741 Garching, Germany
[3] Universitätssternwarte, Scheinerstr. 1, D-86179 München, Germany

E-mail: `norbert.przybilla@sternwarte.uni-erlangen.de`

Abstract. It is generally accepted that the atmospheres of cool/lukewarm stars of spectral types A and later are described well by LTE model atmospheres, while the O-type stars require a detailed treatment of NLTE effects. Here model atmosphere structures, spectral energy distributions and synthetic spectra computed with ATLAS9/SYNTHE and TLUSTY/SYNSPEC, and results from a hybrid method combining LTE atmospheres and NLTE line-formation with DETAIL/SURFACE are compared. Their ability to reproduce observations for effective temperatures between 15 000 and 35 000 K are verified. Strengths and weaknesses of the different approaches are identified. Recommendations are made as to how to improve the models in order to derive unbiased stellar parameters and chemical abundances in future applications, with special emphasis on Gaia science.

1. Introduction

Quantitative stellar spectroscopy employs model atmospheres that are based on many approximations and simplifications. One of the fundamental decisions to be made for the model construction concerns the description of the thermodynamic state of the atmospheric plasma. Either local thermodynamic equilibrium (LTE) may be assumed, or deviations from it may be allowed for (NLTE), i.e. taking the mutual interaction of radiation and matter into account.

It is generally accepted that the atmospheres of cool/lukewarm stars of spectral types A and later are described well by LTE model atmospheres, while O stars require a detailed treatment of NLTE effects due to the high energy density of their radiation field. Both approaches are being followed in the literature for analyses of B-type stars at present, with variable success.

The present work tests the predictive power of common LTE and NLTE model atmosphere and line-formation codes in this transition region, concentrating on main-sequence stars of effective temperatures between 15 000 and 35 000 K (corresponding to spectral types of about B4 to O9). Three approaches are investigated: pure LTE as implemented in the ATLAS9/SYNTHE suite of codes [1, 2], pure NLTE with TLUSTY/SYNSPEC [3], and a hybrid approach based on LTE atmospheres (ATLAS9) and subsequent NLTE line-formation computations with DETAIL/SURFACE [4, 5]. We deviate from the usual 'model vs. model' comparison – which may be misleading as *a priori* it is not clear as to which model is more realistic – by testing observational constraints wherever possible.

Journal of Physics: Conference Series **328** (2011) 012015 doi:10.1088/1742-6596/328/1/012015

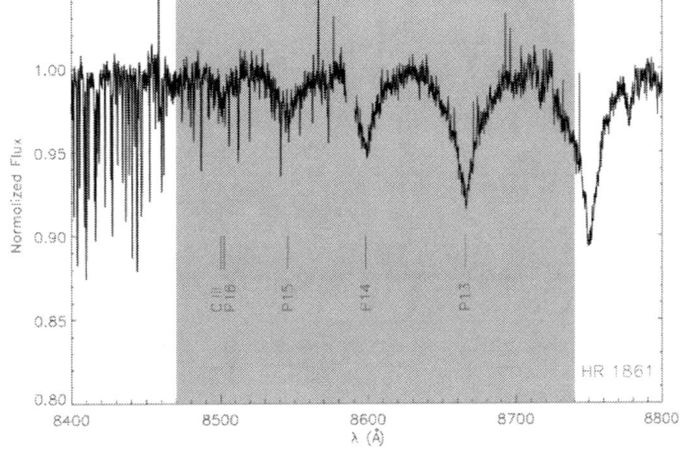

Figure 1. Near-IR spectrum of HR 1861 (B1 IV), observed with FOCES on the Calar Alto 2.2m telescope. The most important lines in the Gaia RVS spectral range (grey region) are identified.

2. The Gaia perspective for investigations of OB-type stars

The legacy of the Gaia mission in terms of OB-type stars will mainly lie in the determination of highly precise parallaxes, proper motions and broad-band space photometry. In contrast to the cooler stars, little information besides radial and projected rotational velocities will be derivable from the spectra obtained with the on-board radial-velocity spectrometer (RVS). The reason for this is the scarcity of spectral lines in the wavelength range of the RVS, which essentially comprises the Paschen series members P_{13} to P_{16} in this kind of star, see Fig. 1. The determination of accurate and precise stellar parameters and elemental abundances will therefore have to rely on ground-based observations, e.g. within the Gaia-ESO survey [6].

3. Models: assumptions & implementations

Quantitative spectroscopy of early-type main-sequence stars and giants in the effective temperature range of interest, $15\,000 \leq T_{\rm eff} \leq 35\,000$ K, can rely on *classical model atmospheres*[1]. Basic assumptions for the model construction are chemical homogeneity, stationarity, plane-parallel geometry and hydrostatic and radiative equilibrium, while the thermodynamic state of the atmospheric plasma can be determined in LTE or NLTE. The former case is conceptually less realistic but attractive since typical computing times per model are a few seconds on modern PCs, while NLTE models may require many CPU hours. A third approach, so-called hybrid NLTE modelling, can combine the strengths of both types of models, performing sophisticated NLTE line-formation computations on LTE atmospheres, at a fraction of the computing cost of full NLTE models. This is particularly attractive in those cases where the atmospheric structure is close to LTE as for the parameter space investigated here which we will show in the following.

For our test we have chosen results from the most common among the many available model atmosphere codes: on the one hand line-blanketed LTE atmospheres and synthetic spectra computed with the ATLAS9/SYNTHE suite of codes [1, 2] and on the other line-blanketed NLTE models calculated with the TLUSTY/SYNSPEC package [3]. Model structures, fluxes and detailed synthetic spectra were adopted from published grids, Castelli's ODFNEW models [8], the Padova spectral library [9], and the OSTAR2002 and BSTAR2006 grids [10, 11].

[1] More luminous stars, either mid/early O-type stars or OB-supergiants close to the Eddington limit, require consideration of their stellar winds, i.e. hydrodynamic model atmospheres (and spherical geometry), see e.g. [7].

Journal of Physics: Conference Series **328** (2011) 012015 doi:10.1088/1742-6596/328/1/012015

Table 1. Model atoms used with our DETAIL/SURFACE computations.

Ion	Ref.	Ion	Ref.	Ion	Ref.	Ion	Ref.
H I	[13]	C II/III	[15, 16]	O I/II	[18, 19]	Si III/IV	[21]
He I/II	[14]	N II	[17]	Mg II	[20]	Fe II/III	[22, 23]

Our own computations rely on ATLAS9 atmospheres and NLTE line-formation computations with updated and extended versions of DETAIL/SURFACE [4, 5]. The former solves the coupled statistical equilibrium and radiative transfer equations, employing the accelerated lambda iteration scheme of [12], while the latter computes the synthetic spectra, using refined line-broadening theories. Model atoms according to Table 1 were employed. More details on our hybrid NLTE approach (labeled 'ADS' henceforth) can be found e.g. in [24].

4. Model tests

In order to facilitate the use of existing grids for the model comparisons, a two-step strategy has to be followed. In the first step, we compare our *tailored* ADS modelling with observation for strategic points in the investigated parameter space to test its ability to reproduce observation. Then, in the second step, ADS models for *generic* parameters are compared to the TLUSTY/SYNSPEC and ATLAS9/SYNTHE models at a grid point closest to the real star's parameters, thus facilitating a meaningful assessment of the predictive power of all these modelling implementations. While the synthetic spectra were convolved with the same rotational and macroturbulent broadening profiles as for the comparison with observation, the Gaussian ('instrumental') profile had to be adjusted to account for the lower resolution of the Padova grid. Due to space restrictions we will discuss the case of the well-studied standard star γ Peg (B2 IV) in detail in the following, and comment on more general aspects of the comparisons briefly[2].

For γ Peg (HD 886) we derived $T_{\rm eff} = 22\,000$ K, a surface gravity of $\log g = 3.95$ (cgs units), microturbulence $\xi = 2$ km/s, solar helium abundance and metal abundances typical for the normal early B-type star population in the solar neighborhood [26, 28]. The nearest parameter combination covered by the published grids coincides with these, except that $\log g = 4.00$ and solar abundances according to [29]. We have adopted the generic grid parameters nevertheless for the ADS model in the second step of the comparison in order to avoid any systematic bias.

Model atmosphere structures and spectral energy distributions. We begin testing the models with a comparison of atmospheric structures in Fig. 2. The temperature structures of the LTE and NLTE models agree very well, except for the outermost layers. Differences amount to up to 1–2% for all depths relevant for the formation of the spectral lines and continua from the far-UV through the optical to the near-IR. Interestingly, the differences are smaller for the low-metallicity comparison at 1/10th of the solar value, despite the fact that NLTE effects should be more pronounced because of increased photon mean-free paths and a harder radiation field. A possible explanation of this unexpected finding could be the differences in the metal line opacities (which matter more at higher metallicity). Many more elements are considered in

[2] Our observational sample comprises a total of 29 apparently slowly-rotating B3 to B0 dwarf and giant stars, for which spectra of high signal-to-noise ratio (S/N $\sim$ 250–800), high resolution ($R = \Delta\lambda/\lambda \simeq$40–48 000) and wide wavelength coverage were obtained using state-of-the-art Echelle spectrographs on 2m-class telescopes, see [25] for a brief overview. For this sample we have shown that combining a sophisticated analysis technique with our ADS modelling allows practically the entire observed spectra in the optical for all these objects [26, 27, 28] to be reproduced to similar or even better quality as shown for γ Peg in the following.

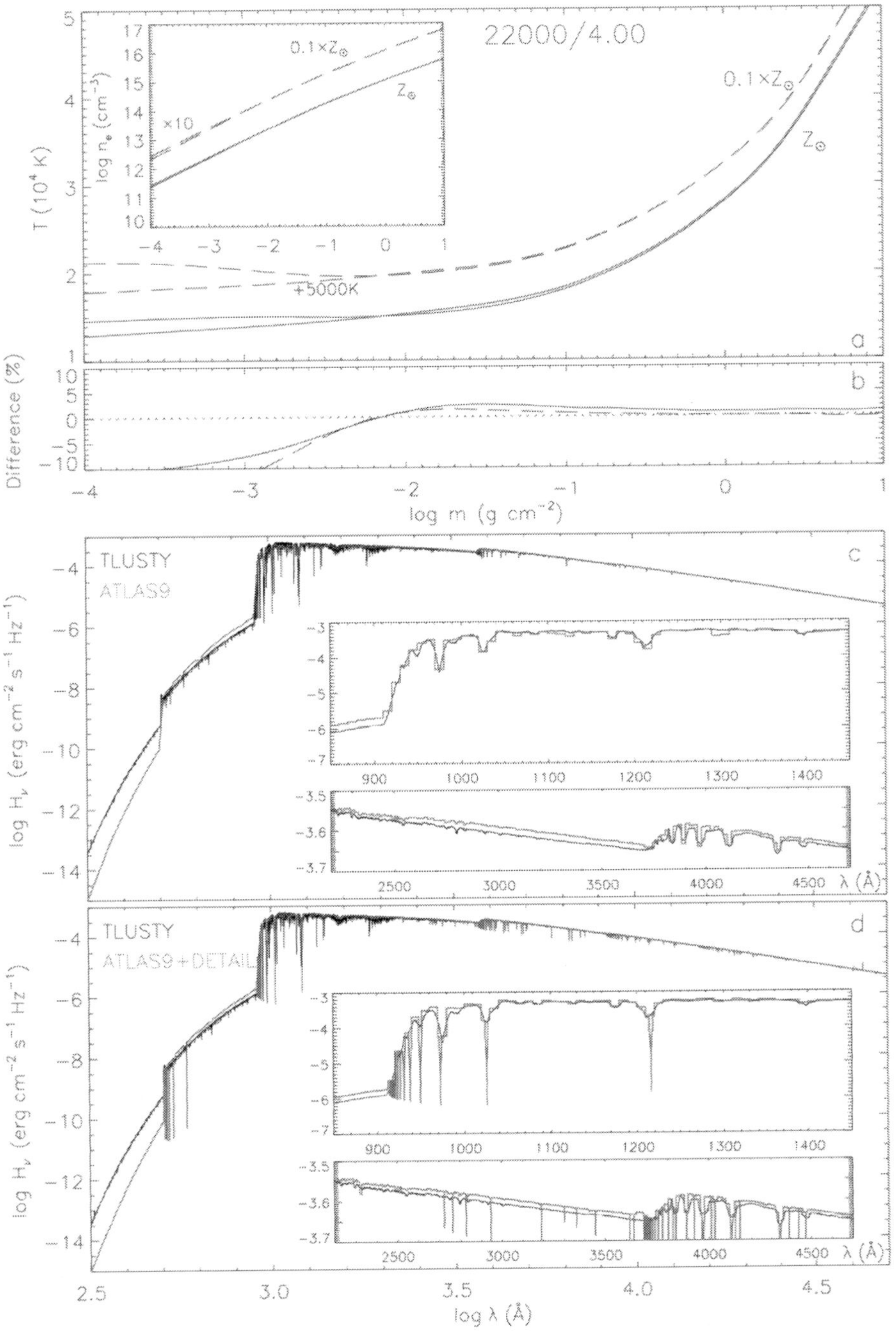

Figure 2. Comparison of model atmosphere structures and resulting spectral energy distributions. Panel (a) presents a comparison of ATLAS9 (red) and TLUSTY (black) temperature structures as a function of mass scale, for models at $T_{\rm eff} = 22\,000$ K, $\log g = 4.00$, $\xi = 2$ km/s at solar (full lines) and 1/10th solar metallicity (dashed lines, shifted by +5000 K). Electron density structures are shown in the inset. Panel (b) displays the resulting differences between the ATLAS9 and TLUSTY temperature structures (in percent). Panel (c) and (d) compare the TLUSTY with ATLAS9 and ATLAS9+DETAIL Eddington fluxes, respectively. The insets highlight the regions around the Lyman and the Balmer jumps. Color coding according to the legends.

Journal of Physics: Conference Series **328** (2011) 012015 doi:10.1088/1742-6596/328/1/012015

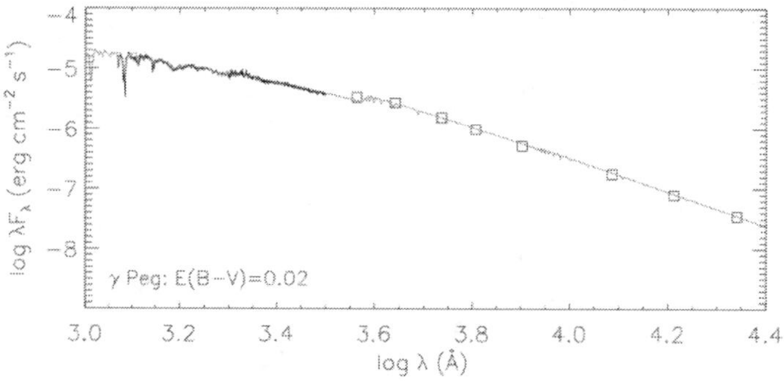

Figure 3. Comparison between our computed SED (red line) and observation for γ Peg (B2 IV): IUE spectrophotometry (black line) and UBVRIJHK photometry (squares).

the ATLAS9 than in the TLUSTY model, resulting in a stronger line-blanketing effect, exactly as indicated in Fig. 2 (a). Very good agreement is found also for LTE and NLTE density structures. Overall, close agreement of the LTE and NLTE structures is found over the entire parameter range considered, for the depths relevant for the formation of the observable line spectra and continua, at $-2 \lesssim \log m \lesssim -0.3$.

As temperature and density stratifications are not directly observable, we verified the effects on line profiles of hydrogen, helium and metal lines by performing line-formation computations with DETAIL & SURFACE on ATLAS9 and TLUSTY atmospheres. Indeed, only small differences are found for the resulting profiles, which are irrelevant for the stellar parameter and elemental abundance determination, see e.g. Fig. 10 of [24]. Maximum effects on metal abundances derived from a few individual lines are about 0.05 dex, and less for mean elemental abundances.

Overall good agreement is also found for the LTE and NLTE spectral energy distributions (SEDs), as exemplified in Fig. 2 (c) and (d). The differences in the continua are because of the temperature differences at the formation depths. Significant discrepancies occur only in the Lyman and helium continua, but it is difficult to test the models in this respect as interstellar absorption prevents observations of the extreme-UV for almost all early-type stars. The model differences in the extreme-UV vary with $T_{\rm eff}$. Very good agreement over the entire spectral range is achieved at higher $T_{\rm eff}$, see Fig. 9 of [24]. Our models can reproduce observed SEDs and therefore the global stellar energy output very well, if the observations are corrected for the appropriate amount of interstellar reddening. An example, for γ Peg, is shown in Fig. 3.

We conclude that *LTE and NLTE model atmospheres are essentially equivalent for dwarf and giant stars over the range* $15\,000 \leq T_{\rm eff} \leq 35\,000$ *K*, for most practical applications.

Line spectra. Resolved line spectra are the most valuable observables as they give information about the temperature and density stratifications within the stars' atmospheres. They are also well-suited for the determination of stellar parameters and elemental abundances.

H & He: Hydrogen and helium provide the deepest features in the spectra of early B-type stars, thus sampling the widest extent of the atmospheric stratification. Our overall best-fit ADS model is compared to observed profiles for many diagnostic lines of γ Peg in Fig. 4. We note explicitly, that *one* synthetic spectrum is displayed in all the individual panels – also for all following comparisons – and not best fits to the individual lines. Overall, an excellent match of observation and theory is obtained. In particular the line wings agree well, being formed in the deeper layers of the atmosphere. Small discrepancies are present for the line cores, in particular for the stronger He I lines, which originate at shallower atmospheric depths. And, apparently the treatment of the He I forbidden components may be improved in some cases.

Journal of Physics: Conference Series **328** (2011) 012015 doi:10.1088/1742-6596/328/1/012015

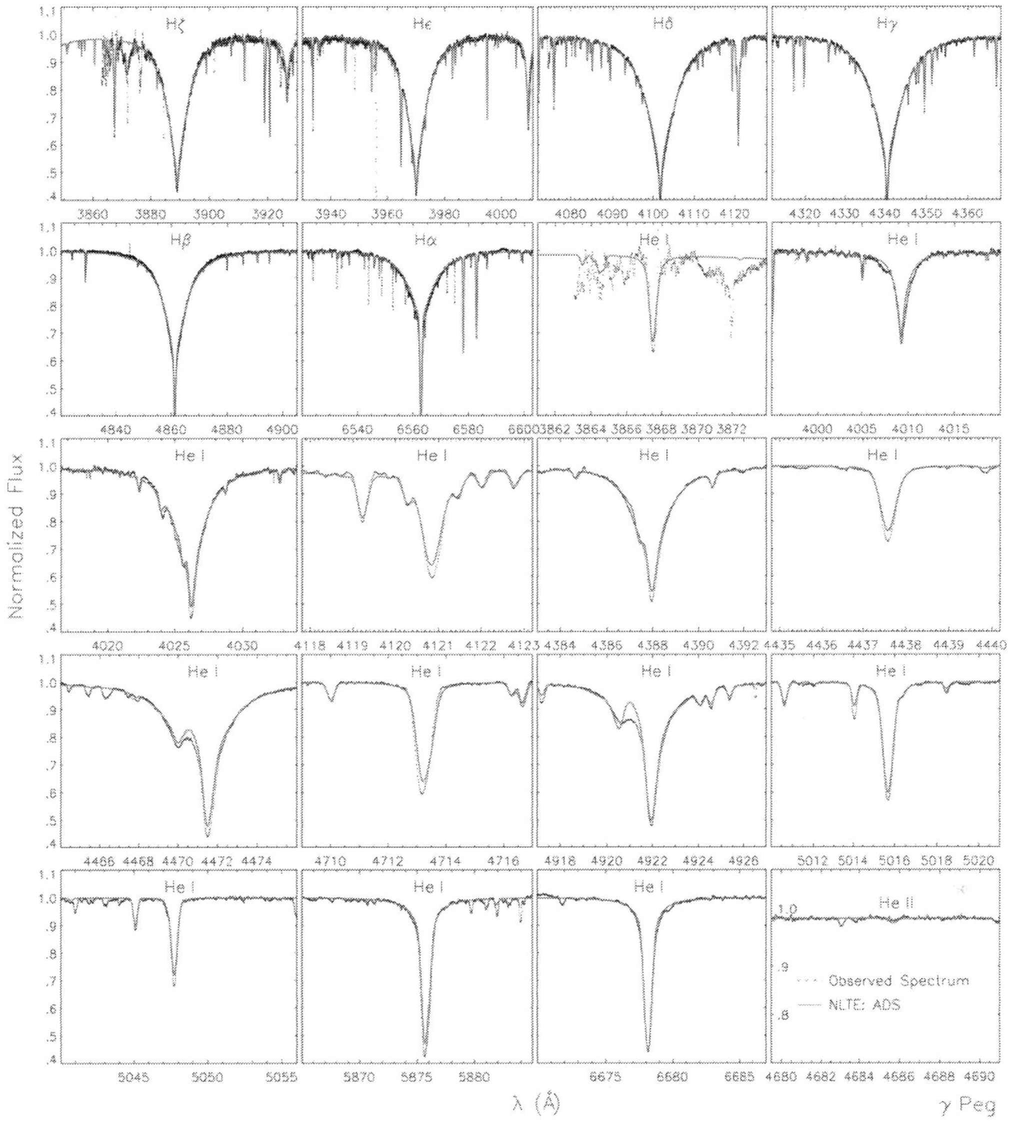

Figure 4. Comparison of strategic H I and He I/II in the observed spectrum of γ Peg (black dots) with our global best-fit ADS model (red line).

About the same fit quality can be achieved throughout the entire parameter space investigated here, with additional minor complications arising at the hotter end of the $T_{\rm eff}$-range, where weak stellar winds start to impact Hα and the He II λ4686 Å line, see [24] for examples.

We conclude that the NLTE modelling with ADS can reproduce the observed H I and He I/II lines reliably, except for some fine details. ADS models thus provide a good reference for the second step of our test, comparison with the TLUSTY/SYNSPEC NLTE and ATLAS9/SYNTHE LTE models, which is shown in Fig. 5. Both NLTE approaches produce closely resembling profiles for this test case, and the good agreement (except for some minor details) holds throughout the entire parameter range considered here for the H I, He II and He I triplet lines. However, at higher temperatures the He I singlet lines computed with TLUSTY/SYNSPEC become significantly weaker than the ADS profiles (and the observations) – the well-known singlet-triplet problem [30].

Journal of Physics: Conference Series **328** (2011) 012015 doi:10.1088/1742-6596/328/1/012015

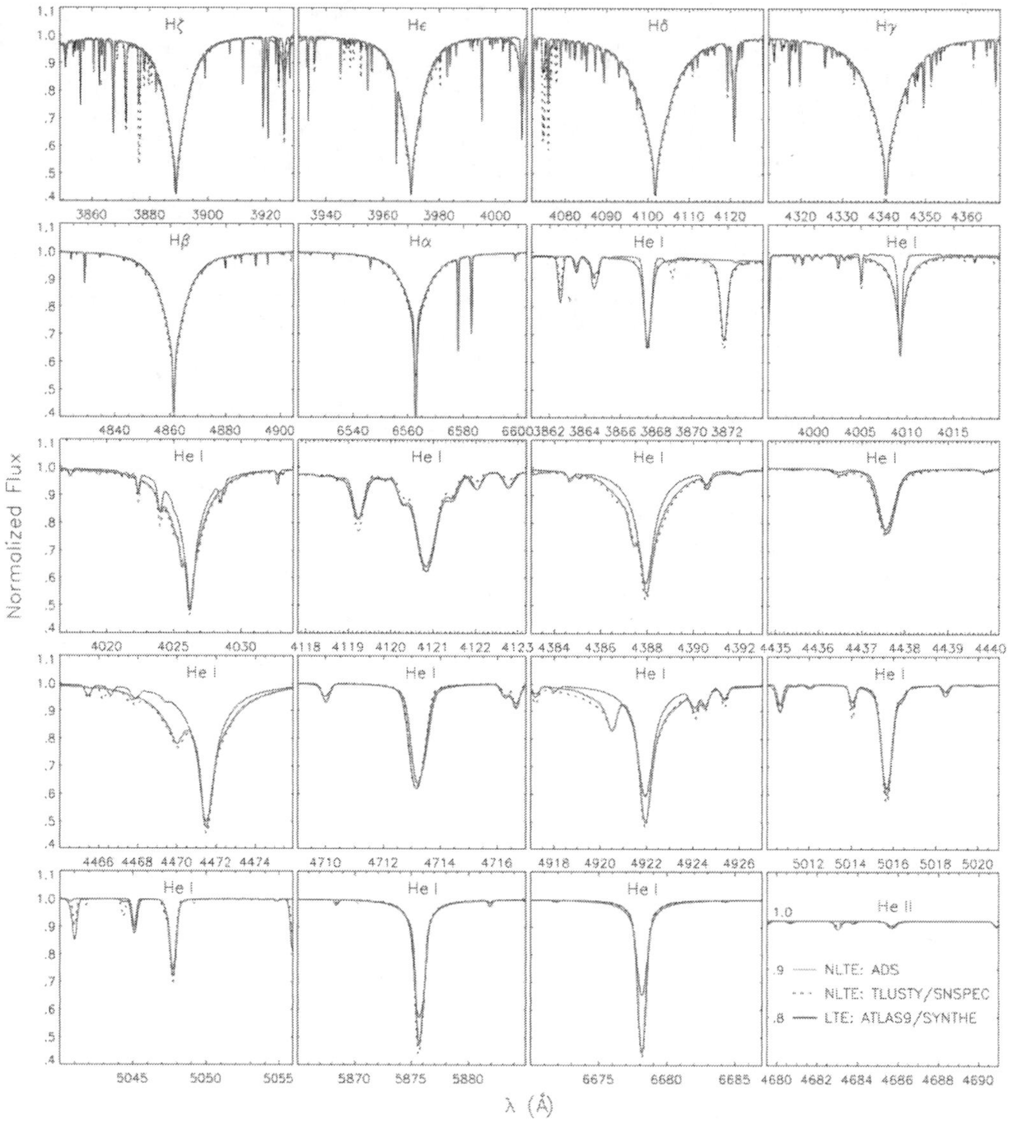

Figure 5. Same as Fig. 4, comparing line profiles computed with ADS, TLUSTY/SYNSPEC and ATLAS9/SYNTHE (line designations according to the legend).

Good agreement between the NLTE models and the ATLAS9/SYNTHE profiles is found for the Balmer lines and a few sharp He I lines, *for this test case*. However, problems with inadequate line broadening – Voigt profiles with constant Stark broadening parameter and no account for forbidden components – are indicated for the majority of the He I features (in particular all diffuse ones) computed with ATLAS9/SYNTHE. This can be resolved simply by implementing the appropriate broadening tables. More fundamental are the discrepancies in the depths of lines such as He I λλ4921, 5875 and 6678 Å, which are genuine NLTE effects.

The comparison over the entire parameter range reveals severe shortcomings of LTE modelling of observed hydrogen and helium lines. Over the $T_{\rm eff}$-range of 22 000 to 25 000 K, NLTE effects quickly set in and affect the cores and wings of the Balmer lines, and subsequently all helium lines. Maximum effects are found for highest $T_{\rm eff}$: the Balmer lines computed in LTE have

Journal of Physics: Conference Series **328** (2011) 012015 doi:10.1088/1742-6596/328/1/012015

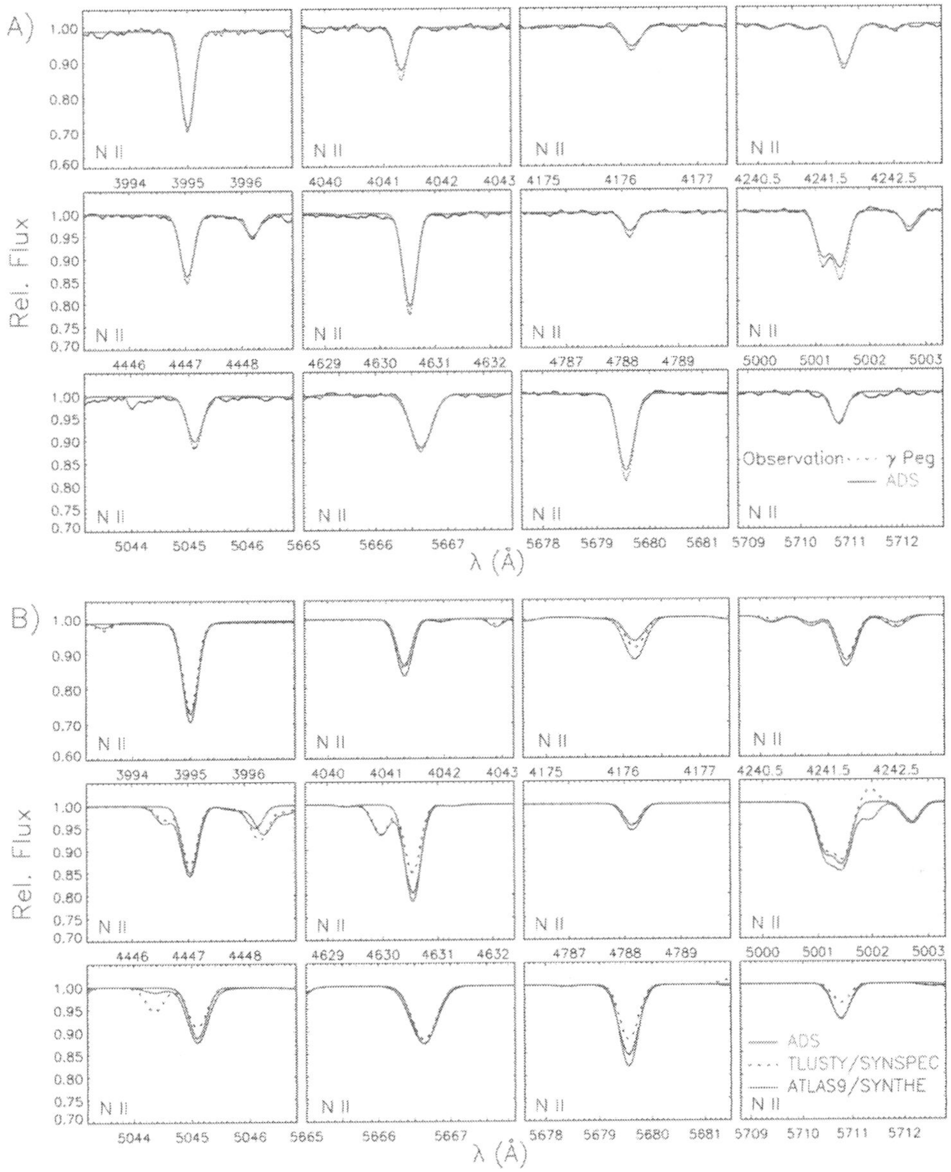

Figure 6. A): comparison of observations for N II lines in γ Peg with our global best-fit ADS model. B): comparison of ADS, TLUSTY/SYNSPEC and ATLAS9/SYNTHE models for the same lines. See figure legends and the text for details.

equivalent widths about 30% lower than in NLTE, and up to a factor of more than 2 for the He I lines. Surface gravities determinations based on Balmer-wing fitting is thus subject to systematic error, e.g. $\log g$ values are overestimated by up to 0.2 dex from LTE analyses of Hγ.

Overall, LTE modelling of the Balmer wings and selected He I lines appears to be largely unbiased by systematic error for main sequence stars up to $T_{\rm eff} \simeq 22\,000$ K. Preference should then be given to the He I $\lambda\lambda$3867, 4121, 4437, 4713, 5016 and 5048 Å transitions, which are least affected by NLTE effects. Either full or hybrid NLTE models have to be used for quantitative analyses of stars at higher $T_{\rm eff}$ when systematic bias is to be avoided. Our ADS approach improves on the TLUSTY OSTAR2002/BSTAR2006 grids by avoiding the He I singlet-triplet problem.

Journal of Physics: Conference Series **328** (2011) 012015 doi:10.1088/1742-6596/328/1/012015

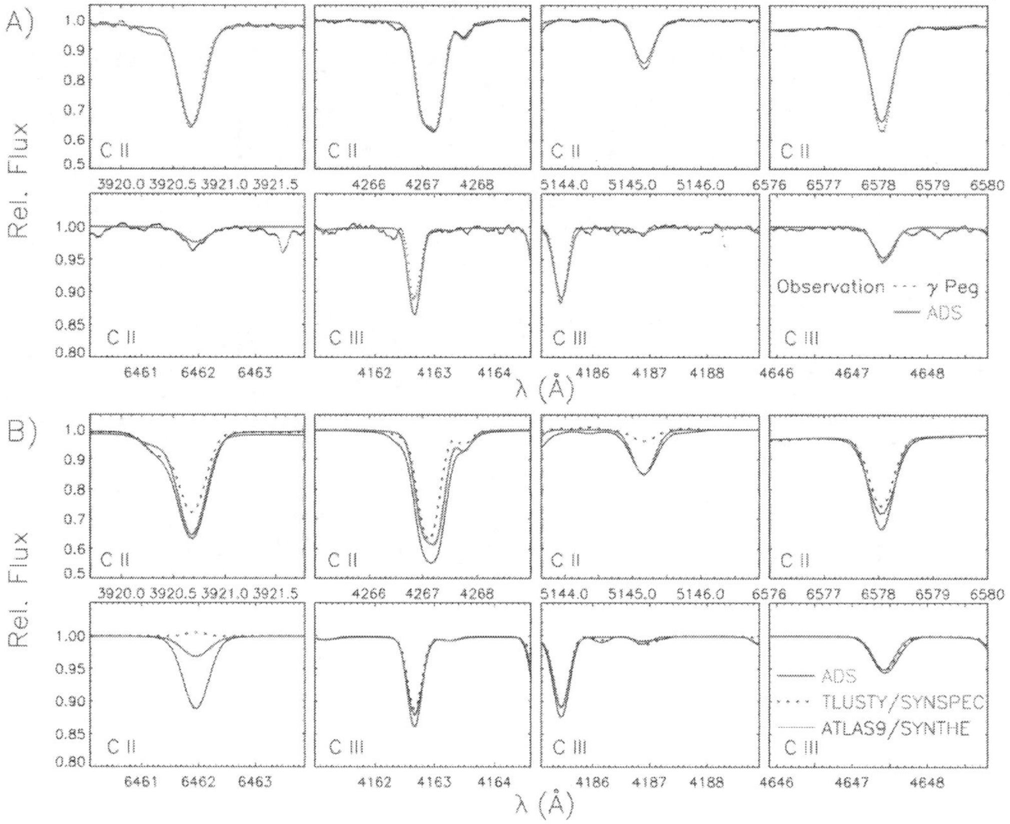

Figure 7. As Fig. 6, for C II/III lines.

Nitrogen: We discuss here only spectral lines of the N II ion, which in contrast to N III lines are observable throughout the entire parameter range. A comparison of important diagnostic lines in the optical spectrum of γ Peg with our global best-fit ADS model is shown in Fig. 6. The fit quality is excellent, in total 40 N II lines were analyzed to derive the mean nitrogen abundance. Therefore, the ADS model provides a good reference for the comparison with the other codes, which is shown in the bottom half of Fig. 6.

In many cases good agreement is found between the TLUSTY/SYNSPEC and ADS NLTE line profiles. However, a significant number of differences exists, e. g. for N II $\lambda\lambda$4630, 5045, 5679 and 5711 Å from the examples shown in Fig. 6, where the TLUSTY/SYNSPEC models are too weak compared to observation. Abundance analyses based on such features would therefore significantly overestimate the nitrogen abundance. Both agreement and discrepancies vary throughout the entire parameter range considered here, so that it is difficult to be quantify them in brief. In general, the differences increase towards lower gravities. The reasons for the differences in the two NLTE approaches result almost entirely from the model atom implementations. We tested this by adopting the N II model atom used in the TLUSTY computations for DETAIL/SURFACE and could tightly reproduce the TLUSTY/SYNSPEC solution.

The comparison of the NLTE ADS reference with the LTE ATLAS9/SYNTHE model shows that NLTE effects are often small, except for lines such as N II λ4176 Å from the examples. In fact, ATLAS9/SYNTHE profiles are closer to the ADS profiles (and therefore observation) than the TLUSTY/SYNSPEC profiles in most cases. The NLTE effects become more pronounced with decreasing surface gravities.

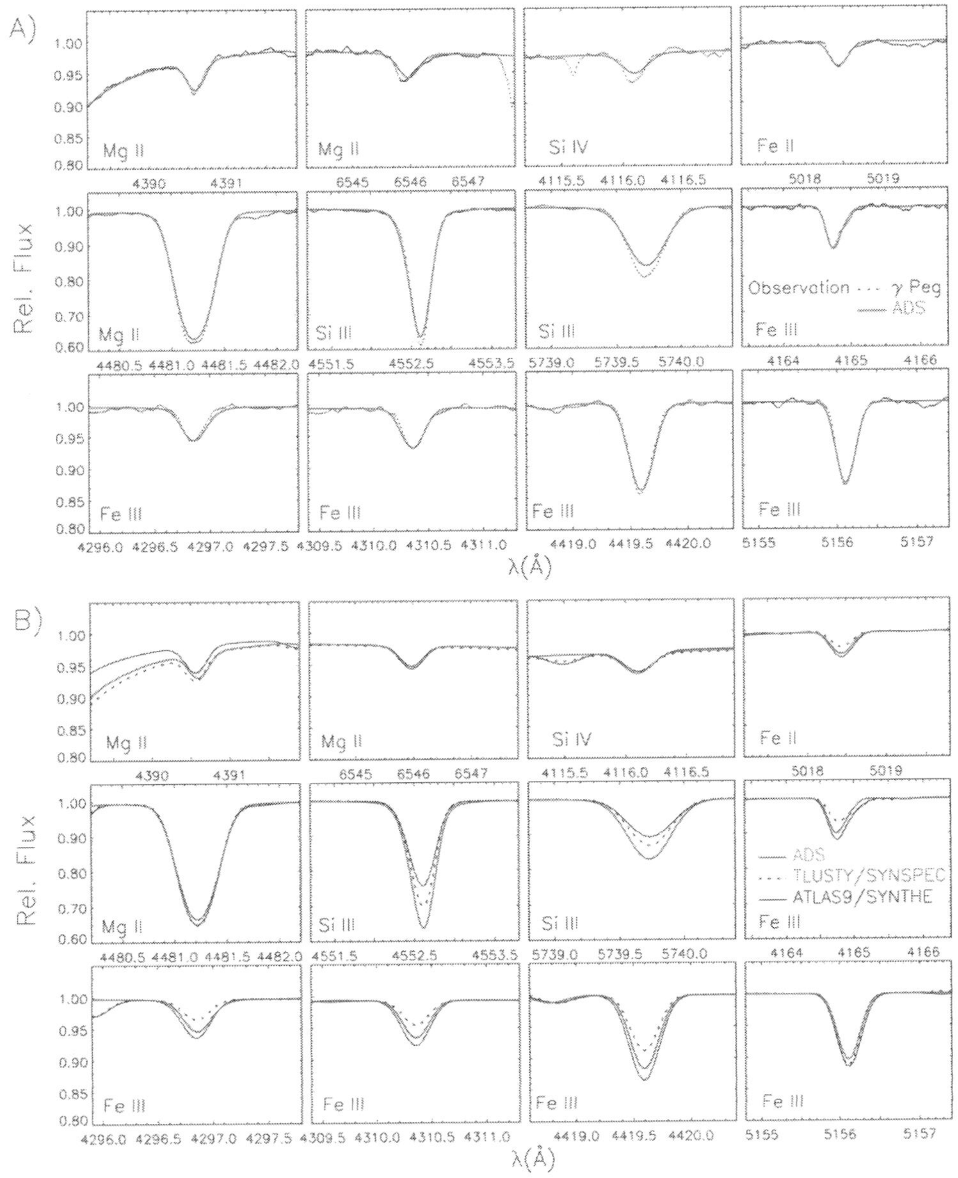

Figure 8. As Fig. 6, for Mg II, Si III/IV and Fe II/III lines.

Oxygen: We investigated the behavior of O I and O II lines, with similar results to those for nitrogen. Because of this we omit a more detailed description here. Noteworthy are the overall relatively small NLTE effects, again with the exception of several multiplets, e.g. the near-IR triplet O I $\lambda\lambda$7771–5 Å. For these lines, large differences between the two NLTE solutions are seen, again probably because of the different model atom implementations.

Carbon: Singly and doubly-ionized carbon were investigated. Again, the ADS model provides a good reference for further comparison with other models, as it reproduces observation well, not only for γ Peg (see examples in Fig. 7) but throughout the entire early B-type main sequence [16].

In particular C II was known as one of the most challenging cases for NLTE line formation computations for decades. The comparison of ADS with the TLUSTY/SYNSPEC solution shows

Journal of Physics: Conference Series **328** (2011) 012015 doi:10.1088/1742-6596/328/1/012015

disagreement in practically every line. Usually the TLUSTY/SYNSPEC profiles are too shallow, and in some cases even turn into (unobserved) emission (C II λ6462 Å). The situation is much better for C III, except for the strong line C III λ4162 Å. The problems with C II persist for other parameter combinations. Certainly, the TLUSTY model atom for carbon needs to be improved to allow for meaningful abundance determinations. For the moment, C III should be preferred for abundance work based on the BSTAR2006 and OSTAR2002 grids.

The comparison with the LTE profiles shows that most of the stronger, but also several of the weak lines show pronounced NLTE effects, throughout the entire parameter range. An exception is C II λ5145 Å (and its other multiplet members), which stays in LTE for all cases of interest. NLTE effects on the C III lines increase significantly for higher $T_{\rm eff}$.

Magnesium, silicon and iron: Examples for the comparison of our global best-fit ADS model with observations for important diagnostic Mg II, Si III/IV and Fe II/III lines in γ Peg are shown in Fig. 8. The Mg II and Fe II/III are reproduced reliably. We are at present incorporating a new silicon model atom, in particular to improve the fits to several lines not shown here, but the old model [21] reproduces lines such as those of the Si III $\lambda\lambda$4552–4575 Å triplet or Si IV λ4116 Å reasonably well, providing a valuable diagnostic for $T_{\rm eff}$-determination via the ionization balance.

Both NLTE solutions, with ADS and with TLUSTY/SYNSPEC, agree well for Mg II, but disagree for almost all Si and Fe lines. This is of concern in the case of silicon as use of the ionization equilibrium would lead to systematically different solutions for $T_{\rm eff}$ and $\log g$ with the BSTAR2006 and OSTAR2002 grids, probably inconsistent with other diagnostics. Moreover, the Si III triplet is often used to constrain the microturbulence velocity, which therefore may also become subject to systematic offsets when based on the TLUSTY grids. The iron lines are systematically shallower in the TLUSTY/SYNSPEC models, implying the derivation of higher iron abundances in applications. Because of the rôle of iron as the main line-opacity source this has further effects on atmospheric structure computations and therefore also on the atmospheric parameter determination. It is highly important to improve these model atoms for TLUSTY.

While the LTE and NLTE solutions for Mg II are very similar for the parameters of γ Peg (Mg II λ4390 Å apparently differs because of its location in the wing of an ill-fitted He I line), the iron lines are weakly and most of the silicon lines strongly NLTE affected. LTE modelling of silicon lines in early B-type stars can lead to systematic errors in atmospheric parameters when using ionization equilibria, and to unreliable elemental abundances. LTE iron abundances are in general somewhat underestimated.

5. Conclusions and Recommendations

We have seen that the data from the Gaia mission will need to be complemented by ground-based (optical) spectroscopy for early-type stars in order to put meaningful constraints on atmospheric parameters and elemental abundances. Three approaches for the modelling of stellar atmospheres in the $T_{\rm eff}$-range of 15 000 to 35 000 K, corresponding to spectral types of about B4 to O9, have been tested here: full LTE (with ATLAS9/SYNTHE), full NLTE (TLUSTY/SYNSPEC) and hybrid NLTE (ADS). The aim of the work was to determine the best means, presently available, to analyze optical spectra of such early-type stars, which is of course of much broader relevance than for Gaia-related science alone.

The comparison showed that pure LTE modelling can yield meaningful results for the low-$T_{\rm eff}$ regime, up to about 22 000 K on the main sequence, *if the right spectroscopic indicators are employed.* An overall good global match between synthetic and observed spectra cannot be expected to be obtained as many lines show NLTE effects.

Pure NLTE modelling with the available TLUSTY/SYNSPEC grids is conceptually superior. However, NLTE atmospheric structures and SEDs turn out to be close to LTE in the parameter space considered. NLTE line spectra of ions with 'simple' atomic structures, like H I, He I/II (with some restrictions at higher $T_{\rm eff}$) or Mg II turn out to be reliable. However, oversimplified model

Journal of Physics: Conference Series **328** (2011) 012015 doi:10.1088/1742-6596/328/1/012015

atoms cause problems for many of the more complex ions/elements. Blind use of the synthetic spectra published in the OSTAR2002 and BSTAR2006 grids should therefore be avoided. *The fact is that NLTE computations based on inappropriate model atoms can potentially introduce larger systematic errors to quantitative analyses than simple LTE modelling.* The TLUSTY model atoms certainly need extensive revision before a good global match with observations can be obtained with this package.

Our NLTE line-formation computations with DETAIL/SURFACE on ATLAS9 atmospheres combine the advantage of low to moderate CPU time requirements with realistic model atoms, facilitating a robust global match of observed spectra to be obtained, see also [27, 28]. It is reassuring that the same model atoms provide as good a match of the synthetic spectra to observations also for other kinds of stars, with drastically different chemical composition such as subdwarf OB stars [31] or extreme helium stars [32, 33, 34]. First applications of our models combined with a sophisticated analysis methodology promise breakthroughs for several astrophysical fields, see e.g. [25].

References

[1] Kurucz R L 1993 *CD-ROM No. 13* (Cambridge, Mass.: SAO)
[2] Kurucz R L and Avrett E H 1981 *Solar Spectrum Synthesis. I. A Sample Atlas from 224 to 300 nm, SAO Spec. Rep.* **391** (Cambridge, Mass.: SAO)
[3] Hubeny I and Lanz T 1995 *ApJ* **439** 875
[4] Giddings J R 1981 *Ph.D. Thesis* (Univ. London)
[5] Butler K and Giddings J R 1985 *Newsletter of Analysis of Astronomical Spectra*, No. 9 (Univ. London)
[6] Blomme R 2011 *J. Phys.: Conf. Series this Volume*
[7] Groh J H 2011 *J. Phys.: Conf. Series this Volume*
[8] Castelli F and Kurucz R L 2003 *Proc. 210th IAU Symposium* ed N Piskunov et al. (San Francisco: ASP) p A20
[9] Munari U, Sordo R, Castelli F and Zwitter T 2005 *A&A* **442** 1127
[10] Lanz T and Hubeny I 2003 *ApJS* **146** 417
[11] Lanz T and Hubeny I 2007 *ApJS* **169** 83
[12] Rybicki G B and Hummer D G 1991 *A&A* **245** 171
[13] Przybilla N and Butler K 2004 *ApJ* **609** 1181
[14] Przybilla N 2005 *A&A* **443** 293
[15] Nieva M F and Przybilla N 2006 *ApJ* **639** L39
[16] Nieva M F and Przybilla N 2008 *A&A* **481** 199
[17] Przybilla N and Butler K 2001 *A&A* **379** 955
[18] Przybilla N, Butler K, Becker S R, Kudritzki R P and Venn K A 2001 *A&A* **359** 1085
[19] Becker S R and Butler K 1988 *A&A* **201** 232
[20] Przybilla N, Butler K, Becker S R and Kudritzki R P 2001 *A&A* **369** 1009
[21] Becker S R and Butler K 1990 *A&A* **235** 326
[22] Becker S R 1998 *ASP Conf. Ser.* **131** 137
[23] Morel T, Butler K, Aerts C, Neiner C and Briquet M. 2006 *A&A* **457** 651
[24] Nieva M F and Przybilla N 2007 *A&A* **467** 295
[25] Nieva M F, Przybilla N and Irrgang A 2011 *J. Phys.: Conf. Series this Volume*
[26] Przybilla N, Nieva M F and Butler K 2008 *ApJ* **688** L103
[27] Nieva M F and Simón-Díaz S 2011 *A&A* **532** A2
[28] Nieva M F and Przybilla N 2011 *A&A* (submitted)
[29] Grevesse N and Sauval A J 1998 *Space Sci. Rev.* **85** 161
[30] Najarro F, Hillier D J, Puls J, Lanz T and Martins F 2006 *A&A* **456** 659
[31] Przybilla N, Nieva M F and Edelmann, H. 2006 *Balt. Astron.* **15** 107
[32] Przybilla N, Butler K, Heber U. and Jeffery C. S. 2005, *A&A* **443** L25
[33] Przybilla N, Nieva M F, Heber U. and Jeffery C. S. 2006, *Balt. Astron.* **15** 163
[34] Kupfer T, Heber U, Przybilla N, Jeffery C S, Behara N T and Butler K 2010 *AIP Conf. Proc.* **1273** 209

Journal of Physics: Conference Series **328** (2011) 012016 doi:10.1088/1742-6596/328/1/012016

Visualization and spectral synthesis of rotationally distorted stars

T H Dall[1] and L Sbordone[2,3]

[1]ESO, Karl-Schwarzschild-Str. 2, 85748 Garching bei München, Germany

[2]Max Planck Institute for Astrophysics, Karl-Schwarzschild-Str. 1, 85741 Garching bei München, Germany

[3]ZAH - Landessternwarte Heidelberg, Königstuhl 12, 69117 Heidelberg, Germany

E-mail: `tdall@eso.org`

Abstract. Simple spherical, non-rotating stellar models are inadequate when describing real stars in the limit of very fast rotation: Both the observable spectrum and the geometrical shape of the star deviate strongly from simple models. We attempt to approach the problem of modeling geometrically distorted, rapidly rotating stars from a new angle: By constructing distorted geometrical models and integrating standard stellar models with varying temperature, gravity, and abundances, over the entire surface, we attempt a semi-empirical approach to modeling. Here we present our methodology, and present simple examples of applications.

1. Introduction

For many years, stellar atmospheric models and spectrum synthesis has been synonymous with the use of standard 1D atmospheric models. Lately, 3-D models have added another dimension of realism to this picture. However, all models in use today are local models, treating only a small part of the stellar surface. Calculation of synthetic spectra then relies on assuming that the stellar atmosphere looks the same everywhere. Real stellar atmospheres are of course not the same everywhere. In this contribution we wish to highlight an aspect of stellar spectral synthesis that is mostly overlooked, namely the deviations from spherical symmetry.

The problems associated with stars that are not spherically symmetric were first addressed by von Zeipel [1; 2], who showed that stars that were flattened due to fast rotation would have a latitude dependent brightness distribution. First modeling of the spectra of rotating distorted stars were started by Collins [3], but the application to real stars were only brought to general attention through later work on Vega [4]. Moreover, the observed spectrum of a star will be influenced by many different additional effects from differential rotation, turbulence, temperature and gravity variations — all latitude dependent. Not to mention effects like spots, plagues, etc. that are both latitude and longitude dependent.

2. Methodology

Ideally, one should be able to take an observed spectrum and perform an inverse analysis, that is, from the spectrum infer the exact distributions of temperature, gravity, abundances, etc., as well as the shape of the star. However, such capabilities are far into the future since we still do

Figure 1. Example of a tiled spherical surface with 22 latitudinal bands ($N = 11$) before the flattening coordinate transformation is applied. Note that the shape of the 1452 triangular tiles varies slightly across the surface.

not understand fully all the effects going into the structure of a star and the formation of its spectrum.

With this work, we are attempting a different approach to learn more about these effects, that determine the structure of geometrically distorted stars and the possibilities of observing those effects. Our approach is:

- We assume a shape of the star – any shape.
- We assume (any) distributions of temperature, gravity, abundances, etc. across the surface.
- We assume any rotation law on the form $v(\theta)$, where θ is the co-latitude.
- At any point on the stellar surface we then compute a resulting intensity spectrum, in the direction of the observer. For this, any kind of local stellar model atmosphere can be used, whether ATLAS, MARCS, or state-of-the-art 3-D models.
- We then sum up all the contributions from the visible part of the star into an 'observed' synthetic spectrum.

This is an *exploratory* approach, aimed at providing more insight into the individual effects involved in shaping the spectra of real stars, rather that an attempt at producing realistic models of real stars, although that is of course the long term aim.

We want to avoid dependence on analytical descriptions of the parameters, which means that we must be able to freely assign any value of any parameter to any point on the surface. For this reason we chose to create a tiled surface, i.e., we cover the entire surface with plane segments to which we can assign any combination of parameters we wish. In principle the only limit to the fine graininess of the surface is the speed of the computer hardware. Since a sphere cannot be tiled with equal-area regular polygons, we chose instead to employ triangular tiling where

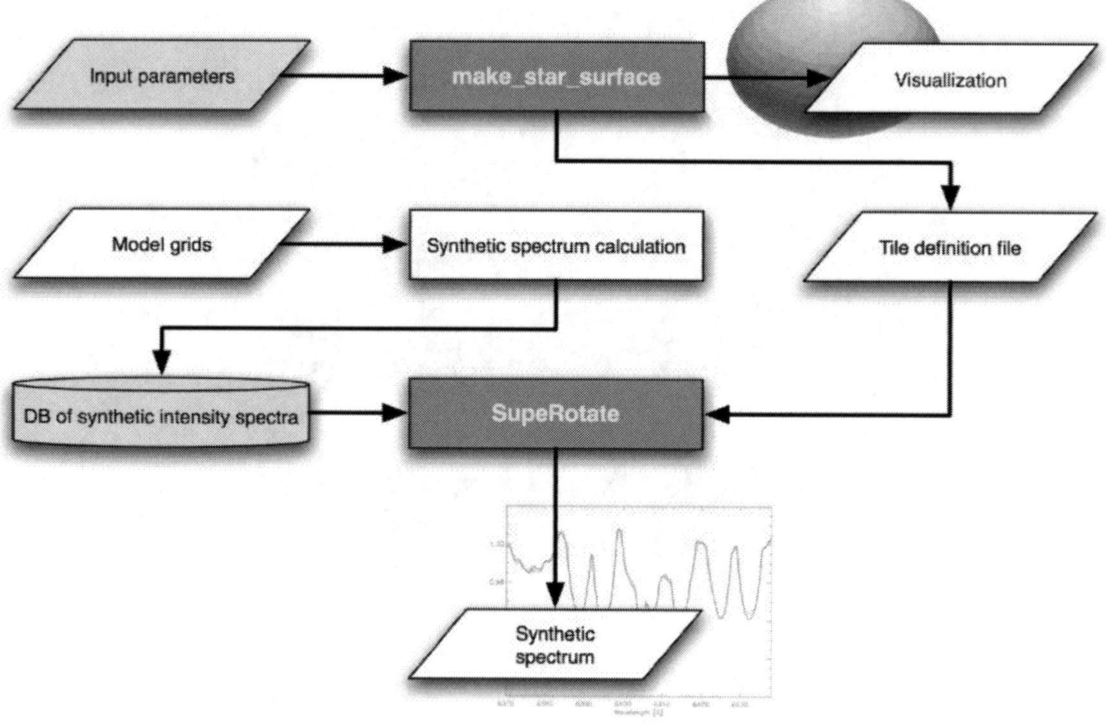

Figure 2. Flow chart describing the process. The blue boxes represent the software modules described in this paper. The box SupeRotate is where the actual spectral synthesis is done. The database of synthetic intensity spectra can be constructed using any current model atmosphere grid.

the individual tiles have slightly different surface areas. The tiles are arranged in latitudinal bands, where N is the number of bands per hemisphere. See Fig. 1 for an example with $N = 11$ resulting in a total of 1452 tiles.

For the sake of demonstration, we have used for this paper only simple z-axis scaling, meaning that the scaling parameter is equal to the ratio between the polar and equatorial radii. In principle any coordinate transformation can be used, whether analytical or numerical. This approach is complimentary to that of calculating equilibrium models of rotating stars [see e.g., 5].

The full procedure of what we have named 'SupeRotate' is shown schematically in Fig. 2. The details will be treated in a later paper.

3. Results

As a zero-order test of the code we first simulated a non-flattened star with no variation of parameters across the surface and compared this to a standard synthetic spectrum derived from the same set of models. We tested this for models with varying rotational velocities, assuming rigid rotation. The results were practically identical for $N \geq 9$.

As more interesting tests we have simulated two cases which resemble well known stars, namely Altair and FK Com. However, these models are intended to test the model sanity rather than produce actual physical results on the two stars.

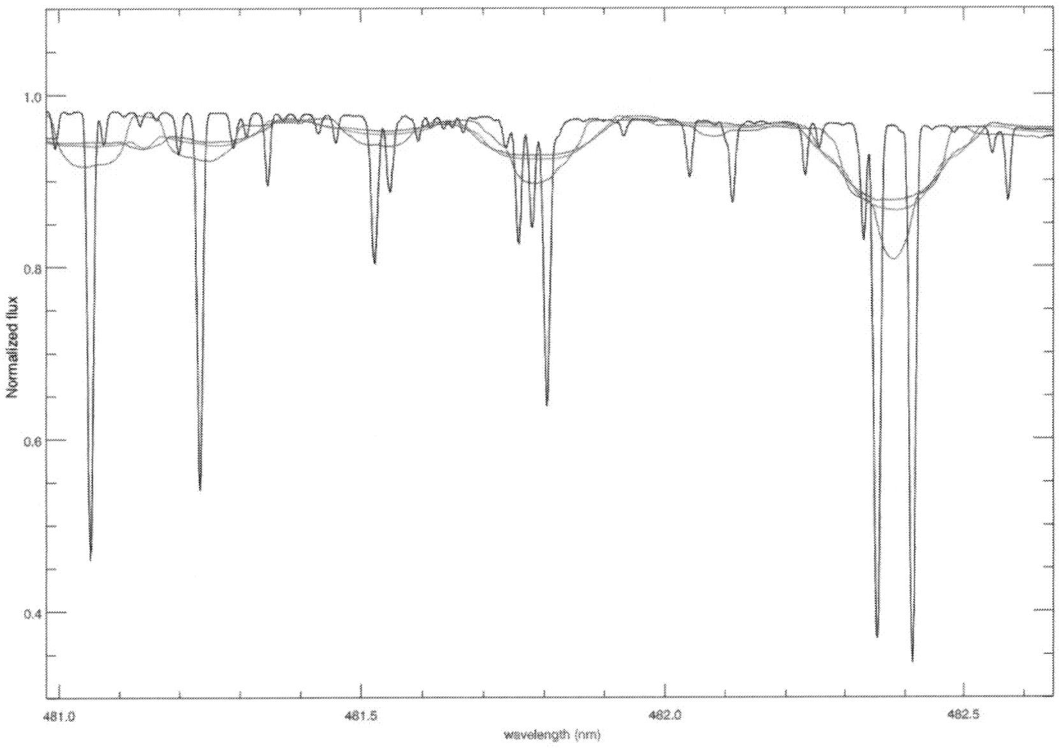

Figure 3. Close up of a section of the resulting synthetic spectrum of 'Altair' with $T_{\rm eff}$ and $\log g$ varying from pole to equator, and at four different values of the inclination; $i = 0°$ (black), $i = 30°$ (blue), $i = 60°$ (magenta), $i = 90°$ (red).

Table 1. Parameters used in the simulations of 'Altair'. The model is then used to calculate synthetic spectra for four values of the inclination; $i = 0°, 30°, 60°, 90°$.

Parameter	Range (pole — equator)
$T_{\rm eff}$	7300 K — 6900 K
$\log g$	4.1 — 3.9

3.1. Models of 'Altair'

Altair is an extensively studied flattened, fast rotating early type star. Its oblateness or flattening of $r_{\rm pol}/r_{\rm eq} = 0.80$ was first demonstrated by van Belle et al. [6] with subsequent studies confirming these findings [7; 8]. For this study, for the purpose of illustration, we are modeling Altair with an equatorial rotational velocity of 80 km s^{-1} as opposed to the true velocity of about 270 km s^{-1}. Thus, we refer to our models as 'Altair' rather than Altair.

We have calculated synthetic spectra for four different inclinations (see Fig. 3) with atmospheric parameters varying linearly in latitude from pole to equator. The parameters used are summarized in Table 1.

Further tests on 'Altair' are planned and will be reported elsewhere.

Journal of Physics: Conference Series **328** (2011) 012016 doi:10.1088/1742-6596/328/1/012016

Table 2. Parameters used in the simulations of 'FK Com' as well as the average parameters from [10]. The spectra are all calculated for the known inclination $i = 60°$.

Parameter	Range (pole — equator)	Standard model
$T_{\rm eff}$	5500 K — 4600 K	5000 K
$\log g$	3.8 — 2.8	3.5

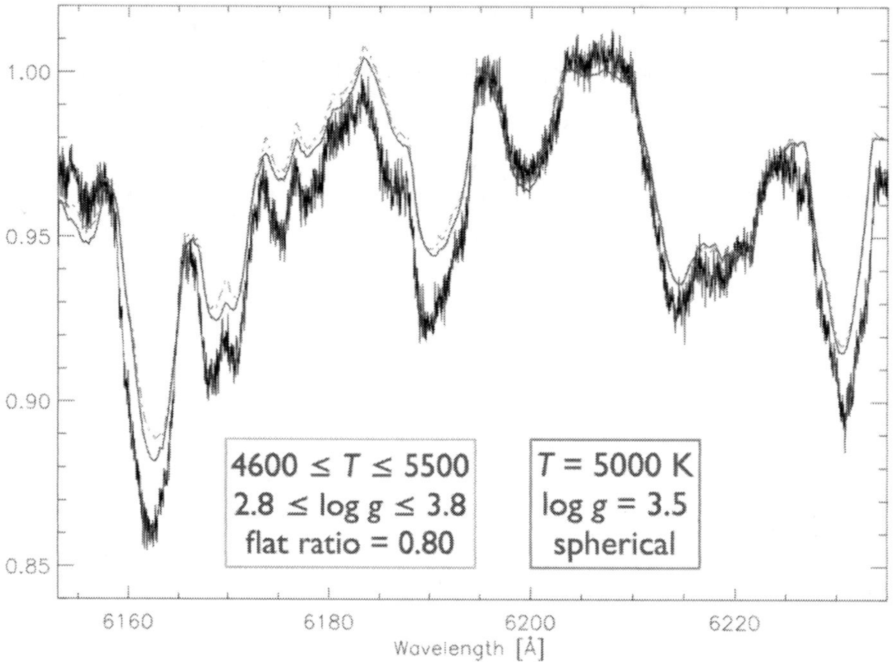

Figure 4. Observed spectrum of FK Com (black) along with our synthetic spectrum from SupeRotate (orange) and from standard synthesis (blue).

3.2. Models of 'FK Com'

FK Com is, with its $v \sin i = 160$ km s^{-1}, the fastest rotating and most active of the FK Comae stars (see [9] and references therein). These are a small group of magnetically very active late-type giants with rotation periods of only a few days. The photometric and spectroscopic characteristics of FK Comae stars are very similar to those of the very active RS CVn stars, but the RS CVn stars are close binary systems while FK Comae stars are single, possibly the result of a merging of the components of a RS CVn system.

The parameters of FK Com are well known and lie within the model grid we have prepared for our testing phase. Though it is believed that temperature, gravity and indeed abundances are varying widely over the surface, we have employed parameters varying smoothly with θ only. Thus, our model is 'FK Com' rather than FK Com. Table 2 lists the parameters used, as well as the standard average stellar model atmosphere consensus for FK Com. In our calculation we used the known $v \sin i$ and inclination $i = 60°$ as well as an assumed flattening of 0.80. The resemblance with the observed spectrum is good as is evident from Fig. 4.

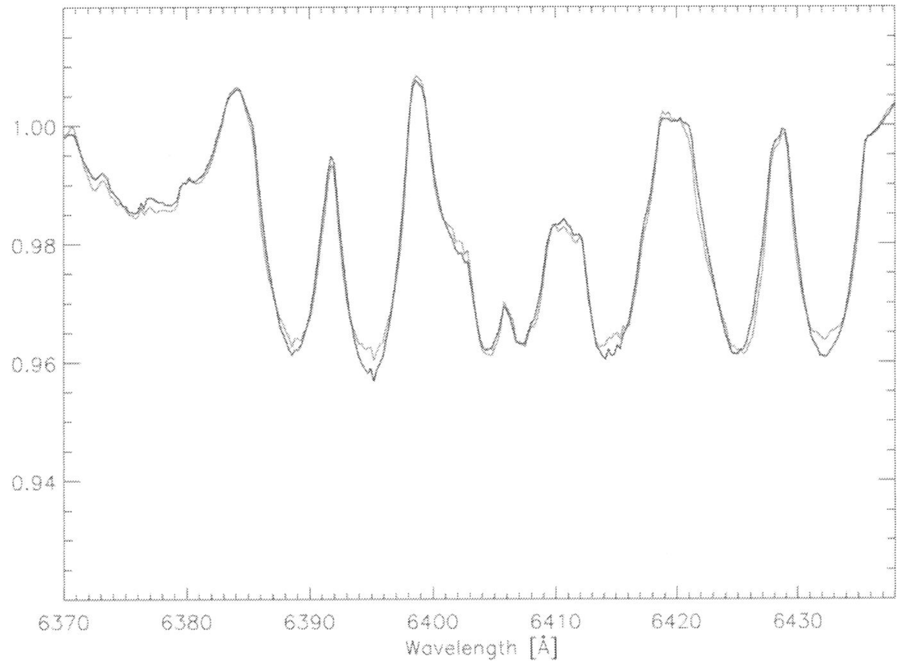

Figure 5. Synthetic spectrum from SupeRotate (orange) and from standard synthesis (blue) in a 70 Å interval.

In Fig. 5 we show the results of the flattened model compared with the standard model for a small spectral range. It is clear that the differences are quite small and possibly too small to be observed, in particular when one considers that the surface of FK Com is a mix of different temperatures, abundance patches, active regions, etc. However, what this code can provide is firm ideas about the relative significance of all these effects and what changes might be induced by variations of any of these parameters.

4. Conclusion

In this contribution we have explored a new modeling approach which is aimed primarily at gaining insight into the underlying individual effects that are shaping the spectra of stars with non-spherical and non-homogeneous surfaces.

Our models are currently at the 'toy model' stage, suited for an exploratory approach. Our models are easy to use, flexible, and highly adaptable. Following further testing, the software will be made freely available to the community. Meanwhile, we welcome requests and suggestions for suitable projects which we will undertake as extended testing.

Acknowledgments

We wish to thank H. Korhonen for kind permission to use the spectrum of FK Com.

References

[1] von Zeipel H 1924 *MNRAS* **84** 684–701

[2] von Zeipel H 1924 *MNRAS* **84** 665–683

[3] Collins II G W 1963 *ApJ* **138** 1134–+

[4] Gulliver A F, Hill G and Adelman S J 1994 *ApJ* **429** L81–L84

[5] MacGregor K B, Jackson S, Skumanich A and Metcalfe T S 2007 *ApJ* **663** 560–572 (*Preprint* 0704.1275)

[6] van Belle G T, Ciardi D R, Thompson R R, Akeson R L and Lada E A 2001 *ApJ* **559** 1155–1164

[7] Peterson D M, Hummel C A, Pauls T A, Armstrong J T, Benson J A, Gilbreath G C, Hindsley R B, Hutter D J, Johnston K J, Mozurkewich D and Schmitt H 2006 *ApJ* **636** 1087–1097 (*Preprint* arXiv:astro-ph/0509236)

[8] Monnier J D, Zhao M, Pedretti E, Thureau N, Ireland M, Muirhead P, Berger J P, Millan-Gabet R, Van Belle G, ten Brummelaar T, McAlister H, Ridgway S, Turner N, Sturmann L, Sturmann J and Berger D 2007 *Science* **317** 342– (*Preprint* 0706.0867)

[9] Korhonen H, Hubrig S, Berdyugina S V, Granzer T, Hackman T, Schöller M, Strassmeier K G and Weber M 2009 *MNRAS* **395** 282–289 (*Preprint* 0812.0603)

[10] Korhonen H, Berdyugina S V, Hackman T, Ilyin I V, Strassmeier K G and Tuominen I 2007 *A&A* **476** 881–891

Journal of Physics: Conference Series **328** (2011) 012017 doi:10.1088/1742-6596/328/1/012017

High-precision atmospheric parameter and abundance determination of massive stars, and consequences for stellar and Galactic evolution

Maria-Fernanda Nieva[1,2], Norbert Przybilla[2] and Andreas Irrgang[2]

[1] Max-Planck-Institut für Astrophysik, Karl-Schwarzschild-Str. 1, D-85741 Garching, Germany
[2] Dr. Karl Remeis-Sternwarte & ECAP, Sternwartstr. 7, D-96049 Bamberg, Germany

E-mail: `nieva@sternwarte.uni-erlangen.de`

Abstract. The derivation of high precision/accuracy parameters and chemical abundances of massive stars is of utmost importance to the fields of stellar evolution and Galactic chemical evolution. We concentrate on the study of OB-type stars near the main sequence and their evolved progeny, the BA-type supergiants, covering masses of ~6 to 25 solar masses and a range in effective temperature from ~8000 to 35 000 K. The minimization of the main sources of systematic errors in the atmospheric model computation, the observed spectra and the quantitative spectral analysis play a critical role in the final results. Our self-consistent spectrum analysis technique employing a robust non-LTE line formation allows precise atmospheric parameters of massive stars to be derived, achieving 1σ-uncertainties as low as 1% in effective temperature and ~0.05–0.10 dex in surface gravity. Consequences on the behaviour of the chemical elements carbon, nitrogen and oxygen are discussed here in the context of massive star evolution and Galactic chemical evolution, showing tight relations covered in previous work by too large statistical and systematic uncertainties. The spectral analysis of larger star samples, like from the upcoming Gaia-ESO survey, may benefit from these findings.

1. Introduction

Massive stars in the Galaxy are vastly outnumbered by lower-mass stars, like e.g. solar-type stars. Yet, massive stars are key drivers of the dynamics and evolution of the interstellar medium (ISM). As sources of ionizing radiation and momentum – via stellar winds and supernovae – they heat the ISM, initiate turbulence and contribute to the enrichment of the ISM with heavy elements. From the observer's perspective, massive stars are Galactic beacons throughout the UV to near-infrared wavelengths. They are observable not only nearby but out to large distances, tracing the star-forming regions within the spiral arms of the Milky Way. Massive stars are therefore key objects to understand the history and evolution of our Galaxy, and their study is of utmost relevance in the framework of the Gaia mission.

We concentrate here on the study of OB (late-O and early-B) main sequence stars and their progeny, the BA-type supergiants (BA-SGs) in order to trace their evolutionary history from the zero-age main sequence to the late stages of their life. The present sample of 64 stars covers a range in mass of ~6 to 25 solar masses and a range in effective temperature from ~8000 to 35 000 K. Highly accurate and precise stellar parameters and chemical abundances have been derived using robust spectral synthesis that accounts for deviations from the assumption of local thermodynamic equilibrium (non-LTE) and a self-consistent spectral analysis that utilizes multiple ionization equilibria. While the method is superior to standard techniques – which employ photometric temperature indicators or the ionization equilibrium of only one element –, it is also more time consuming. We have therefore expanded our efforts to implement first a semi-automatic and now an almost fully-automatic analysis procedure in order to facilitate the study of larger star samples at similar accuracy and precision. As it will be shown here, the efforts to obtain high-quality results from high-resolution observations at high S/N-ratio are extremely valuable for the progress of several fields, like the evolution of the chemical elements in the Galaxy, the evolution of stars under different conditions and also the evolution of the ISM.

In the following, we briefly describe the star sample (Sect. 2), the codes used to compute the synthetic models in non-LTE (Sect. 3), as well as the self-consistent spectral analysis based in simultaneous fits of individual lines of hydrogen, helium and several metals (Sect. 4). Two main results of this work are presented, addressing the fields of stellar evolution (Sect. 5) and Galactic chemical evolution (Sect. 6). Finally, the conclusions are drawn in Sect. 7.

2. The star sample and observations

The star sample has been carefully chosen in order to avoid observationally induced systematic biases, see [1] for a discussion of selection criteria. The OB main sequence star sample contains 29 single, sharp-lined and chemically inconspicuous stars. The stars are located within a radius of 400 pc from the Sun. The BA supergiant sample is composed of 35 stars located at distances up to 4 kpc from the Sun. For the whole sample of OB dwarfs and BA-SGs observations have been collected with several high-resolution spectrographs, e.g. the Fiber-fed Extended Range Optical Spectrograph (FEROS) and the Fibre Optics Cassegrain Echelle Spectrograph (FOCES, at the 2.2m telescopes at La Silla and Calar Alto, respectively) and the FIbre-fed Echelle Spectrograph (FIES, at the Nordic Optical Telescope, La Palma) at high S/N-ratio (250 to 800). Details of the observations can be found in [1, 2] for the OB sample and in [3, 4, 5] for the BA-SG sample.

3. Spectrum synthesis in non-LTE

The non-LTE line-formation computations for the OB dwarfs follow the methodology discussed in detail in our previous studies for H and He [6], for C [7, 8], and for N, O, Ne, Mg, Si and Fe [9]. For the BA-SGs the methodology is described in [3], and references therein. In brief, a non-LTE approach is employed to solve the restricted non-LTE problem on the basis of prescribed LTE atmospheres. This technique provides an efficient way to compute realistic synthetic spectra in all cases where the atmospheric structure is close to LTE, like for the stars analysed here [6]. The computational efforts can thus be focused on robust non-LTE line-formation calculations.

3.1. Models and programs

The model atmospheres were computed with the ATLAS9 code [10] which assumes plane-parallel geometry, chemical homogeneity, and hydrostatic, radiative and local thermodynamic equilibrium. Line blanketing was realized here by means of opacity distribution functions (ODFs) from [11]. Solar abundances of [12] were adopted in all computations. The model atmospheres were held fixed in the non-LTE calculations. Non-LTE level populations and model spectra were obtained with recent versions of DETAIL [13] and SURFACE [14]. The coupled radiative transfer and statistical equilibrium equations were solved with DETAIL, employing an accelerated lambda iteration scheme of [15]. This allowed even complex ions to be treated in a realistic way. Synthetic spectra were calculated with SURFACE, using refined line-broadening theories. Continuous opacities due to hydrogen and helium were considered in non-LTE and line blocking was accounted for in LTE via Kurucz' ODFs. Updates of some of the published model atoms for non-LTE computations were carried out introducing improved oscillator strengths and collisional data from *ab-initio* computations, see [1, 3].

4. Spectral analysis

Our analysis method is based on the simultaneous reproduction of all spectroscopic indicators via an iterative line-fitting procedure aiming to derive atmospheric parameters and chemical abundances self-consistently. In contrast to common strategies in stellar spectroscopy, this analysis technique takes full advantage of the information encoded in the *line profiles* at different wavelength ranges simultaneously. Integrated quantities like equivalent widths W_λ are not used in this approach. The stellar parameters primarily derived here are the effective temperature $T_{\rm eff}$, surface gravity $\log g$, microturbulent velocity ξ, (radial-tangential) macroturbulent velocity ζ, projected rotational velocity $v \sin i$ and elemental abundances $\varepsilon(X) = \log(X/{\rm H}) + 12$.

4.1. Stellar parameter and abundance determination

Special emphasis was given to use multiple indicators in order to minimize the chance of the stellar atmospheric parameters and chemical abundance determination being biased by residual systematic errors. The following spectroscopic indicators were utilized in the analysis:

- $T_{\rm eff}$: all available hydrogen and helium lines, and multiple independent ionization equilibria; confirmation via spectral energy distributions (SEDs);
- $\log g$: wings of all available hydrogen lines and multiple ionization equilibria; for many of the OB dwarfs also confirmation via HIPPARCOS distances [16];
- ξ: several elements with spectral lines of different strength enforcing no correlation between $\varepsilon(X)$ and the strength of the lines (equivalent to $\varepsilon(X)$ being independent of W_λ);
- $v \sin i$ and ζ: metal line profiles;
- $\varepsilon(X)$: a comprehensive set of metal lines.

Usually in stellar analyses, once the stellar parameters are fixed one commences with the abundance determination, treating this as an essentially independent step. In our approach the abundance and stellar parameter determination are tightly related because of the use of ionization equilibria. In consequence, only few species are left to finalize the analysis. Another difference to typical literature studies is the large number of spectral lines evaluated by us per species (typically, ~150 to 250 lines are analysed in total per star), see e.g. [1, 3], and the consistency achieved from the different ionization stages of the various elements. All the various improvements in observations, modelling and analysis methodology facilitated results at much higher accuracy to be achieved than possible in standard works, e.g. [17] to [29]. The high quality of the results could be retained over a large parameter space, spanning nearly 25 000 K in $T_{\rm eff}$ and ranging from close to the zero-age main sequence (ZAMS) to the late stages of evolution. Consequently, an excellent match of the computed and the observed spectra is achieved globally and in the details.

4.2. A semi-automatic analysis

We have computed a comprehensive grid of models (in total of the order ~100 000 synthetic spectra) to perform the analysis. A powerful fitting routine for the semi-automatic comparison of observed and theoretical spectra, SPAS[1], provides the means to interpolate between model grid points for up to three parameters simultaneously and allows to apply instrumental, rotational and (radial-tangential) macrobroadening functions to the resulting theoretical profiles. Interactive work in some decisive steps on the analysis with SPAS paid off as much more accurate results could be obtained. Crucial was the selection of the appropriate spectroscopic indicators for the parameter determination which may vary from star to star upon availability of specific spectral lines (depending on stellar temperature, spectrum quality and the observed wavelength

[1] Spectrum Plotting and Analysing Suite, SPAS [30].

coverage). All spectral lines unsuited for analysis because of e.g. blends, low S/N, uncorrectable normalization problems, incomplete correction of cosmics, or known shortcomings in the modelling needed to be excluded. Also a verification and, possibly, correction of the automatic continuum normalization lead to a gain in precision. Every element was analysed independently and some interactive iterations for fine-tuning the parameter determination were needed in order to find a unique solution that reproduces all indicators simultaneously. This facilitated also to derive realistic uncertainties for the stellar parameters. The standard deviations around the average parameter values were adopted, as derived from the various independent spectral indicators. Likewise, uncertainties of elemental abundances were determined from the line-to-line scatter found from the analysis of the individual features. Finally, it was thus possible to derive a simultaneous, self-consistent solution for atmospheric parameters and chemical abundances, and also to quantify their statistical uncertainties. The novel approach [1] provides results meeting the same quality standard as our previous work, e.g. [8, 9].

4.3. A fully-automatic analysis

All pros and cons of the spectral analysis that have been learnt in the previous work have been recently encoded by one of us into an even more efficient and automatic procedure, allowing the whole previous analysis to be done even faster and at similar precision. The independent analyses per element and several interactive iterations until convergence of all parameters are hereby replaced by one simultaneous fit of the entire suitable spectrum (freed from features not included in our models such as missing lines, cosmics, telluric lines, reduction artifacts, among others). Interpolating in pre-calculated model grids, the technique applied is capable to sample the whole multi-parameter space spanned by atmospheric parameters and elemental abundances at the same time. For instance, line blends due to macroscopic effects (i.e. instrumental, rotational, and macroturbulent broadening) are thus implemented absolutely correct. With this method, atmospheric parameters and chemical abundances are objectively, simultaneously, and self-consistently constrained by exploiting all available spectroscopic indicators.

5. Implications for massive star evolution

The availability of atmospheric parameters and elemental abundances for larger star samples facilitates to stellar evolution models to be tested. Significant improvements in the accuracy and precision of the observational results are in particular valuable, as they allow such tests to be pursued in greater detail.

Energy production in massive stars is governed by the CNO cycles throughout most of their lifetime. The nuclear-processed material may reach their surface layers through rotational mixing already during their main-sequence phase [31] opening up a very powerful diagnostic to test models of stellar evolution. The changes of the surface abundances reflect the actions of the dominating CN-cycle initially, following a well-defined nuclear path.

Numerous studies of CNO abundances in massive stars of the Milky Way are available from the literature, mostly for early B-type stars close to the main sequence and for BA-SGs. We illustrate the results of several key publications in the N/O–N/C diagrams of Fig. 1. Some of the more recent studies [17, 18, 19] are based on non-LTE model atmospheres, while the bulk of the data were obtained from non-LTE line-formation computations on LTE model atmospheres – which is equivalent to the full non-LTE approach in the cases under consideration [6].

The main-sequence stars show overall a wide range of N/O–N/C combinations, with the deviations from the predictions increasing in the supergiants. This picture is difficult to be interpreted – even more so if only one of the elements (like nitrogen) is considered. Moreover, several observational data points pose a challenge for the evolution models. On the other hand, the abundance uncertainties are very large, with a typical statistical 1σ-error in abundance per element of about a factor $\sim$2, while systematic uncertainties are often largely underestimated

Journal of Physics: Conference Series **328** (2011) 012017 doi:10.1088/1742-6596/328/1/012017

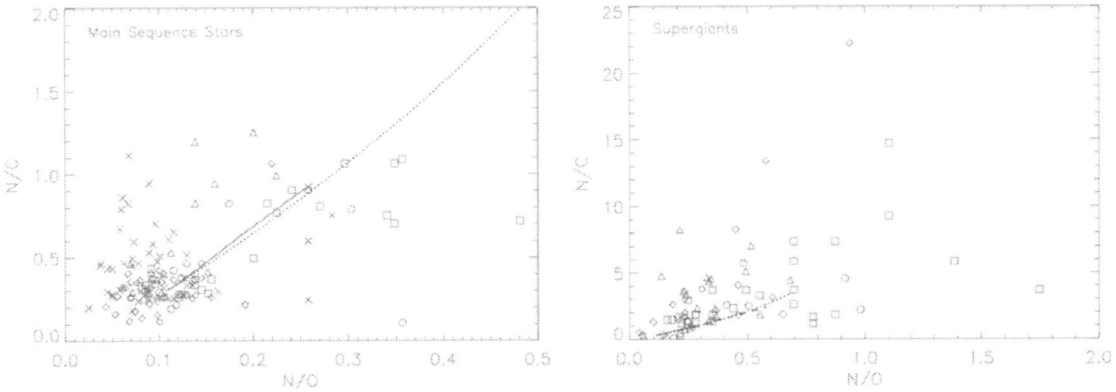

Figure 1. Observational constraints on the mixing of CNO-burning products in massive stars from non-LTE analyses in the literature. Mass ratios N/C over N/O are displayed. Left-hand panel: main-sequence stars. Circles: [20]; triangles: [21]; diamonds: [22, 23, 24, 25]; squares: [26]; crosses: [19]. Right-hand panel: BA-type supergiants. Triangles: [27], [28]; circles: [29]; squares: [17]; diamonds: [18]. Error bars can be larger than the plotting range. The lines represent predictions from evolution calculations, for a rotating 15 $M_{\odot}$ star with $v_{\rm rot}^{\rm ini} = 300\,{\rm km\,s^{-1}}$ until the end of the main sequence: solid red line, until the end of He burning: dashed blue line [32], and for a star of the same parameters that in addition takes the interaction of rotation and a magnetic dynamo into account [33], until the end of the MS: dotted line, respectively. The predicted trends are similar for the entire mass range under investigation.

(for a discussion of this see [34]) or even unaccounted for. The error bars in Fig. 1 are larger than the entire plotting range in many cases. In consequence, no *definite* conclusions can be drawn on the quality of the stellar evolution models from these data.

The behaviour of CNO abundances, previously presented for a sub-sample of 20 stars in [35], is now shown for the whole star sample of 64 stars in Fig. 2. In contrast to the literature values (Fig. 1), a clear and tight trend is found, confirming the predicted locus of N/O–N/C abundance ratios. Most of our main-sequence objects cluster around the pristine Cosmic Abundance Standard (CAS) values [1], i.e. they are unmixed, while about 1/3 of the stars show a mixing signature of varying magnitude, following the predicted nuclear path with $d(N/C)/d(N/O) = 4.6$ (for initial CAS abundances) tightly. Stellar evolution models based e.g. on the solar values by [36] would predict a different nuclear path (with slope $\sim$3.0).

However, as already indicated above, the models of [32] for rotating stars with mass loss evolving towards the red supergiant stage (solid line in Fig. 2) predict mixing that is too low (see [35]), in particular for most of the supergiants. Five reasons may provide an explanation.

I) Higher than average rotation velocities in the progenitor stars of these supergiants on the main sequence may reconcile the situation for some objects.

II) Evolution models for rotating stars that also account for the interaction of rotation and a magnetic dynamo [33] predict enhanced mixing signatures of the amount required (dotted line).

III) Some stars may have evolved in a close binary, which can also lead to enhanced mixing associated with mass transfer.

IV) Some objects may have been siblings to τ Sco on the main sequence, climbing up the N/O–N/C relation even further in their further evolution.

V) Supergiants may already have evolved through the red supergiant phase (e.g., on a blue loop) to expose first dredge-up abundance ratios, which could quantitatively also explain the observations (dashed line).

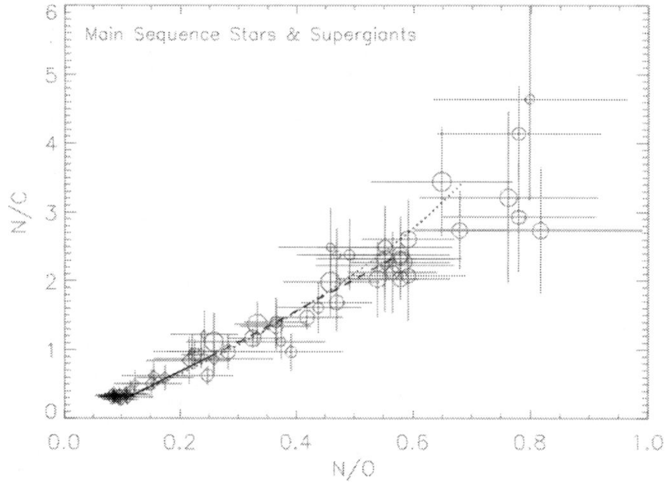

Figure 2. N/C vs. N/O abundance ratios (by mass) of our non-LTE analysis for the sample of 64 stars. B-type main-sequence stars are displayed as diamonds, BA-type supergiants as circles. The symbol size encodes the stellar mass and error bars give 1σ-uncertainties. The different lines describe evolutionary model predictions for $15\,M_\odot$ stars identical to those in Fig. 1.

6. Implications for Galactic chemical evolution

Nucleosynthesis in successive generations of stars has enriched the cosmic matter with heavy elements ever since the first Population III stars were born. Studies of various objects like Galactic stars, planetary nebulae and H II regions allow the cosmic enrichment history to be traced and the specific production sites of individual elements to be constrained. The Cosmic Abundance Standard (CAS), derived from our sample of unmixed B-stars, provides valuable input for the comparison of models with observations, as it marks the *present-day endpoint of Galactochemical evolution*, in particular for a typical spiral galaxy like the Milky Way. Here, we concentrate on the role of the CAS values in the context of the Galactic evolution of the light elements CNO as traced by stellar analyses, see Fig. 3.

Our achievements in improving the precision and accuracy of the B-star analyses over previously published work, relative to the trends in the evolution of the CNO abundances as derived from solar-type stars, become obvious in the comparison of the left- and right-hand panels of Fig. 3. While the enormous scatter in the B-star abundances was difficult to be reconciled with the data from solar-type stars in the past, our data for the B-stars indicate a high degree of chemical homogeneity, at absolute abundance values that may differ from the solar standard.

While the investigations of the cosmic chemical evolution are currently focusing on the early phases at low metallicity, an accurate knowledge of the present-day endpoint of the evolution is nonetheless essential. The reason for this is that the interpretation of the data is based on comparisons with Galactic chemical evolution models, which have to (but not always do) match the present-day composition as a boundary condition. It is therefore important for the entire modelling which reference values are used, solar or CAS abundances. In particular the differences in the C/O ratio are appreciable, also with respect to the majority of nearby solar-type stars, amounting to almost 50%. Taken at face value, this difference indicates that the C/O enrichment of the interstellar medium in the present-day solar neighborhood occurred slower than at the

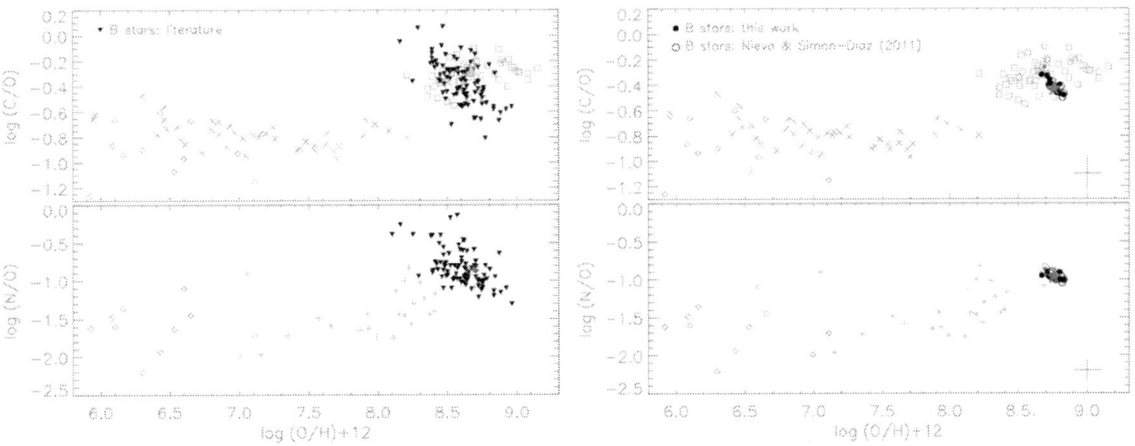

Figure 3. Observational constraints on the chemical evolution of Galactic CNO abundances: abundance ratios log (C/O) and log (N/O) vs. O abundance. *Left-hand panels*: black triangles: early B-stars from the literature, like in Fig. 1, except for [19]. Data for low-mass stars are displayed as green symbols – squares: solar-type dwarfs [37]; crosses: solar-type dwarfs and subgiants [38]; diamonds: unmixed cool giants [39]; plus signs: unevolved solar-type stars [40]. Solar abundance ratios of [36] are also indicated (⊙). *Right-hand panels*: black dots: unmixed B-type stars from the present work, black circles: unmixed stars from [2]. Data from the literature like in the left-hand panels. The Cosmic Abundance Standard is also indicated (red star). Error bars typical for individual stars in the present study are shown.

Sun's place of birth.

Note that the Sun may be viewed as an extreme but still compatible case in terms of the distribution of the stars from the present sample in the log O–log C/O diagram, but the absolute values for the carbon abundance differ significantly. The solar and CAS data on N/O are rather compatible on the other hand. A systematic investigation of nitrogen abundances in high-metallicity solar-type stars would be desirable for further comparisons.

Overall, it is astonishing how different and at the same time how similar the young and old star populations in the solar neighborhood are. It is for the first time that this is elaborated, as the lack of high precision and accuracy in many previous studies of early B-type stars prevented any meaningful conclusions to be drawn.

7. Conclusions

Over the past few years, we have made great efforts to reduce uncertainties in quantitative spectral analyses of OB-type stars near the main sequence and their evolved progeny, BA-type supergiants. Self-consistent analyses that bring all spectroscopic indicators – Balmer, helium lines and metal ionization equilibria – into match simultaneously yield drastically reduced systematic errors in the determination of atmospheric parameters and chemical abundances. More recent work focused on the development of a fully-automatic fitting routine for the analysis of much larger samples of stars than feasible at present. We are confident that future studies of early-type stars, like within the ESO-Gaia survey, will largely benefit from these developments.

This leap in the accuracy and precision of early-type star analyses has already produced some exciting results. We gave concrete examples on the behaviour of the light elements CNO in the context of stellar and Galactic chemical evolution, which will stimulate further investigations by theorists that will give meaningful answers to several of today's fundamental astrophysical questions.

Journal of Physics: Conference Series **328** (2011) 012017 doi:10.1088/1742-6596/328/1/012017

References

[1] Nieva M F and Przybilla N *A&A*, submitted
[2] Nieva M F and Simón-Díaz S 2011 *A&A* **532** 2
[3] Przybilla N, Butler K, Becker S R and Kudritzki R P 2006 *A&A* **445** 1099
[4] Schiller F and Przybilla N 2008 *A&A* **479** 849
[5] Firnstein M 2010 *Ph.D. Thesis* (Univ. Erlangen-Nuremberg)
[6] Nieva M F and Przybilla N 2007 *A&A* **467** 295
[7] Nieva M F and Przybilla N 2006 *ApJ* **639** L39
[8] Nieva M F and Przybilla N 2008 *A&A* **481** 199
[9] Przybilla N, Nieva M F and Butler K 2008 *ApJ* **688** L103
[10] Kurucz R L 1993b *CD-ROM No. 13* (Cambridge, Mass.: SAO)
[11] Kurucz R L 1993a *CD-ROM No. 2–12* (Cambridge, Mass.: SAO)
[12] Grevesse N and Sauval A J 1998 *Space Sci. Rev.* **85** 161
[13] Giddings J R 1981 *Ph.D. Thesis* (Univ. London)
[14] Butler K and Giddings J R 1985 in *Newsletter of Analysis of Astronomical Spectra, No. 9* (Univ. London)
[15] Rybicki G B and Hummer D G 1991 *A&A* **245** 171
[16] van Leeuwen F 2007 *Hipparcos, the new reduction of the raw data* (Dordrecht: Springer)
[17] Crowther P A, Lennon D J and Walborn N R 2006 *A&A* **446** 279
[18] Searle S C, Prinja R K, Massa D and Ryans R 2008 *A&A* **481** 777
[19] Hunter I *et al.* 2009 *A&A* **496** 841
[20] Kilian J 1992 *A&A* **262** 17
[21] Gies D R and Lambert D L 1992 *ApJ* **387** 673
[22] Cunha K and Lambert D L 1994 *ApJ* **426** 170
[23] Daflon S, Cunha K and Becker S R 1999 *ApJ* **522** 950
[24] Daflon S, Cunha K, Becker S R and Smith V V 2001a *ApJ* **552** 309
[25] Daflon S, Cunha K, Butler K and Smith V V 2001b *ApJ* **563** 325
[26] Morel T, Hubrig S and Briquet M 2008 *A&A* **481** 453
[27] Venn K A 1995 *ApJ* **449** 839
[28] Venn K A and Przybilla N 2003 in *ASP Conf. Ser.* **304** 20
[29] Takeda Y 2000 *PASJ* **52** 113
[30] Hirsch H 2009 *Ph.D. Thesis* (Univ. Erlangen-Nuremberg)
[31] Maeder A and Meynet G 2000 *ARA&A* **38** 143
[32] Meynet G and Maeder A 2003 *A&A* **404** 975
[33] Maeder A and Meynet G 2005 *A&A* **440** 1041
[34] Nieva M F and Przybilla N 2010 *ASP Conf. Ser.* **425** 146
[35] Przybilla N, Firnstein M, Nieva M F, Meynet G and Maeder A 2010 *A&A* **517** A38
[36] Asplund M, Grevesse N, Sauval A J and Scott P 2009 *ARA&A* **47** 481
[37] Gustafsson B, Karlsson T, Olsson E, Edvardsson B and Ryde N 1999 *A&A* **342** 426
[38] Fabbian D, Nissen P E, Asplund M, Pettini M and Akerman C 2009 *A&A* **500** 1143
[39] Spite M *et al.* 2005 *A&A* **430** 655
[40] Israelian G, Ecuvillon A, Rebolo R, García-López R, Bonifacio P and Molaro P 2004 *A&A* **421** 649

Journal of Physics: Conference Series **328** (2011) 012018 doi:10.1088/1742-6596/328/1/012018

Tests of simulated Gaia BP/RP spectra with LDS (Low Dispersion Spectroscopy) photographic sky surveys

René Hudec[1,2] and Lukáš Hudec[2]

[1] Astronomical Institute, Academy of Sciences of the Czech Republic, CZ–251 65 Ondřejov, Czech Republic

[2] Czech Technical University in Prague, Faculty of Electrical Engineering, Technicka 2, CZ–166 27 Prague 6, Czech Republic

E-mail: rhudec@asu.cas.cz

Abstract. The LDS (Low Dispersion Spectroscopy) performed in various extended sky surveys with optical telescopes using objective prism and photographic plates offers an interesting opportunity to test simulated low-dispersion spectra for the Gaia BP/RP photometers and to compare them with real data, especially for objects with strong emission lines. We present a review of astrophysics with LDS performed in the past, as well as an overview of existing extended sky surveys (with photographic plates) providing LDS data. Some of them provide almost complete coverage of the northern or southern hemisphere (e.g. the Northern and Southern Mt Wilson - Michigan Hα surveys or the German La Paz Bolivia Southern Spectral Sky Survey). We show examples of these data and discuss a comparison of existing LDS plate data with expected/simulated Gaia BP/RP data. We show examples of real data for objects with very strong and wide emission features confirming that such features will be detectable with Gaia BP/RP. We also discuss the importance of Gaia RP/BP low-dispersion spectroscopy for astrophysical studies.

1. Introduction

ESA-Gaia is an ambitious mission to chart a three-dimensional map of the Milky Way, in the process revealing the composition, formation, and evolution of the Galaxy. Gaia will provide unprecedented positional and radial velocity measurements with accuracies required to produce a stereoscopic and kinematic census of about one billion stars in our Galaxy and throughout the Local Group. It amounts to ~1 percent of the Galactic stellar population. Combined with astrophysical information for each star provided by onboard multi-color photometry, these data will have the necessary precision for quantifying the early formation, and subsequent dynamical, chemical, and star formation evolution of the Milky Way. The Gaia payload consists of a single integrated instrument, the design of which is characterized by a dual telescope concept, with a common structure and a common focal plane (`sci.esa.int/gaia/`). Both telescopes are based on a three-mirror anastigmatic (TMA) design. Beam combination is achieved in image space with a small beam combiner. Silicon-carbide (SiC) ultra-stable material is used for the mirrors and telescope structure. There will be a large common focal plane with an array of 106 CCDs. The large focal plane also includes areas dedicated to the spacecraft's metrology

Journal of Physics: Conference Series **328** (2011) 012018 doi:10.1088/1742-6596/328/1/012018

and alignment measurements. Three instrument functions/modes are designed: (i) Astrometric mode for accurate measurements, even in densely populated sky regions of up to 3 million stars deg^{-2}, (ii) Photometric mode based on low-resolution, dispersive, spectro-photometry using the Blue and Red Photometers (BP and RP) for continuous spectra in the 320–1000 nm band for astrophysics and chromaticity calibration of the astrometry, and (iii) Spectroscopic (RVS) mode for high-resolution spectra, using a grating, and covering the narrow 847–874 nm band. The expected limiting magnitude is 20 in photometric mode (`sci.esa.int/gaia`).

In this paper we focus on the 'photometric mode' RP/BP of Gaia. In reality the use of dispersive elements (prism) produce ultra low-dispersion spectra. One disperser called BP (Blue Photometer) operates in the wavelength range of 330–660 nm; the other is called RP (Red Photometer) and covers the wavelength range 650–1000 nm. The dispersion increases at shorter wavelengths, and ranges from 4 to 32 nm/pixel for BP, and from 7 to 15 nm/pixel for RP (`sci.esa.int/gaia`). However, we note that the photometric CCDs are located at the edge of the focal plane where the quality of the images is more sensitive to aberrations than the astrometric images [1].

Evidently the Gaia BP/RP images (ultra low-dispersion spectra) can be simulated by LDS ground-based surveys with photographic plates. Hence, the survey data can be used to test the simulated LDS for Gaia. For Gaia, most important are the full northern or southern sky surveys, as they can provide for any object in the sky, above a certain magnitude threshold, the real spectrum, and in many cases, several or even many spectra at various times.

The BP and RP spectra will be binned on-chip in the across-scan direction, whereas along-scan binning is not foreseen. The RP and BP will be able to reach object densities on the sky of at least 750 000 objects deg^{-2}. The observed complex images can be simulated by the GIBIS simulator (Fig. 3). GIBIS is a pixel-level simulator of the Gaia mission intended to simulate how the Gaia instruments will observe the sky based on realistic simulations of the astronomical sources and instrumental properties. It is a research branch of the global Gaia Simulator (GaiaSimu) currently under development within Gaia Coordination Unit 2: Data Simulations.

2. Astrophysics with LDS in the past

The LDS (Low-Dispersion Spectroscopy) astrophysics was developed and performed at numerous observatories (many in the USA) between about 1909 and nearly 1980. Numerous LDS plates were observed with Schmidt telescopes (plates with objective prism, also called slitless spectra), but there were also other instruments using a prism. The LDS plates were used for various projects e.g., QSO, blue and emission-line galaxy searches, emission line and Hα surveys, stellar classification, determination of radial velocities in galactic fields, etc., but they were seldom used after about 1980. As a consequence current knowledge about LDS among astronomers (especially young astronomers) is very limited.

The LDS using astronomical photographic plates dates back more than 100 years, which allows us to investigate very long-term spectral changes. For example, LDS observed from 1909 at Carnegie Observatories, Pasadena, CA, USA, are shown in Fig. 2. It is important because certain types of variable stars (VS) such as Miras, Cepheids, and a few cases of other stars, mostly peculiar variables, exhibit large variations of their spectral types. The research field is, however, poorly exploited, as the studies were very laborious (plates mostly required visually inspection) and rather limited, and hence there is no review of the spectral variability of VS.

3. Low Dispersion Spectral survey plate databases

Before Gaia, low dispersion spectra were frequently observed in the last century with various photographic telescopes using the objective prism. Some of them are listed here (the dispersion data are given in the next section; we also note that many other similar surveys exist):

Journal of Physics: Conference Series **328** (2011) 012018 doi:10.1088/1742-6596/328/1/012018

(1) Schmidt Camera Sonneberg. Sky survey (selected fields) with the 50/70 cm Schmidt telescope. Online access not yet available but the scans can be provided upon request.

(2) Bolivia Expedition Spectral Plates. These plates represent the coverage of the southern sky with spectral and direct plates, directed by the Potsdam Observatory. The plates are stored at Sonneberg Observatory and were observed during the years 1924–1928; in total ~70 000 prism spectra were estimated and published in Potsdam Publ. 26–19 in 1930.

(3) Hamburg Quasar Survey. A wide-angle objective prism survey searching for quasars with $B<17^{\mathrm{m}}.5$ in the northern sky. The survey plates have been observed with the former Hamburg Schmidt telescope located at Calar Alto/Spain since 1980. Online access is available.

(4) Byurakan Survey. The Digitized First Byurakan Survey (DFBS) is the digitized version of the First Byurakan Survey (FBS). It is the largest spectroscopic database in the world, providing low-dispersion spectra for 20 000 000 objects on 1139 FBS fields which equals 17056 deg^2. Online access is available. Sky coverage: DEC>$-15°$, and all RA (except for the Milky Way). The prism spectral plates were observed by the 1 m Schmidt telescope. Limiting magnitude: $17^{\mathrm{m}}.5$ in V. Spectral range: 340–690 nm, with a spectral resolution of 5 nm.

(5) Spectral survey plates at PARI (Pisgah Astronomical Research Institute), NC, USA. The total list of LDS plates at PARI is as follows: HK Survey 4° prism, A-Stars Survey 1.8° prism, He Survey UV filters, High Luminosity Survey, IR Survey 4° prism, Red Survey 4° prism, OB Survey 4 °prism, QSO Survey 1.8° prism, SGH Survey 1.8° prism, All-Sky Survey 4° prism, Blue Survey 10° prism, and anti center Survey 4° prism. Possibly the best suited for Gaia LDS simulations is the Case QSO-Survey. The telescope is the 61/91 cm Burrell Schmidt at Kitt Peak with 1.8° prism, plate FOV of 5° by 5°, limiting B magnitude of 18, the emulsion is IIIaJ Baked, and the spectral range from 330 nm to 530 nm. The Burrell Schmidt-type telescope of the Warner and Swasey Observatory at its new location on Kitt Peak in Arizona was used for a spectroscopic survey of the region with galactic latitude $\geq +30°$. and declination $\geq +30°$. The plates covering 5° by 5°, were observed with the 1.8° prism, providing a dispersion of 135 nm/mm at Hγ. The categories of objects that were cataloged are blue and/or emission-line galaxies, probable H II regions, blue and/or emission-line stellar objects, known and probable blue stars, main-sequence late B- and A-type stars, suspected field horizontal-branch stars of types A and F including RR Lyrae variables, suspected F- and G-type subdwarfs showing UV excess, faint carbon and late M suspected halo giants, and peculiar objects [2].

(6) The Henize MtWilson-Michigan Southern Hα Sky Survey. 290 high-quality plates of 15 × 15 inches observed in 1950–1952 in South Africa using a dedicated telescope by Karl Henize (for his Dissertation). A few examples of prominent emission spectral features found by K. Henize are shown in Fig. 5. They were found in objective prism spectral sky surveys. The estimated rate of strong emission and normal stars is 1: 10 000 in the Michigan-Mt Wilson Southern Hα Survey. There are hints that at least some of these strong emission features are variable. The used dedicated telescope was D25 cm, with a dispersion of 45 nm/mm at Hα. Note that filters were used in this program. A full-size plate is shown in Fig. 4.

(7) MtWilson–Michigan Northern Hα Sky Survey recently stored at Carnegie Observatories, Pasadena, CA, USA. This collection contains nearly 1000 LDS survey plates of various sizes and represents a northern analogy to the Henize MtWilson-Michigan Southern Hα Sky Survey.

4. The Ultra-Low Dispersion Spectroscopy by Gaia

4.1. Ultra LDS provided by Gaia RP/BP

We have identified the following astrophysical aspects of the LDS provided by Gaia BP/RP.

- Strong emission lines (also variable)
- Prominent spectral variability
- Possibility of spectroscopic Gaia alerts

Journal of Physics: Conference Series **328** (2011) 012018 doi:10.1088/1742-6596/328/1/012018

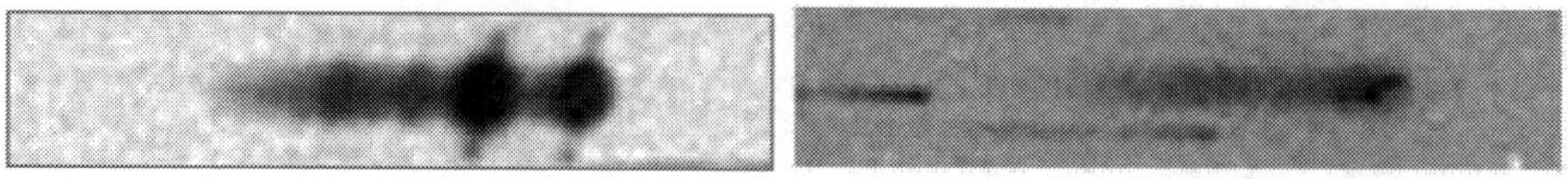

Figure 1. Examples of objects with emission lines observed with spectral dispersion similar to Gaia BP/RP: planetary nebula (left-hand panel) and a UV excess galaxy (right-hand panel). Source: Digitized Byurakan Survey.

- Follow-up by ground-based RTs with LDS
- Plate Sky Surveys can serve as real LDS simulator
- Confirm simulated spectra for any object brighter than ∼18 mag., even at various epochs.

4.2. Algorithms

The algorithms for automated analyses of digitized spectral plates were developed by students of informatics [3]. The main goals are as follows: the automated classification of spectral classes, searches for spectral variability (both in the continuum and lines), searches for objects with specific spectra, correlation of spectral and brightness variability, searches for transients, and the application to Gaia. The archival spectral plates observed with the objective prism offer the possibility to simulate the Gaia low-dispersion spectra and related procedures such as searches for spectral variability and variability analyses based on spectrophotometry. We focus on the sets of spectral plates in the same region of the sky over long-term intervals with good sampling that allow us to simulate Gaia BP/RP output. The main task is the automated classification of the stellar objective prism spectra from digitized plates, and a simulation and feasibility study for low-dispersion Gaia spectra.

4.3. Comparison of the Gaia Low Dispersion Spectra versus spectral plates

The comparison of low-dispersion spectra expected from the Gaia RP/RP photometers with those recorded in various astronomical photographic spectral sky surveys is an important task. The motivation of these studies is as follows: (1) Comparison of the simulated Gaia BP/RP images with those obtained from digitized Schmidt spectral plates (both using dispersive elements) for selected test fields, and (2) Feasibility study for the application of algorithms developed for the plates to Gaia. The dispersion is an important parameter and is briefly further discussed: (1) Gaia BP: 4–32 nm/pixel i.e., 400–3200 nm/mm, 9 nm/pixel i.e., 900 nm/mm at Hγ, RP: 7–15 nm/pixel i.e., 700–1500 nm/mm. A PSF FWHM∼2 px i.e., the spectral resolution is ∼18 nm. (2) Plates Schmidt Sonneberg (typical mean value): the dispersion for the 7° prism 10 nm/mm at Hγ, and 23 nm/mm at Hγ for the 3° prism. (3) Bolivia Expedition plates: 9 nm/mm, with calibration spectrum. (4) Hamburg QSO Survey: 1.7° prism, 139 nm/mm at Hγ, spectral resolution of 4.5 nm at Hγ. (5) Byurakan Survey: 1.5° prism, 180 nm/mm at Hγ, resolution 5 nm at Hγ, and (6) PARI prism dispersion: 150 nm/mm at 450 nm. We find that the Gaia BP/RP dispersion is ∼5 to 10 times smaller than the dispersion of typical digitized spectral prism plates, and the spectral resolution of Gaia is ∼3 to 4 times less compared to plates. Note that for plates the spectral resolution is seeing-limited, hence, the values represent the best values. For plates affected by bad seeing the spectral resolution is only ∼2 times better than compared to Gaia BP/RP.

It confirms that in the already available sky plate archives there are photographic spectral records with low-dispersion spectra very similar to what Gaia BP/RP will provide.

Journal of Physics: Conference Series **328** (2011) 012018 doi:10.1088/1742-6596/328/1/012018

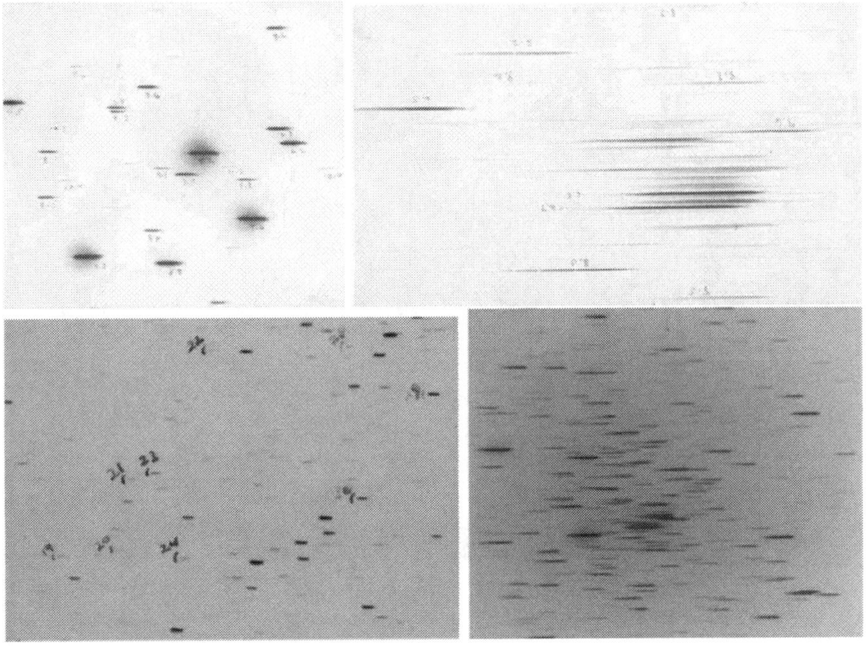

Figure 2. Probably the oldest astronomical objective prism plates of Mt Wilson Observatory at Carnegie Observatories in Pasadena, CA, USA, were observed in 1909. These data are not only of historical interest because they can be used to study long-term spectral evolution and/or changes of particular objects.

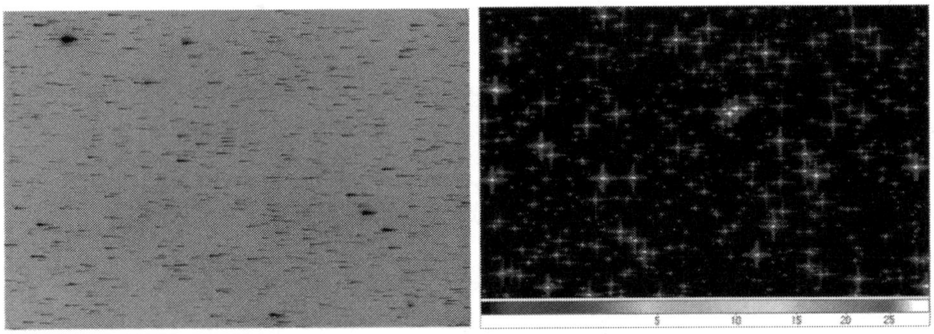

Figure 3. Comparison of the Case sky survey LDS plate (left-hand panel) and simulated (Gibis simulator) Gaia BP output (right-hand panel) for similar dispersions (but different sky regions).

4.4. Most important LDS plate surveys/databases for simulated Gaia LDS

We have identified the following plate surveys as the most promising for the application to Gaia (simulations, algorithm tests, tests of simulated LDS).

- German La Paz Bolivia Expedition: Southern Sky Coverage D
- Hamburg Quasar Spectral Survey D
- Digitized Byurakan Spectral Survey D
- Northern H alpha MtWilson-Michigan Sky Survey PD
- Southern H alpha MtWilson-Michigan Sky Survey PD

Journal of Physics: Conference Series **328** (2011) 012018 doi:10.1088/1742-6596/328/1/012018

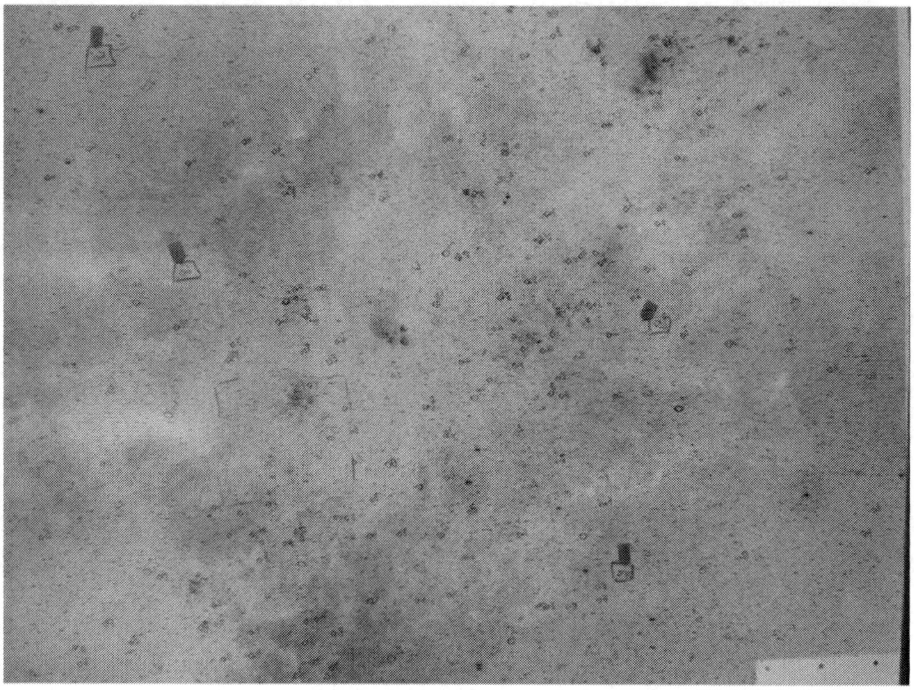

Figure 4. Example of Henize Mt Wilson-Michigan Southern Sky Hα Survey plate, manually evaluated and marked by K. Henize.

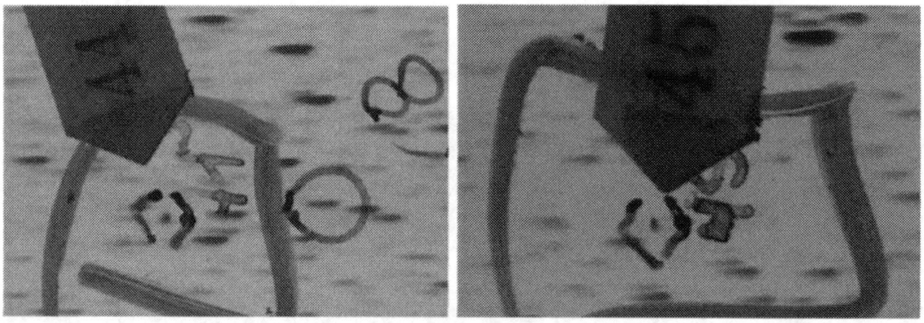

Figure 5. Examples of marked objects with very strong emission lines with spectral dispersion similar to the Gaia BP/RP. Southern Michigan-Mount Wilson Hα sky survey by K. Henize with plates located at PARI, NC, USA. Note that the marked objects show no continuum, only strong emission features.

Note: The listed surveys represent an almost full-sky hemisphere coverage. D = Digitized, PD = Partly Digitized.

5. The Spectrophotometry

Despite the very low dispersion discussed above, the major strength of Gaia data for many fields of science will be in spectrophotometry, because the low-dispersion spectra may be converted to numerous well-defined color filters. For example, Optical Afterglows (OAs) of Gamma-Ray Bursts (GRBs) are known to exhibit rather specific color indices that distinguish them from other types of the astrophysical objects [4,5]. A reliable classification of optical transients will

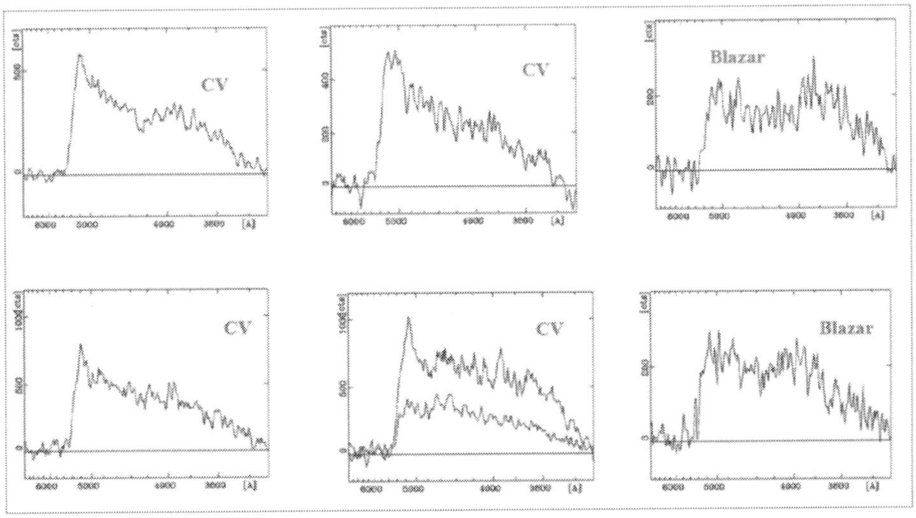

Figure 6. The flux profiles of LDS of selected objects (cataclysmic variables and blazars from the digitized Hamburg QSO survey.

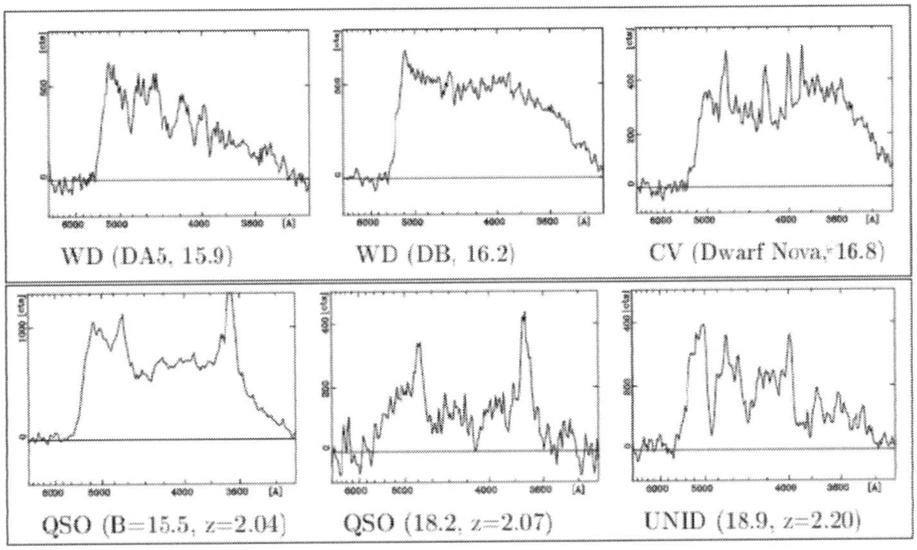

Figure 7. The Hamburg QSO digitized plates were used for the identification of ROSAT X-ray sources. Here we show a few examples. For more details see [9].

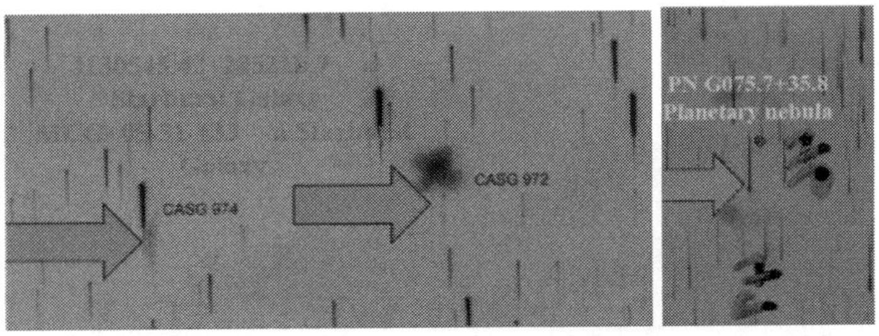

Figure 8. The prominent spectral features marked on LDS plates at PARI, NC, USA.

hence, in principle, be possible with this method. Colors of microquasars may serve as another example; they display blue colors with a diagonal trend formed by individual objects. This method can be used for optically faint, hence distant, objects as well. Apart from these examples color studies may improve our current understanding of various types of astrophysical objects, including cataclysmic variables and various types of high-energy sources.

We note however that the color indices cannot be calculated without the decontamination of BP/RP spectra [6,1]. The energy redistribution effect in the Gaia BP and RP spectra arising from contamination by the wings of the image profiles has been mentioned and investigated by [6], [7], and [8]. According to them, Gaia spectra may be used for stellar classification either after applying contamination corrections, or by using standard calibration stars with known physical parameters and observed by the Gaia spectrophotometers. According to [6], in the latter case, there is no possibility to calculate real spectral energy distributions, magnitudes, color indices, color excesses, or other photometric quantities. The classification must be performed by matching the observed pseudo-energy distributions of target and standard stars, or by using pattern-recognition algorithms (or template matching) over the entire spectrum to estimate the astrophysical stellar parameters.

6. Conclusion

The ESA-Gaia satellite will provide ultra-low dispersion spectra with the BP and RP photometers and present new challenges for astrophysics and informatics. The nearest analogy is the digitized prism (slitless) spectral plates of LDS sky surveys. Digitized surveys can be used for simulations and tests of Gaia data and algorithms. Some Gaia algorithms have been already proposed and tested. Some types of variable stars are known to exhibit large changes of spectral type. The research field is however poorly exploited and perhaps more discoveries can be expected with Gaia data because the Gaia mission will allow us to investigate the spectral behavior of a huge amount of objects observed during 5 years with good spectroscopic sampling. However, the Gaia data must be first decontaminated for scientific applications, as we discuss in this paper. We note that for investigations of (possible) prominent spectral changes with Gaia BP/RP data we will mostly rely on the overall spectral continuum flux profile and on bright and wide emission spectral features (and/or color changes) instead of absorption lines. The latter spectral lines will be hardly visible with the Gaia photometers due to the very small spectral dispersion.

7. Acknowledgments

We thank the referee Dr. L. Houziaux (Royal Academy of Belgium, Brussels) for carefully reading the paper and for several comments that have helped to improve its clarity. We appreciate the support provided by the PARI Institute, NC, USA, especially by Thurburn Baker and Michael Castelaz. The Czech participation in the ESA Gaia project is supported by the PECS project 98058. The scientific part of the study is related to the grants 205/08/1207 and 102/09/0997 provided by the Grant Agency of the Czech Republic. Some aspects of the project described here represent a natural continuation of the Czech participation to the ESA INTEGRAL (ESA PECS 98023). The analyses of digitized low-dispersion spectral plates at the PARI Institute are supported by MSMT KONTAKT Project ME09027.

8. References

[1] Straizys V 2010 *personal comm.*
[2] Pesch P and Sanduleak N 1984 in *Astronomy with Schmidt-type Telescopes*, Proceedings of the 78th IAU Coll. in Asiago Italy Aug 30 – Sep 2 1983, ed. M Capaccioli (Reidel Publ. Dordrecht) p 53
[3] Hudec L 2007 *Algorithms for spectral classification of stars, BSc. Thesis Charles University Prague*
[4] Simon V, Hudec R, Pizzichini G and Masetti N 2001 *A&A* **377** 450

[5] Simon V, Hudec R, Pizzichini G and Masetti N 2004 in *Gamma-Ray Bursts: 30 Years of Discovery* Gamma-Ray Burst Symp. (AIP Conf. Proc.) **727** 487
[6] Straizys V *et al* 2006 *Baltic Astronomy* **15** 449
[7] Montegriffo P *et al* 2007 GAIA-C5-TN-OABO-PMN-001-1, Gaia Livelink
[8] Montegriffo P 2009 GAIA-C5-TN-OABO-PMN-002, Gaia Livelink
[9] Zickgraf F J *et al* 2003 *A&A* **406** 535

Journal of Physics: Conference Series **328** (2011) 012019 doi:10.1088/1742-6596/328/1/012019

Hot stars in the Gaia-ESO Survey

Ronny Blomme

Royal Observatory of Belgium, Ringlaan 3, B-1180 Brussel, Belgium

E-mail: Ronny.Blomme@oma.be

Abstract. The Gaia-ESO Survey will produce spectra of about 100 000 stars, using the VLT FLAMES instrument. This includes hot, massive stars in a number of selected clusters. I describe the on-going cluster selection as well as the work package responsible for analyzing the hot-star spectra.

1. Large surveys

Astronomy has recently entered the era of large surveys. Many photometric surveys, as well as a few spectroscopic ones, are currently underway or are planned for the near future. A (non-exhaustive) list of these surveys is given in Table 1.

Astronet, the consortium of the largest funding agencies for astronomy in Europe, recently reviewed the situation of the European 2- to 4-meter class telescopes (ETSRC 2010). One of the recommendations of the Review Committee was that some of these telescopes be used for complementary programmes in support of European space missions. Where needed, new instrumentation will be developed to efficiently achieve these goals. As a consequence, these telescopes will no longer be available for individual astronomers, or small groups of them, but only for larger consortia of astronomers.

Finally, the upcoming Gaia mission promises to deliver high-accuracy positions, distances, proper motions and photometry of 1 billion stars, with each star being observed (on average) 80 times. Spectroscopy of 150 million stars will also be obtained, on average 40 times for each star.

All the above makes it clear that we have entered the age of 'industrial' astrophysics. This will imply considerable changes in the way that astronomers do their research.

2. Gaia and massive stars

2.1. The Gaia satellite

The Gaia satellite[1] will provide astrometry and photometry down to magnitude $G{\approx}20$ (for hot stars, the Gaia-band G magnitude is approximately equal to the Johnson V magnitude). Radial velocities will be determined for all stars down to $G_{\rm RVS} \approx 17$, and stellar atmosphere parameters down to $G_{\rm RVS} \approx 13$ (again, for hot stars, the magnitude in the Radial Velocity Spectrometer filter ($G_{\rm RVS}$) is approximately equal to the Johnson V magnitude). However, for massive early-type stars the spectroscopic estimates may be too optimistic, as the spectral windows are not

[1] For the expected science performance of Gaia, see:
http://www.rssd.esa.int/index.php?project=GAIA&page=Science_Performance

Journal of Physics: Conference Series **328** (2011) 012019 doi:10.1088/1742-6596/328/1/012019

Table 1. Non-exhaustive list of current and near-future surveys.

acronym	name	distinguishing features
SDSS	Sloan Digital Sky Survey	images of 350 million objects, spectra of 1.5 million
UKIDSS	UKIRT Infrared Deep Sky Survey	JHK photometry over 7500 square degrees down to K=18.3
IPHAS	INT/WFC Photometric H-alpha Survey of the Northern Galactic Plane	photometry in Hα, r and i filters
UVEX	UV-Excess Survey of the Northern Galactic Plane	photometry in U, g, r and He I 5876
VPHAS+	VST/OMEGACAM Photometric H-alpha Survey of the Southern Galactic Plane	southern counterpart to IPHAS and UVEX
SkyMapper	SkyMapper	photometry southern sky in $uvgriz$ filters, down to mag 22
Pan-STARRS	Panoramic Survey Telescope & Rapid Response System	photometry down to mag 24; whole sky will be observed 3 times in each lunar cycle
VVV	VISTA Variables in the Via Lactea	deep IR atlas in 5 passbands
LSST	Large Synoptic Survey Telescope	photometry of billions of objects; entire visible sky every 3 nights for 10 years; down to mag 24-25
RAVE	RAdial Velocity Experiment	spectra of up to 1 000 000 stars. Outside the Galactic Plane, so no significant number of hot stars are observed.
LEGUE	LAMOST Experiment for Galactic Understanding and Evolution	multi-fibre spectroscopy with 4000 fibres. Will observe Galactic halo stars, stars in the anticentre, and bright Galactic disk stars. This includes a complete census of OB stars in the Galactic plane ($20° < l < 230°$).
GALAH	GALactic Archaeology with HERMES	high resolution spectra of over a million Galactic disk stars. Uncertain if many OB stars will be included.

Journal of Physics: Conference Series **328** (2011) 012019 doi:10.1088/1742-6596/328/1/012019

optimized for those stars. The expected precision in radial velocity for a B1 V star is 1 $\mathrm{km\,s^{-1}}$ at V=7 and 9 $\mathrm{km\,s^{-1}}$ at V=12, and the precision will quickly degrade for fainter stars.

2.2. Number of massive stars

To better understand the important contribution that Gaia will make to massive-star research, we can estimate how many currently unknown massive stars will be detected by Gaia. There are various ways to obtain a rough values for this number. Shara et al. (2011) estimate the number of Wolf-Rayet stars to be in the range 1000 – 6500. Taking a factor of 10 more O-type stars than W-R stars, there should be 10 000 – 65 000 O-type stars. An alternative estimate starts from the viewpoint that there should be about 100 clusters of the 30 Dor type in the Galaxy (Hanson & Popescu 2008). Based on that we estimate that there should be of the order of 100 000 massive stars. A third possibility is to look to the Besançon Universe model, which predicts that Gaia will observe about 900 000 B-type stars (Jordi and Carrasco 2007).

However, the number of O and early B-type stars currently known is quite small compared to these predictions. The most extensive catalogue is the one maintained by Reed[2] (2003). It is based on spectral type determinations that have been published in the literature. It lists 19766 stars of type O and early-B (the catalogue is known to be incomplete for main-sequence stars later than spectral type B2).

Figure 1 plots the histogram of the Reed catalogue as a function of V magnitude. It is clear that the catalogue becomes incomplete around V=10. If we require also some spectroscopic information (red line), we lose another magnitude.

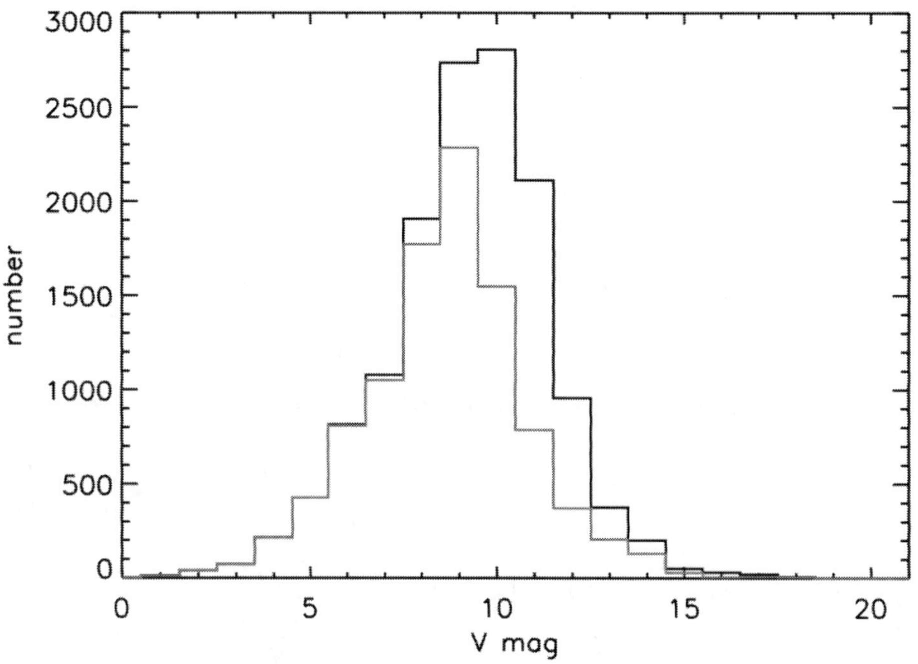

Figure 1. Histogram of the Reed catalogue, as a function of V-magnitude (black line). The red line indicates the histogram limited to those stars for which spectroscopic information is available.

[2] See also: `http://othello.alma.edu/~reed/obnotes.doc`

2.3. Gaia contribution to hot-star research

The good news therefore is that Gaia will provide at least one order of magnitude more hot stars than are currently known. Furthermore, we will know the distance to all of them with considerably higher accuracy than before. This improved census of massive stars will allow us to address a number of scientific questions, such as the organization of the Galactic Plane and the spiral arms. It will allow crucial quantitative tests of massive star evolution, and it will improve the number statistics of various evolutionary phases, including the better detection and statistics of the important, short-lived phases (LBVs, binary mergers, ...).

The bad news is that we will have little or no spectroscopic information for most of these stars. The Gaia spectroscopy does not reach the same magnitude limits as the astrometry and the photometry. Furthermore the spectroscopic domain is limited to 847 – 874 nm; for hot stars we will mainly detect a number of Paschen lines in that wavelength range, but little else.

Recognizing the massive stars will therefore have to be done using the Gaia photometry. That, however, will be complicated by the so-called $A_V - T_{\rm eff}$ degeneracy. Due to noise in the data, a unique determination of effective temperature ($T_{\rm eff}$) and interstellar extinction (A_V) is never possible. But, in addition to that, the confidence regions of the best-bit solution do not show a nice error-ellipse in the $T_{\rm eff}$, A_V plane, but rather an extended "ridge" (Bailer-Jones 2010). This effect is especially serious for the hotter, more massive stars (Blomme et al. 2011). One possible way to reduce the impact of this problem is to use the fact that massive stars frequently are found in clusters. By assuming a constant interstellar extinction in that cluster, the effective temperatures will be better determined (although that assumption may not always be valid).

3. Gaia-ESO Survey (GES)

3.1. Overview of the survey

To obtain more spectroscopic information for all types of Galactic stars, the Gaia-ESO Survey (GES) was proposed. The project is led by G. Gilmore and S. Randich and includes more than 250 Co-Investigators. It will use the VLT FLAMES instrument (on the UT2 telescope), with both the Giraffe and UVES fibres. It will require 300 nights, spread over 5 years. We expect that of order 10^5 Giraffe spectra will be taken, and of order 10^4 UVES spectra. The proposal was recommended for implementation by the ESO Observing Programmes Committee (OPC) and the Survey Management Plan is currently (Summer 2011) being developed.

The main goal of the GES is to study the formation and evolution of the Milky Way and its stellar populations. This includes determining the substructure of the halo, the nature of the bulge, studying the formation of the thick and thin disks, as well as the dynamical and stellar evolution of open clusters. The GES will therefore sample all these major components of the Milky Way. As part of the survey, about 60 old clusters will be studied (these are defined as having an age larger than 100 Myr) and 40 young clusters. Among those young clusters, 13 were chosen specifically for their massive-star content.

3.2. Massive stars in the GES

Our specific interest is in the O, B and A-type stars in these clusters. A number of scientific issues can be addressed with the spectra that will be collected for these types of stars. We will determine the stellar parameters and compare their position in the Hertzsprung-Russell diagram with theoretical evolutionary tracks and isochrones. This will lead to critical tests of stellar evolution modelling. The data will also help to constrain the upper part of the Initial Mass Function (IMF). For those stars where the Hα emission is strong enough, we will determine the mass-loss rates. The radiatively driven winds of these stars are known to be clumped, and the large number of stars studied will allow us to get a handle on this clumping issue.

Further analysis will also lead to the determination of abundances for a number of elements. From evolutionary calculations, it is known that the Nitrogen abundance is strongly influenced by rotation. While these predictions hold well in the Small and Large Magellanic Clouds, the situation is more confused in our Galaxy (Hunter et al. 2009).

The hot stars can also be used to study Galactic abundance gradients. Because of their high luminosity, they can be seen over larger distances than the cooler stars, thus providing a larger baseline. Because they are also much younger, they provide abundance values much closer to the present-day ones. By comparing the gradient based on young hot stars with that inferred from older populations, one can derive the time evolution of the gradient; this is one of the most powerful constraints on models of thin disc formation.

The data will also be relevant to Be stars. By systematically comparing Be stars to normal B stars, we can better understand the origin of the Be phenomenon. A higher fraction of Be stars is predicted for either higher masses (Zorec et al. 2005), or for more evolved main-sequence stars (Ekström et al. 2008). A large sample of high-resolution spectra will allow a much more accurate age determination for more stars, and thus will disentangle the mass and age effects.

3.3. Observation strategy

We expect spectra of approximately 1500 O, B, A stars to be observed as part of the GES.

The observation strategy for hot stars is different from that for cooler stars. The latter are observed in Giraffe gratings that are more in the red part of the spectrum (HR10, HR15N, and HR21). For the hot stars we use gratings in the blue part of the spectrum, as well as one that covers the Hα line (see Table 2). The table also lists the spectral lines that are of specific relevance to hot stars either to determine the stellar parameters or the abundances.

Besides the Giraffe gratings, we also intend to use the six fibres of the UVES spectrograph with the CD3 520 nm setting. To position all fibres, we will use astrometry from the 2MASS survey. We aim to achieve a S/N = 100 in O, B stars: this will result in a radial velocity accuracy of 0.5 km s^{-1} (for slow and medium-fast rotators). With a per-pixel S/N of 50 for UVES, good abundance determinations will be possible.

3.4. Cluster selection

A preliminary selection of the massive star clusters to be studied has already been done. A major selection criterion is of course the science interest of the cluster. But a number of technical issues play a role as well. We need to be reasonably sure about the cluster membership of the stars. While this may be less of a problem for the earliest spectral types, it does become more problematic when we consider the A-type stars.

Furthermore, we also need a sufficient number of stars within a limited magnitude range, so we can efficiently use the 131 FLAMES fibres to make a simultaneous integration, resulting in spectra with usable S/N. An example is shown in Fig. 2, where we plot the histogram of the cluster stars in bins of B-magnitude (we prefer to use the B-magnitude, because most of our gratings are in the blue part of the spectrum). To indicate what spectral types these bins correspond to, we correct the observed B-magnitude for interstellar extinction and distance, and convert it into spectral type, assuming that the stars are on the Zero-Age Main Sequence (ZAMS). The upper part of the figure shows the three proposed pointings of the FLAMES instrument: one for the magnitude range 10.0 to 13.0 (with $n = 57$ targets), one for magnitude 13.0 – 15.0 ($n = 104$) and one for magnitude 15.0 – 17.0 ($n = 98$).

Taking into account the procedures outlined above, we made a preliminary list of massive-star clusters for the GES (see Table 3). This list will most probably undergo changes and extensions, before it becomes final.

Table 2. FLAMES settings to be used in GES for hot star observations.

Giraffe gratings		
grating	wavelength range (Å)	spectral lines (wavelengths in Å)
HR03	4033–4201	Hδ 4102
		He I 4121, He I 4144, Si IV 4089, Si IV 4116, Si II 4129
		O II 4075, 4133, 4157, 4185
HR05A	4340–4587	He I 4387, He I 4471, He II 4542
		Si III 4552, Si III 4568, Si III 4575, Mg II 4481
		N II 4447, O II 4350, 4367, 4396, 4415, 4452
HR06	4538–4759	He I 4713, He II 4541, He II 4686 (wind line)
		Si IV 4631, Si IV 4654, Si III 4560
		C III 4650, N II 4601, 4614, 4631, 4643, O II 4595, 4640-70, 4700
HR14A	6308–6701	Hα 6563 (wind line)
		He I 6678, He II 6527, He II 6683, Si II 6347
		C II 6580
UVES settings		
CD3, 520 nm	4140–6210	overlaps with HR03, 05A, 06. Also includes Hγ.
		UVES fibres will be used for the brightest stars only.

Table 3. Preliminary list of GES clusters selected for their massive-star content. Also listed is the number of proposed pointings.

cluster	# pointings	cluster	# pointings
NGC 3293	3	NGC 4755	3
Trumpler 16	4	Dolidze 25	3
NGC 3766	2	Collinder 228	2
Trumpler 27	2	Westerlund 2	1
NGC 6649	5	IC 2944	2

4. Work package on hot stars

One of the work packages of the GES is responsible for the analysis of the O, B and A-type stars. It is led by the author and its task is to determine stellar parameters and abundances from the observed spectra of O, B, A stars.

First of all, the effective temperature and surface gravity will be determined. As these stars are part of a cluster, we will also have a good estimate of the distance (and another work package will be responsible for refining that value). Given the distance, we can also derive the luminosity, radius and mass. By comparing these values to evolutionary tracks, we can also determine the age and initial mass. Where appropriate, we will also measure the mass-loss rate (assuming no clumping) from the observed Hα line profile.

Journal of Physics: Conference Series **328** (2011) 012019 doi:10.1088/1742-6596/328/1/012019

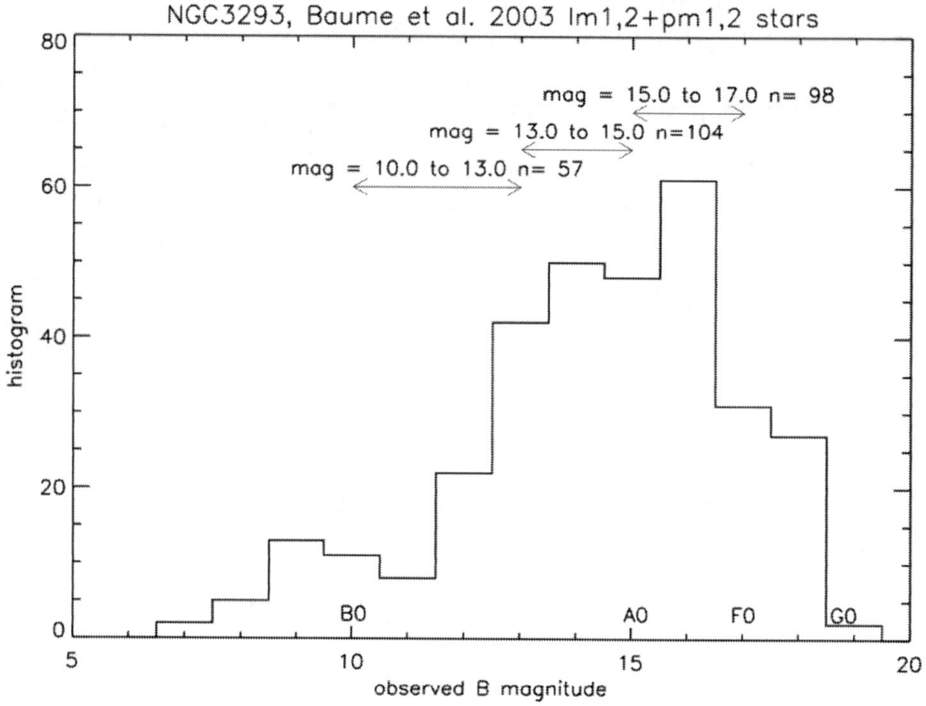

Figure 2. Histogram of the stars in the cluster NGC 3293, in bins of B-magnitude. The interpretation as spectral types assumes that the star is on the ZAMS. The top of the figure shows the three proposed pointings for this cluster, each pointing covering a specific magnitude range.

The techniques we will use for this are the classical ones where the observed spectra (or equivalent widths of the lines) are compared to synthetic spectra. A number of groups have volunteered to help in the effort needed for this work package.

5. Conclusions

A large number of astronomical surveys are currently running or are planned for the near future. The Gaia satellite will add substantial information to the ever growing amount of data. The Gaia-ESO project will complement all this with a large number of high-resolution spectra. Specifically for hot stars, we expect to make a substantial addition to the number of stars studied, which will lead to interesting results in this field of research.

Acknowledgments

I am grateful to Dr. S. Randich (INAF, Arcetri Astrophysical Observatory, Italy), whose referee report helped to improve the clarity of this paper.

References

Bailer-Jones C A L 2010 *MNRAS* **403** 96

Blomme R, Frémat Y, Lobel A and Martayan C 2011 *Gaia: At the Frontiers of Astrometry* EAS Publications Series vol 45, ed C Turon et al (Les Ulis: EDP Sciences) p 373

Ekström S., Meynet G., Maeder A and Barblan, F. 2008 *A&A* **478** 467

ETSRC 2010 *Report by the European Telescope Strategic Review Committee on Europe's 2-4m telescopes over the decade to 2020* `http://htdocs.factory02.com/astronet/IMG/pdf/PlaquetteT2_4m-final-2.pdf`

Journal of Physics: Conference Series **328** (2011) 012019 doi:10.1088/1742-6596/328/1/012019

Hanson M M and Popescu B 2008 *Massive Stars as Cosmic Engines* IAU Symposium 250, ed F Bresolin et al (Cambridge: Cambridge University Press) p 307

Hunter I et al 2009 *A&A* **496** 841

Jordi C and Carrasco J M 2007 *The Future of Photometric, Spectrophotometric and Polarimetric Standardization* ASP Conference Series vol 364, ed C Sterken (San Francisco: Astronomical Society of the Pacific) p 215

Reed B C 2003 *AJ* **125** 2531

Shara M M, Faherty J K, Zurek D, Moffat A F J, Gerke J, Doyon R, Artigau E and Drissen L 2011 A Near-Infrared Survey of the Inner Galactic Plane for Wolf-Rayet Stars II. Going Fainter: 72 More New WR Stars *Preprint* astro-ph/1106.2196

Zorec J, Frémat Y and Cidale L 2005 *A&A* **441** 235

Journal of Physics: Conference Series **328** (2011) 012020 doi:10.1088/1742-6596/328/1/012020

Modeling the wind and photosphere of massive stars with the radiative transfer code CMFGEN

Jose H Groh

Max-Planck-Institut für Radioastronomie, Auf dem Hügel 69, D-53121 Bonn, Germany

E-mail: jgroh@mpifr.de

Abstract. Massive stars are extremely luminous and characterized by mass loss through radiation-driven stellar winds. The radiation emitted at the stellar surface may interact with the wind, making the analysis of the emerging spectrum a very challenging task. In addition to the luminosity, effective temperature, and surface gravity, several other stellar parameters impact the spectral morphology of these objects, remarkably the mass-loss rate and the wind terminal velocity. That is generally the case for OB supergiants, Luminous Blue Variables, and WR stars. CMFGEN [1] comprises the state-of-the-art in non-LTE radiative transfer and has been successfully applied to the above classes of objects over the last decade. The code assumes spherical symmetry, stationary outflow, and both photospheric and wind lines can be treated in non-LTE. Full line blanketing due to hundreds of thousands of spectral lines is included, as well as wind clumping. Here I discuss the assumptions behind CMFGEN and present examples of models and spectroscopic analyses.

1. Introduction

Massive stars are rare but essential constituents of a stellar population. They are the main contributors to the input of ionizing photons, energy, and momentum into the interstellar medium and playing a key role in the chemical enrichment of their host galaxy. Recent observational studies indicate that stars with initial masses well above 100 $M_{\odot}$ – and sometimes perhaps as much as 300 $M_{\odot}$ – are formed in the Milky Way and the Local Group [2]. Theoretical studies based on numerical simulations suggest that stars with high initial masses ($\sim 100\ M_{\odot}$; e.g., [3]) were routinely formed in the early Universe.

Current evolutionary models indicate that very massive stars, with initial masses greater than roughly 40 $M_{\odot}$, begin their lives as O-type stars burning hydrogen in their cores. After several millions of years, they unavoidably go through an unstable, short-lived stage, known as the Luminous Blue Variable (LBV) phase, when a significant fraction of the stellar mass and angular momentum are lost through winds and giant eruptions. In the current standard picture of stellar evolution, LBVs are progenitors of hydrogen-free Wolf-Rayet (WR) type, helium-core burning stars (see e.g., [4, 5, 6]).

Before going any further, we would like to emphasize the reasons for analyzing spectra of stars in general, as this provides a bridge to compare observations to theoretical evolutionary models. Spectroscopic analysis provides a way to maximize the amount of information extracted from the observations, helping to properly understand massive stars. While photometric observations provide essentially magnitudes and colors, spectroscopy allows one to measure line equivalent widths and ratios, radial velocities, and determined line profiles. However,

Journal of Physics: Conference Series **328** (2011) 012020 doi:10.1088/1742-6596/328/1/012020

to obtain temperatures, luminosities, chemical composition, among other stellar properties, a careful modeling of the observed spectrum has to be undertaken.

In contrast to low-mass stars like the Sun, massive stars lose mass during their whole lifetime through radiation-driven stellar winds. This occurs because the radiation field of these stars is extremely strong, which causes the star to lose mass at enormous rates (10^{-7} to 10^{-3} $M_{\odot}\text{yr}^{-1}$). The stellar wind may become so dense that the stellar photosphere becomes extended and is formed in a moving layer. Numerous emission lines could then be formed in the wind, veiling (sometimes completely) the underlying photospheric spectrum. Because the radiation emitted by the stellar surface interacts with the optically-thick, dense wind, the analysis of the stellar spectrum is a very challenging task. In addition to the classical stellar qualities such as the temperature, luminosity, and effective gravity, several other quantities such as the mass-loss rate and wind terminal velocity may affect the morphology of the spectrum. Complex radiative transfer models, which include the necessary physics to study the radiation transport across the atmosphere and wind, are needed to obtain realistic parameters of massive stars.

Here we discuss the state-of-the-art code non-LTE radiative transfer code CMFGEN, its assumptions, advantages, and limitations, as well as present illustrative examples of models and spectroscopic analyses. Rather then being an exhausting review on the code itself or on the extensive science done with CMFGEN, the goal of this paper is to present, from a user's standpoint, the essential properties of the code and the basics of spectroscopic analysis of massive stars using photospheric and wind diagnostics. The reader is referred to [1, 7, 8] for a comprehensive description of CMFGEN and to [9] for a detailed description of diagnostics commonly used in the spectroscopic analysis of OB and WR stars.

2. CMFGEN in a nutshell

CMFGEN has been developed and maintained by Dr. D. John Hillier (University of Pittsburgh, USA), and the code and the associated atomic data are publicly available[1]. Analyses with CMFGEN can potentially provide insights into many subfields of Stellar Astrophysics, and below we highlight the main applications of CMFGEN (related subfields are in parenthesis):

- derive accurate abundances, stellar and wind parameters (*Stellar Evolution*);
- derive accurate extreme-UV ($\lambda < 912$ Å) radiation fields (*Nebular Photoionization*);
- provide fundamental data for the study of starbursts (*Star Formation*);
- better understanding of the hydrodynamics of winds (*Radiation Hydrodynamics*);
- testbed of approximate methods that can be used in more complex geometries and in inhomogeneous media (*Multi-dimensional Radiative Transfer*).

Recently, CMFGEN has been adapted to model type II Supernovae, making it possible to obtain distances and diagnostics of the SN progenitor. The reader is referred to [10] and references therein for further information about spectroscopic analysis of SNe with CMFGEN, and hereafter we focus exclusively on massive stars.

2.1. The code suite

CMFGEN comprises not only a single code, and actually a suite of codes are used in the process of analyzing stellar spectra. While the temperature, atmospheric and wind structures, and level populations are obtained with CMFGEN itself, the observed spectrum is computed with a separate code (CMF_FLUX, see [8]). Many accompanying routines for plotting and analysis (DISPGEN and PLT_SPEC) and computation of Rosseland opacities (MAIN_LTE), among others, are provided.

[1] http://kookaburra.phyast.pitt.edu/hillier/web/CMFGEN.htm

Journal of Physics: Conference Series **328** (2011) 012020 doi:10.1088/1742-6596/328/1/012020

2.2. The atomic data

As we heard in this Workshop, the quality and self-consistency of atomic data plays a key role in the spectroscopic analysis and determination of stellar parameters. The atomic data contained in CMFGEN comes from a variety of sources, which are mainly the Opacity Project [11, 12], the NIST database[2], and from several individuals such as Robert Kurucz[3], Keith Butler (personal comm.), Sultana Nahar & Anil Prandham[4], and Gary Ferland (personal comm.).

CMFGEN stores atomic data in ASCII files in unique directory structures, with separate files containing collisional cross sections, oscillator strengths and energy levels, auto ionization rates, superlevels designation, and photoionization cross sections from the ground and excited states. As also pointed out by N. Przybilla (in this Volume), one should not assume the atomic data of a non-standard spectral feature to be correct.

The atomic data format is unique to CMFGEN, and conversion from published data into CMFGEN is time consuming. Unfortunately, this hampers the portability of atomic data and test models using atomic datasets from different groups has been done on a very limited basis.

2.3. Transfer options

Depending on the problem to be solved, the user has five options to compute the radiative transfer in stellar atmospheres with CMFGEN:

- Plane parallel geometry for solving the formal and moment equations;
- Plane parallel co-moving frame for solving the formal and moment equations. Includes the zeroth-order terms in v/c (i.e., we only retain v/c terms that multiply $\partial/\partial v$ terms). With this option, one can treat monotonic *vertical* velocity fields in the atmosphere, and the radiation transfer is computed in the frame moving with the gas. This has the advantage that opacities and emissivities are isotropic.
- Spherical co-moving frame. This option solves the formal and moment equations taking in spherical geometry, including the zeroth-order terms in v/c. A radial monotonic velocity field has to be specified.
- Spherical co-moving frame relativistic. Similar to option 3, but can treat fully relativistic (but stationary) monotonic outflows.
- Spherical time-dependent co-moving frame fully relativistic. Similar to option 4, but including time dependence in the moment equation for monotonic outflows. This option is still under tests and development as of September 2011.

2.4. Pros and cons of using CMFGEN

CMFGEN includes many physical processes that were neglected in previous codes, allowing the user to perform a spectroscopic analysis as realistically as possible nowadays. As described above, CMFGEN can handle spherically-symmetric geometries that is mandatory for modeling the majority of massive stars (with the exception of B dwarfs, see N. Przybilla, this Volume), since the presence of a wind makes a plane-parallel approximation invalid. CMFGEN computes continuum and line formation in the non-LTE regime, and a realistically hydrostatic structure can be computed until just below the sonic point. This allows the simultaneous treatment of spectral lines formed in the atmosphere, wind, and in the transition region between the two regimes, making CMFGEN applicable to the study of massive stars from various evolutionary states, such as OB, LBV, and WR stars.

[2] http://nist.gov/pml/data/asd.cfm
[3] http://kurucz.harvard.edu
[4] http://www.astronomy.ohio-state.edu/~nahar/nahar_radiativeatomicdata/index.html

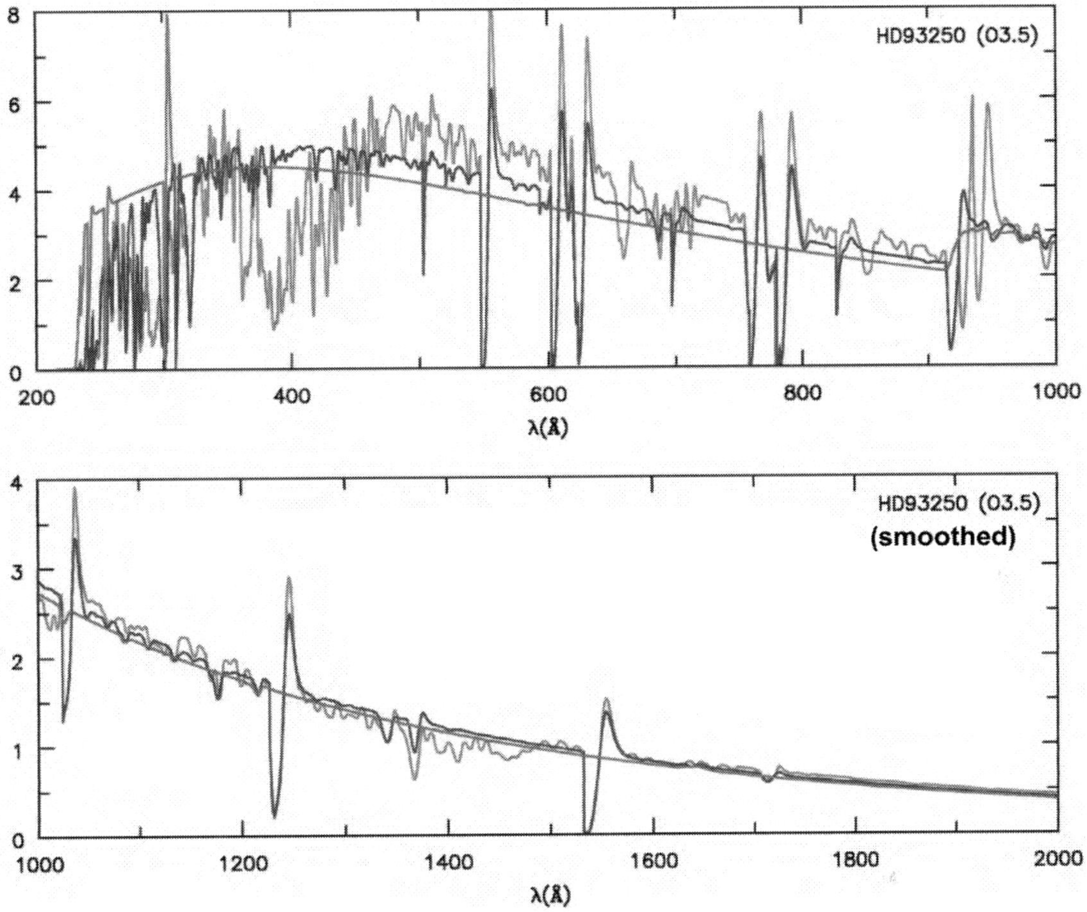

Figure 1. Comparision between CMFGEN synthetic spectra appropriate for the O 3.5 star HD 93250 including H and He only (red), H, He, and CNO (blue), and H, He, CNO, Fe, and other Fe group elements (green). The model parameters are: $L_\star = 1.3 \times 10^6\ L_\odot$, $T_{\rm eff} = 45700$ K, $R_\star = 18.3\ R_\odot$, $\log g = 4.0$, $\dot{M} = 5.6 \times 10^{-7}\ M_\odot {\rm yr}^{-1}$, and $v_\infty = 3000\ {\rm km\,s}^{-1}$. The spectra were smoothed for better visualization. Figure courtesy of D. John Hillier.

As noticed in the early 1990s [13], wind clumping may have a significant impact in obtaining the mass-loss rate of massive stars. CMFGEN allows for the presence of clumping within the wind using a volume-filling factor approach, assuming that material close to the star is unclumped and reaches maximum clumpiness at very large distances. More complex clumping laws can also be taken into account [14].

Line overlap and full line blanketing are also consistently included in CMFGEN through the concept of superlevels, which groups similar energy levels into one single superlevel to be accounted for in the statistical equilibrium equations [15], reducing the computational demand. CMFGEN allows the user to include energy levels and associated transitions from H, He, C, N, O, F, Ne, Na, Mg, Al, Si, P, S, Cl, Ar, K, Ca, Ti, Cr, Mn, Fe, Co, and Ni. Line blanketing plays a major role in the spectroscopic analysis of massive stars, as illustrated in Figure 1. The effects of x-rays are included in an approximate way, as these are crucial in determining the ionization structure of weak winds such as those from O dwarfs [16, 17].

However, it is important that the user is aware of the limitations of CMFGEN. At the moment,

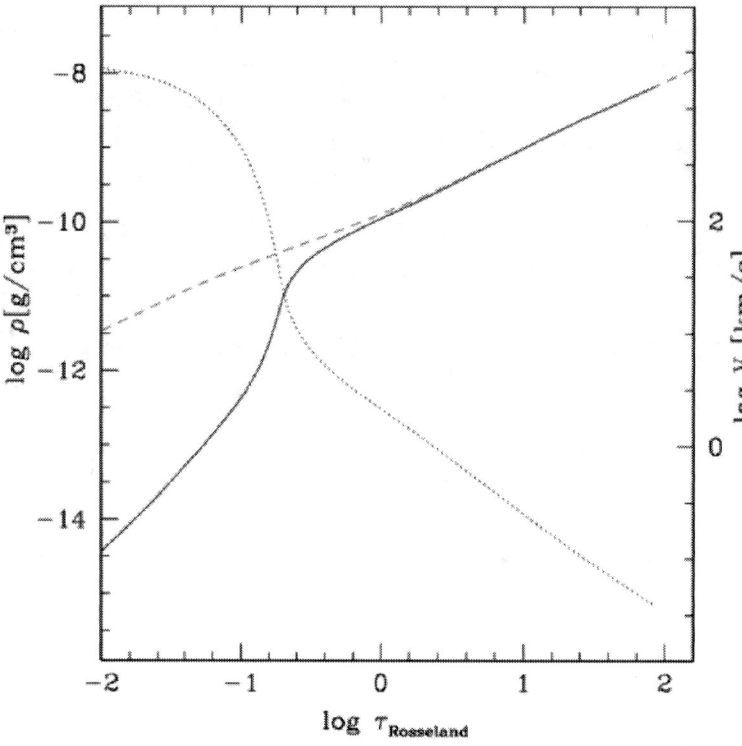

Figure 2. Typical hydrodynamical structure of a CMFGEN model, showing density (left y axis) and velocity (right y axis) as a function of optical depth for a model with $T_{\rm eff}$ = 40000 K, log g =3.50 and $\dot{M} = 1.4 \times 10^{-5}\ M_{\odot}{\rm yr}^{-1}$. The hydrostatic structure computed with the TLUSTY code (red dashed) or, alternatively, with CMFGEN itself, is connected to a beta-type law (dotted), producing the final CMFGEN hydrodynamical structure (solid line). From [18].

the code does not solve the hydrodynamic equations of the wind, and therefore, a velocity law has to be assumed a priori. In CMFGEN, the velocity law is parameterized by a beta-type law in the wind (or double-beta law), which is joined to the hydrostatic structure just below the sonic point (Fig. 2). Parameters from the velocity law, such as the wind terminal velocity and β, can be constrained from the observed line profiles though.

In addition, the computational time of a CMFGEN model (typically 10–20 hours) may become a severe limitation, specially in the analysis of large samples of objects. Usually, given the large parameter space to be explored, a grid approach to obtain the best fit model is not feasible, and the user has often to rely on physical insights.

3. Spectroscopic analysis of massive stars with CMFGEN

Here we provide the basic principles behind the comparison between synthetic and observed spectra, and refer the reader to [9] for a more detailed description of some of the diagnostics. In CMFGEN, each model is defined by the effective temperature, luminosity, mass-loss rate, wind volume filling factor, wind terminal velocity, effective gravity (or stellar mass), and by the abundances of the chemical species. This means that, depending on the parameter regime and the amount of data available, one could be able to obtain the above stellar and wind properties by comparison between models and observations.

3.1. Stellar parameters
As opposed to using bolometric corrections, fitting the observed spectral-energy distribution (SED) has become the standard method to constrain the bolometric luminosity of massive stars (assuming the distance is known). The errors associated with this procedure vary between 0.05 and 0.15 dex, depending on the spectral range used for the fit and the quality of the photometry (see e.g., [19, 20, 21, 22, 2, 9]). This technique has the advantage of providing the reddening law towards the target as a byproduct (and thus being less prone to uncertainty in the reddening law), and is particularly well suited for determining the relative changes in the bolometric luminosity in the case of LBVs [20, 21], as shown in Figure 3.

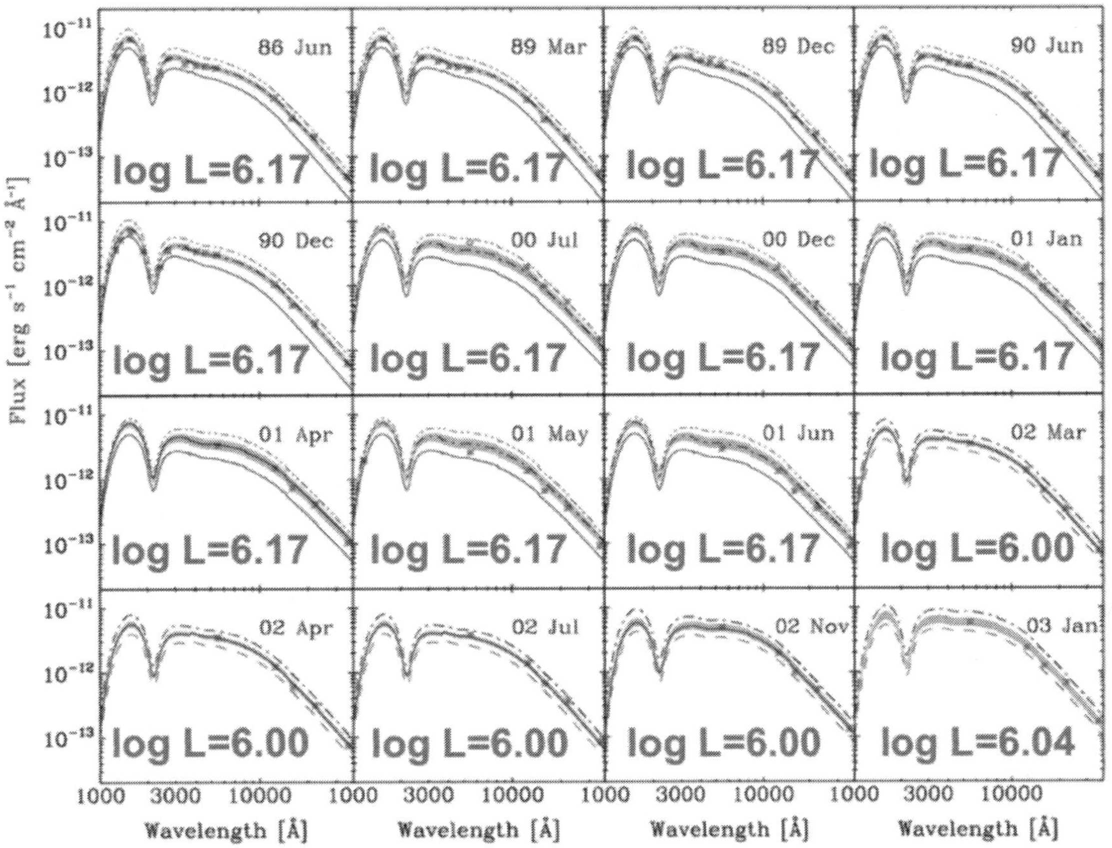

Figure 3. Comparison between the observed spectral energy distribution of AG Car and CMFGEN models with different bolometric luminosities. The panels are ordered chronologically, and in decreasing effective temperature, from left to right, top to bottom. For each panel we show selected models with $L_\star = 2.2 \times 10^6\ L_\odot$ (dark green dash-triple dotted line), $L_\star = 1.5 \times 10^6\ L_\odot$ (black dash-dotted line), $L_\star = 1.1 \times 10^6\ L_\odot$ (light green solid line), $L_\star = 1.0 \times 10^6\ L_\odot$ (blue solid line), and $L_\star = 0.7 \times 10^6\ L_\odot$ (purple dashed line). For a given epoch, models with different $L_\star$ had their $R_\star$ and $\dot{M}$ scaled in order to still fit the observed spectral lines. The gray region corresponds to the uncertainty in the model flux which comes from the luminosity. In all panels, the CMFGEN models were reddened using the Galactic extinction law of [23] with $R_V = 3.5$ and $E(B-V) = 0.65$. Note the trend towards lower luminosity as the effective temperature (radius) decreases (increases) during 2001–2003. The uncertainty in the observed photometry is smaller than the size of the symbols in the UV, optical, and in the near-IR. From [20].

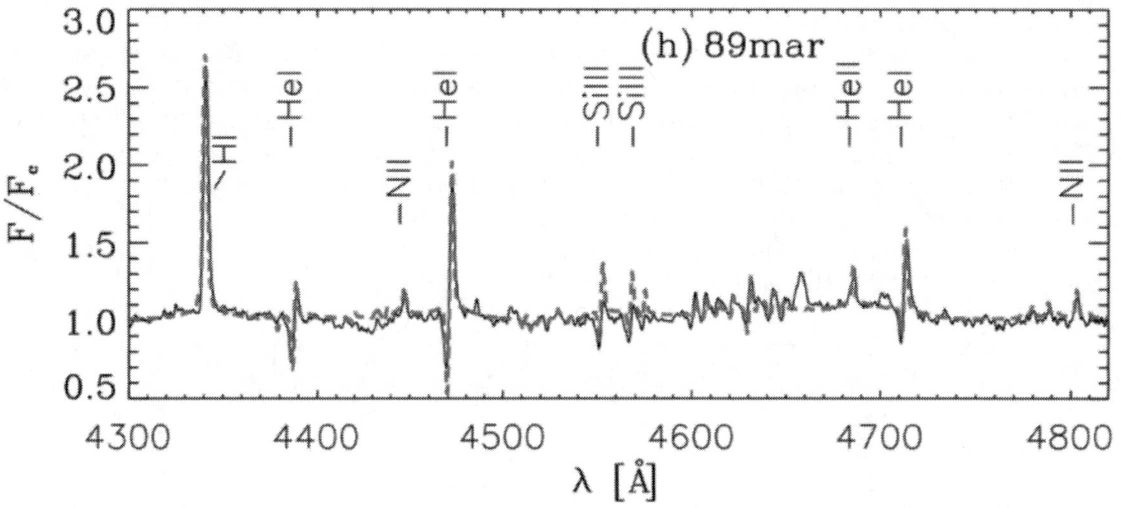

Figure 4. Optical spectra of AG Car taken at 1989 March (black solid line) compared with the best-fit CMFGEN model (red dashed line). The model parameters are: $L_\star = 1.5 \times 10^6\ L_\odot$, $T_{\rm eff} = 22800$ K, $R_\star = 58.5\ R_\odot$, $M = 70\ M_\odot$, $\dot{M} = 1.9 \times 10^{-5}\ M_\odot {\rm yr}^{-1}, f = 0.10$, and $v_\infty = 300\ {\rm km\,s^{-1}}$. From [20].

Once the bolometric luminosity has been obtained, one can use the aid of stellar evolution models to obtain the *current* and *initial* stellar masses. Indeed, this method has been applied to suggest that the most massive stars in the Local Group, located at the center of the R136 cluster in the Large Magellanic Cloud, had initial masses between 200–300 $M_\odot$[2].

The effective temperature is usually obtained from the ionization balance, which can be probed using spectral lines from different ionization stages of the same chemical species. The classical optical diagnostic for O stars is the pair He I 4471/ He II 4542 [9], although lines from Fe IV/V/VI in the ultraviolet and He I/ He II in the K-band [9] can also be used. In the case of LBVs, He II 4542 is extremely weak and thus He II 4686, together with He I 4471, He I 4713/ 5876/6678, are preferred to obtain the He ionization balance [24, 20]. Lines from Si II/Si III/Si IV, N I/ N II, C II/C III, and O I/ O II can also be used as temperature diagnostics in LBVs [20], as well as Fe I/ Fe II and Mg I/ Mg II optical lines in cooler LBVs. Figure 4 shows the best fit model of the LBV AG Carinae around 4500 Å, illustrating how both He I and He II lines can be matched simultaneously for $T_{\rm eff}$=22 800 K.

In OB stars, the effective gravity can be constrained from the pressure-broadened wings of Hydrogen lines. The Balmer Hδ and Hγ lines are the classical diagnostics, since the influence of the stellar wind on these lines is minimized. However, in massive stars with dense winds, such as LBVs and WRs, even Hδ, Hγ, or the Paschen lines have emission components caused by the wind. Notice in Figure 4, in the case of AG Car, how Hγ shows a P-Cygni profile and thus cannot be used for constraining $\log g$. Usually, $\log g$ (and thus masses) of WRs and LBVs cannot be obtained from spectroscopic analysis, with the exception of cool LBVs with thin winds, such as W243 in the massive cluster Westerlund 1 [25].

3.2. Wind parameters

The presence of a stellar wind affects significantly the spectral morphology. Constraints on the mass-loss rate are usually obtained from non-saturated lines with high optical depth. While UV resonance lines of N V, C IV, Si IV, N IV can be employed in the case of massive stars with

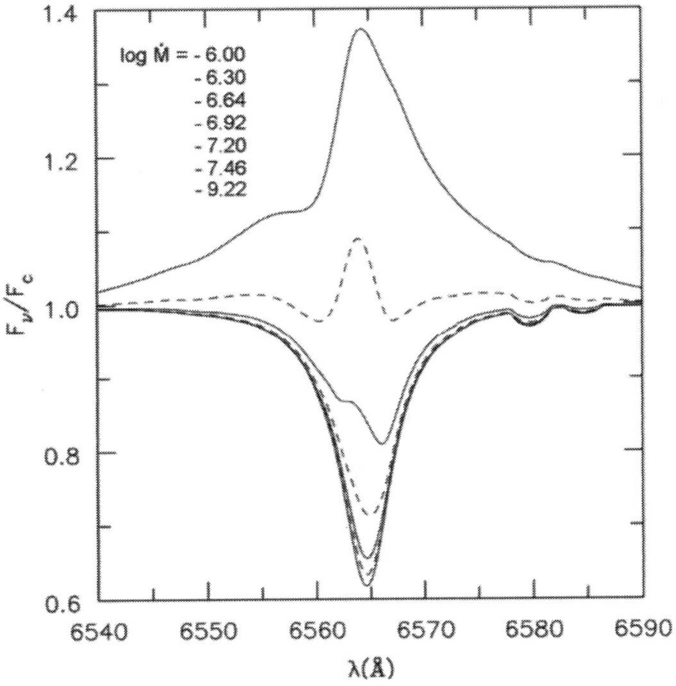

Figure 5. Variations in the Hα morphology as a function of the adopted $\dot{M}$. All synthetic spectra were computed with CMFGEN, and the one with the strongest emission has $\log(\dot{M}/M_\odot \mathrm{yr}^{-1}) = -6.00$, while the one with strongest absorption has $\log(\dot{M}/M_\odot \mathrm{yr}^{-1}) = -9.22$. Models with intermediate values of $\dot{M}$ are shown alternatively with solid and dashed lines for clarity. From [17].

relatively weak winds, recombination lines such as Hα (or another strong H line) or He II 4686 are the classical $\dot{M}$ diagnostics for stars with dense winds such as LBVs and WRs. Figure 5 illustrates the effects of $\dot{M}$ on the morphology of Hα in the case of an O dwarf, showing that this line turns from an absorption to emission profile as $\dot{M}$ increases [17].

Since there is overwhelming evidence for massive star winds being clumped (at least) on small scales (see contributions in [26]), recombination lines actually constrain the ratio $\dot{M}f^{-0.5}$, where f is the wind volume filling factor. Diagnostics that are not dependent on the density squared, such as the electron scattering wings [13] or non-saturated UV resonance lines (e.g, O V 1371 in O stars, [27]), can be used to constrain f (Figure 6).

The wind terminal velocity is best constrained using metallic resonance lines, which are usually found in the UV. In O stars and WRs, the classical diagnostics are N V 1240, Si IV 1393–1403, C IV 1548–50, and N IV 1718 [9], while in LBVs, these are C II 1334–1335, C III 1175, and Mg II 2800–2802 [20]. Whenever UV spectroscopic is unavailable, for instance for stars with large amounts of reddening, a *lower* limit to v_∞ can be obtained either using the width of the emission or the velocity of the maximum absorption of H and He lines in the optical and near-IR.

4. CMFGEN and Gaia

As described above, an accurate spectroscopic analysis of massive stars usually relies on mutli-wavelength diagnostics, or at least makes use of a large fraction of the UV, optical, or near-IR bands. Since Gaia will cover a relatively narrow spectral region in medium spectral

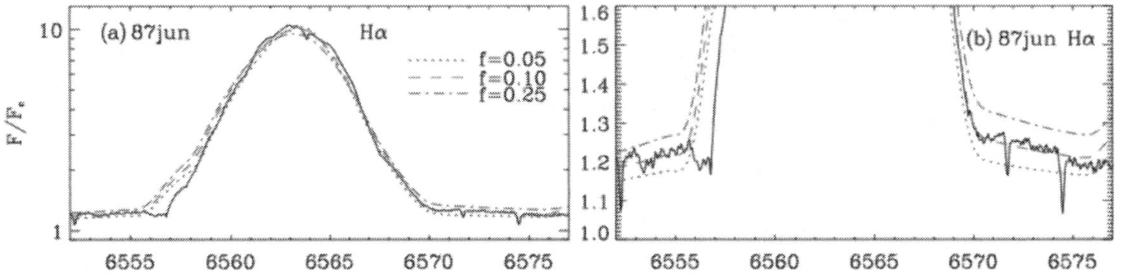

Figure 6. Optical spectra of AG Car taken around Hα in 1987 June (black line) compared with CMFGEN models assuming different wind volume-filling factors f and with $\dot{M}$ scaled to produce a similar amount of Hα emission (i.e., all models have the same $\dot{M}f^{-0.5} = 4.7 \times 10^{-5}\ M_{\odot}\mathrm{yr}^{-1}$). We display models with $f = 0.05$ (dotted blue line), $f = 0.10$ (dashed red line), and $f = 0.25$ (dot-dashed green line). Notice that the model with $f = 0.25$ produces too strong electron scattering wings.

resolution around 8500 Å, an unambiguous determination of stellar and wind parameters of individual targets will likely not be possible. However, the broad spectral range of the Gaia spectrophotometric mode (3300–11500 Å) will provide spectral energy distributions that could be used to roughly constrain the evolutionary stage of individual targets by determining whether these are OB-type, LBV, or WR stars. In this context, follow-up ground-based surveys, such as the Gaia-ESO public survey (see Blomme, this Volume), will play a key role in maximizing the amount of massive star science that will be obtained out of the Gaia survey.

Acknowledgements

I thank the referee, Dr. Jean-Claude Bouret (Laboratoire d'Astrophysique de Marseille – France), for the careful reading and comments on the manuscript. I would like to thank the organizers, in particular Alex Lobel, for the invitation to talk about CMFGEN and massive stars in the context of Gaia, and for setting up a very nice Workshop. I also acknowledge travel support provided by the organizers, and D. John Hillier for making CMFGEN available, for the continuous support, and for sharing figures and viewgraphs that certainly increased the quality of my presentation.

5. References

[1] Hillier D J and Miller D L 1998 *Astrophysical Journal* **496** 407
[2] Crowther P A, Schnurr O, Hirschi R, Yusof N, Parker R J, Goodwin S P and Kassim H A 2010 *Monthly Notices of the Royal Astronomical Society* **408** 731–751 (*Preprint* 1007.3284)
[3] Bromm V, Coppi P S and Larson R B 2002 *Astrophysical Journal* **564** 23–51 (*Preprint* arXiv:astro-ph/0102503)
[4] Maeder A and Meynet G 2000 *Annual Review of Astronomy and Astrophysics* **38** 143–190
[5] Meynet G and Maeder A 2003 *Astronomy and Astrophysics* **404** 975–990 (*Preprint* arXiv:astro-ph/0304069)
[6] Langer N, Norman C A, de Koter A, Vink J S, Cantiello M and Yoon S 2007 *Astronomy and Astrophysics* **475** L19–L23 (*Preprint* 0708.1970)
[7] Hillier D J and Lanz T 2001 *Spectroscopic Challenges of Photoionized Plasmas* (*Astronomical Society of the Pacific Conference Series* vol 247) ed G Ferland & D W Savin pp 343–+
[8] Busche J R and Hillier D J 2005 *Astronomical Journal* **129** 454–465
[9] Martins F 2011 *Bulletin de la Societe Royale des Sciences de Liege* **80** 29–41 (*Preprint* 1010.5347)
[10] Dessart L, Hillier D J, Livne E, Yoon S C, Woosley S, Waldman R and Langer N 2011 *Monthly Notices of the Royal Astronomical Society* **414** 2985–3005 (*Preprint* 1102.5160)
[11] Seaton M J 1987 *Journal of Physics B Atomic Molecular Physics* **20** 6363–6378

[12] Hummer D G, Berrington K A, Eissner W, Pradhan A K, Saraph H E and Tully J A 1993 *Astronomy and Astrophysics* **279** 298–309
[13] Hillier D J 1991 *Astronomy and Astrophysics* **247** 455–468
[14] Najarro F, Figer D F, Hillier D J, Geballe T R and Kudritzki R P 2009 *Astrophysical Journal* **691** 1816–1827 (*Preprint* 0809.3185)
[15] Anderson L S 1989 *Astrophysical Journal* **339** 558–578
[16] Martins F, Schaerer D, Hillier D J, Meynadier F, Heydari-Malayeri M and Walborn N R 2005 *Astronomy and Astrophysics* **441** 735–762 (*Preprint* arXiv:astro-ph/0507278)
[17] Marcolino W L F, Bouret J C, Martins F, Hillier D J, Lanz T and Escolano C 2009 *Astronomy and Astrophysics* **498** 837–852 (*Preprint* 0902.1833)
[18] Martins F, Schaerer D and Hillier D J 2005 *Astronomy and Astrophysics* **436** 1049–1065 (*Preprint* arXiv:astro-ph/0503346)
[19] Groh J H, Oliveira A S and Steiner J E 2008 *Astronomy and Astrophysics* **485** 245–256 (*Preprint* arXiv:0804.1766)
[20] Groh J H, Hillier D J, Damineli A, Whitelock P A, Marang F and Rossi C 2009 *Astrophysical Journal* **698** 1698–1720 (*Preprint* 0904.2363)
[21] Clark J S, Crowther P A, Larionov V M, Steele I A, Ritchie B W and Arkharov A A 2009 *Astronomy and Astrophysics* **507** 1555–1565 (*Preprint* 0909.4160)
[22] Crowther P A and Evans C J 2009 *Astronomy and Astrophysics* **503** 985–990 (*Preprint* 0907.2539)
[23] Fitzpatrick E L 1999 *Publications of the Astronomical Society of the Pacific* **111** 63–75 (*Preprint* astro-ph/9809387)
[24] Najarro F, Hillier D J and Stahl O 1997 *Astronomy and Astrophysics* **326** 1117–1134
[25] Ritchie B W, Clark J S, Negueruela I and Najarro F 2009 *Astronomy and Astrophysics* **507** 1597–1611 (*Preprint* 0910.0529)
[26] Hamann W R, Feldmeier A and Oskinova L M (eds) 2008 *Clumping in hot-star winds*
[27] Bouret J C, Lanz T, Hillier D J, Heap S R, Hubeny I, Lennon D J, Smith L J and Evans C J 2003 *Astrophysical Journal* **595** 1182–1205 (*Preprint* arXiv:astro-ph/0301454)

Journal of Physics: Conference Series **328** (2011) 012021 doi:10.1088/1742-6596/328/1/012021

The IACOB project: A grid-based automatic tool for the quantitative spectroscopic analysis of O-stars

S Simón-Díaz[1,2], N Castro[3], A Herrero[1,2], J Puls[4], M Garcia[1,2] and C Sabín-Sanjulián[1,2]

[1]Instituto de Astrofísica de Canarias, 38200 La Laguna, Tenerife, Spain
[2]Departamento de Astrofísica, Universidad de La Laguna, 38205 La Laguna, Tenerife, Spain
[3]Inst. of Astronomy & Astrophysics, National Obs. of Athens, 15236 Athens, Greece
[4]Universitätssternwarte München, Scheinerstr. 1, 81679 München, Germany

E-mail: ssimon@iac.es

Abstract. We present the IACOB grid-based automatic tool for the quantitative spectroscopic analysis of O-stars. The tool consists of an extensive grid of FASTWIND models, and a variety of programs implemented in IDL to handle the observations, perform the automatic analysis, and visualize the results. The tool provides a fast and objective way to determine the stellar parameters and the associated uncertainties of large samples of O-type stars within a reasonable computational time.

1. Introduction

The era of large spectroscopic surveys of massive stars has already begun, providing us with a huge amount of high-quality spectra of Galactic and extragalactic O-type stars. The *IACOB spectroscopic survey of Northern Galactic OB-stars* is one of them. This long-term observational project is aimed at building a multi-epoch, homogeneous spectroscopic database of high-resolution, high signal-to-noise ratio spectra of Galactic bright OB-stars[1]. Associated with this spectroscopic dataset are several working packages aimed at its scientific exploitation. Within the framework of the Working Package *WP3: Quantitative Spectroscopic Analysis*, we have developed a powerful tool for the automatic analysis of optical spectra of O-type stars. The tool provides a fast and objective way to determine the stellar parameters and the associated uncertainties of large samples of O-stars within a reasonable computational time. Initially developed to be used for the analysis of spectra of O-type stars from the *IACOB spectroscopic database*, the tool is now also being applied in the context of the *VLT-FLAMES Tarantula Survey* (VFTS) project [3], and other studies of stars of this type.

2. The IACOB grid-based tool

Apart from the already mentioned characteristics (automatic, objective, and fast), we also aimed at a tool that is portable, versatile, adaptable, extensible, and easy to use. As shown throughout the following text, this philosophy has guided the whole development of the tool.

[1] Some notes on the present status of *IACOB spectroscopic database*, the IACOB project, and the various working packages can be found in [1] and [2].

IOP Publishing
doi:10.1088/1742-6596/328/1/012021

One of the key-stones of any automatic quantitative spectroscopic analysis is the computation of large samples of synthetic spectra (using a stellar atmosphere code), to be compared with the observed spectrum. In contrast to other possible alternatives (the *genetic algorithm* – GA – employed by [4], or the *principal component analysis* – PCA – approach followed by [5]), we decided to base our automatic tool on an extensive, precomputed grid of stellar atmosphere models and a line-profile fitting technique (i.e., a grid-based approach[2]) This option was possible thanks to the fast performance of the FASTWIND code[3], and the availability of a cluster of ~300 computers at the Instituto de Astrofísica de Canarias (IAC) connected through the CONDOR[4] workload management system.

The grid is complemented by a variety of programs implemented in IDL, to handle the observations, perform the automatic analysis, and visualize the results. The IDL-package has been built in a modular way, allowing the user, e.g., to easily modify how the mean values and uncertainties of the considered parameters are computed, or which evolutionary tracks will be used to estimate the evolutionary masses.

2.1. *The IACOB FASTWIND grid*

In the following we outline the main characteristics of the FASTWIND grid used for the analysis of Galactic O-type stars that is presently incorporated within the IACOB grid-based tool.

Input stellar parameters considered in the FASTWIND models:

- **Effective temperature ($T_{\rm eff}$) and gravity ($\log g$):** Table 1 indicates the ranges of $T_{\rm eff}$ and $\log g$ considered in the grid. Basically, the grid points were selected to cover the region of the $\log T_{\rm eff}$ – $\log g$ diagram where the O-type stars are located.
- **Helium abundance ($\epsilon_{\rm He}$) and microturbulence ($\zeta_{\rm t}$):** The grid includes six values of helium abundance — $\epsilon_{\rm He} = {\rm N_{He}}/({\rm N_H}+{\rm N_{He}})$ —, indicated in Table 1. For all models, a microturbulence $\zeta_{\rm t,mod} = 15\,{\rm km\,s^{-1}}$ was adopted in the computation of the atmospheric structure, and four values of the microturbulence were considered in the formal solution.
- **Radius (R):** Computing a FASTWIND model requires an input value for the radius. This radius has to be close to the actual one (which will be derived in the final step of the analysis, and hence is not know from the beginning on). We assumed a radius for each ($T_{\rm eff}$, $\log g$) pair following the calibration by [8]. The grid is hence divided in 20 regions in which a different radius is considered. For example, for the case of $\log g = 4.0$ and 3.5 dex, the radii range from 7 to 12 $R_\odot$ and from 19 to 22 $R_\odot$, respectively.
- **Wind parameters ($\dot{M}$, v_∞, β):** As conventional for grid computations of stellar atmosphere models for the optical analysis of O-stars, the mass loss rate ($\dot{M}$) and terminal velocity (v_∞) of the wind have been compressed, together with the radius, into the wind strength parameter (or optical depth invariant), $Q = \dot{M}/(v_\infty R)^{1.5}$ (see [7]). Ten $\log Q$-planes were considered for the grid. For each FASTWIND model, $\dot{M}$ and v_∞ need to be specified. First, a terminal velocity $v_\infty = 2.65\,v_{\rm esc}$ was adopted, following [9]; then a mass loss rate was computed for the given $\log Q$, v_∞, and R. Finally, the exponent of the velocity law, β, was assumed as a free parameter ranging from 0.8 to 1.8.

[2] A similar approach has been recently applied by [6] for the analysis of a sample of 12 low resolution spectra of B supergiants in NGC55.
[3] Among the various available modern stellar atmosphere codes incorporating wind and line-blanketing effects, FASTWIND [7] is the fastest.
[4] http://www.cs.wisc.edu/condor/

 doi:10.1088/1742-6596/328/1/012021

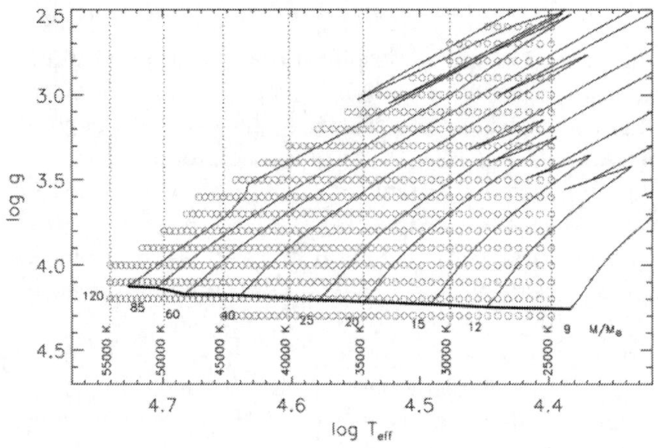

Figure 1. Location of ($T_{\rm eff}$, $\log g$)-pairs considered in the FASTWIND grid incorporated into the present version of the IACOB grid-based tool. Evolutionary tracks from [11].

Table 1. Range of values considered for the six free parameters in the FASTWIND grid.

$\log g$:	[2.6 − 4.3] dex (with step size 0.1 dex)
$T_{\rm eff}$:	≥ 25 000 K (step size 500 K), upper limit defined by the 120 $M_\odot$ track (see Fig. 1)
$\zeta_{\rm t}$:	Model: 15 km s^{-1}, Formal solution: 5, 10, 15, 20 km s^{-1}
$\epsilon_{\rm He}$:	0.06, 0.09, 0.13, 0.17, 0.20, 0.23
$\log Q$:	−15.0, −14.0, −13.5, −13.0, −12.7, −12.5, −12.3, −12.1, −11.9, −11.7
β:	0.8, 1.0, 1.2, 1.5, 1.8

- **Metallicity**: A solar metallicity (following [10]) was assumed for the whole grid[5].

Synthetic lines: The following (optical) lines where synthesized in the formal solution: $H_{\alpha-\epsilon}$, He I λλ4026, 4120, 4143, 4387, 4471, 4713, 4922, 5015, 5048, 5875, 6678, and He II λλ4026, 4200, 4541, 4686, 5411, 6406, 6527, 6683.

In order to optimize the size and the read-out time of the grid, only part of the output from the FASTWIND models is kept and stored in IDL xdr-files. This includes the input parameters, the H/He line profiles and equivalent widths, the information about the stellar atmosphere structure and the emergent flux distribution, and the synthetic photometry resulting from the computed emergent flux distribution. Using these xdr-files, the size of the grid can be reduced to a 10 % of the original one. IDL can restore each xdr-file and compute the corresponding X^2_T quantity (see Sect. 2.2) in 0.02 − 0.1 s per model, i.e., the tool can pass through 80 000 models in 30 min − 1 hour. Following the main philosophy of the IACOB grid-tool, only the reduced grid is used by the tool, whilst the original grid is safely kept on hard-disk at the IAC. This allows for an easy transfer of the grid to an external disk, hence satisfying our constraint for portability.

The FASTWIND grid presently incorporated into the IACOB grid-based tool consists of ∼ 192 000 models (∼ 32 000 models per He-plane). The reduced size is ∼ 34 Gb. This predefined grid can be easily updated and/or extended if necessary using appropriate scripts implemented in IDL, CONDOR, and LINUX. For example, a new He-plane can be computed and prepared for use in ∼ 3 days.

[5] We also calculated similar grids for other metallicities ($Z = 0.5$, 0.4, and 0.2 $Z_\odot$).

IOP Publishing
doi:10.1088/1742-6596/328/1/012021

2.2. The IACOB grid-based tool IDL package

The guideline of the automatic analysis is based on standard techniques applied within the quantitative spectroscopic analysis of O-stars using optical H/He lines that have been described elsewhere (e.g. [12], [13]). The whole spectroscopic analysis is performed by means of a variety of IDL programs following the steps indicated below. In brief, once the observed spectrum is processed, the tool obtains the quantity X_T^2 (i.e. an estimation of the goodness of fit) for each model within a subgrid of models selected from the global grid, and determines the stellar parameters and their associated uncertainties by interpreting the obtained X_T^2 distributions.

Step 1: The input

In this first step, the user has to provide the observed spectrum, to indicate the corresponding resolving power (R), and to give some pre-determined information about the star – the projected rotational velocity ($v \sin i$), the size of the extra line-broadening ($\Theta_{\rm RT}$), and the absolute visual magnitude ($M_{\rm V}$).
Concerning the grid, the user must select the appropriate metallicity, and indicate the range of values for the various free parameters (defining the subgrid of models to be considered within the analysis). This latter option allows the tool to be faster (using optimized ranges for the various free parameters), or the user to perform a preliminary quick analysis by fixing some of the parameters. For example, one can obtain a quick estimate on $T_{\rm eff}$, $\log g$, and $\log Q$ in less than 5 min, by fixing the other three parameters ($\zeta_{\rm t}$, $\epsilon_{\rm He}$, and/or β) .
Finally, the H/He lines which shall be considered in the analysis and the corresponding weights need to be specified.

Step 2: Processing of the observed spectrum

In many cases, the observed spectra need to be processed before launching the automatic analysis, because of, e.g., nebular contamination affecting the cores of the H and He I lines, the need to improve the normalization of the continuum adjacent to the line, and the presence of cosmic rays.
The possibility to correct the observed spectrum for these effects has been incorporated into corresponding IDL procedures. The implemented options include (for each of the considered lines): local renormalization, selection of the wavelength range of the line, clipping/restoring of certain parts of the line, and computation of the signal-to-noise ratio (S/N) in the adjacent continuum.
Once finalized, the processed spectrum is stored. This spectrum and all associated information is used in *Step 3*, and also each time the analysis of the same star/spectrum is (re-)launched. This way the user can easily reconsider his/her decisions after a first analysis.
In this step, the user can select a model from the grid to help him/her with the processing of the observed spectrum. One can, for example, make a quick preliminary analysis (see above) fixing some parameters and including only few lines, and then use the resulting best model to check the processed lines and find constraints for the processing of the others.

Step 3: Estimation of the goodness of fit for each considered model

This step, together with *Step 4*, are the most important ones, constituting the core of the program. In the present version of the IACOB grid-based tool we have considered a procedure as described below; however, we are aware that this procedure can be subject to discussion/improvements. Having this in mind, and following the philosophy of the tool (regarding versatility and adaptability), the corresponding IDL modules have been implemented with the

IOP Publishing
doi:10.1088/1742-6596/328/1/012021

possibility to be easily modified. The fast performance of the IACOB grid-based tool makes it very powerful to investigate various alternative strategies.

In our present version, the tool computes, for every model in the subgrid, the quantity X_L^2 for each considered line

$$X_L^2 = \frac{1}{N_\lambda} \sum_{\lambda=1}^{N_\lambda} \frac{(F_{m,\lambda} - F_{o,\lambda})^2}{\sigma_L^2} \tag{1}$$

where $F_{m,\lambda}$ and $F_{o,\lambda}$ are the normalized fluxes corresponding to the synthetic and observed spectrum, respectively; $\sigma_L = (\mathrm{S/N})^{-1}$ accounts for the S/N of the line; and N_λ is the number of frequency points in the line. Under ideal conditions (e.g., for a perfect model, but see below), X_L^2 corresponds to a reduced χ^2.

In a second step, the X_L^2 values for each model are corrected for possible deficiencies in the synthetic lines (due, for example, to deficiencies in the model, an incorrect characterization of the noise of the line, bad placement of the continuum, or bad characterization of the line-broadening). To this aim we compute, for each line, the standard deviation of the residuals $\sigma_c = \sigma(F_{m,\lambda} - F_{o,\lambda})$ from that model that results in the minimum X_L^2 for the given line. Then, the following correction is applied:

$$X_{L,c}^2 = X_L^2 \frac{\sigma_L^2}{\sigma_c^2} \tag{2}$$

Using these individual $X_{L,c}^2$ values and the weights assumed for each of the considered lines (w_L, e.g., [4]), a global X_T^2 is obtained:

$$X_T^2 = \sum_{L=1}^{N_L} w_L \, X_{L,c}^2 \tag{3}$$

where N_L is the number of lines. For a large number of frequency points per line, X_T^2 should be normally distributed.

Thus, *Step 3* provides the values of the quantity X_T^2 associated with each of the models included in the considered subgrid. As indicated in Section 2.1, this step can last from a few seconds to less than about 1 hour, depending on the number of models in the subgrid. An example of X_T^2-distributions (actually, $p_{X_T^2} = \mathrm{e}^{-0.5 X_T^2}$) with respect to the various stellar parameters is presented in Figure 2.

Step 4: Computation of mean values and uncertainties

The previous step provides the X_T^2-distributions for each of the six parameters derived through the spectroscopic analysis ($T_{\rm eff}$, $\log g$, $\epsilon_{\rm He}$, $\zeta_{\rm t}$, $\log Q$, and β). If the absolute visual magnitude is provided, the tool automatically determines R, $\log L$, and $M_{\rm sp}$ (spectroscopic mass) for each model, and hence the corresponding X_T^2-distributions for these parameters are available as well. If also the terminal velocity is provided, a similar computation is performed for the mass-loss rate (using Q, R, and v_∞). Finally, one of the IDL modules computes the evolutionary masses ($M_{\rm ev}$) of the models from an interpolation in the ($\log T_{\rm eff}$, $\log g$)-plane using the tracks provided by a stellar evolution code. Since the X_T^2-distributions for $T_{\rm eff}$ and $\log g$ are computed and stored in *Step 3*, the computation of the corresponding distributions for R, $\log L$, $M_{\rm sp}$, $M_{\rm ev}$ can be easily repeated in a few seconds, in case a different $M_{\rm v}$ and/or evolutionary tracks need to be considered.

Journal of Physics: Conference Series **328** (2011) 012021 doi:10.1088/1742-6596/328/1/012021

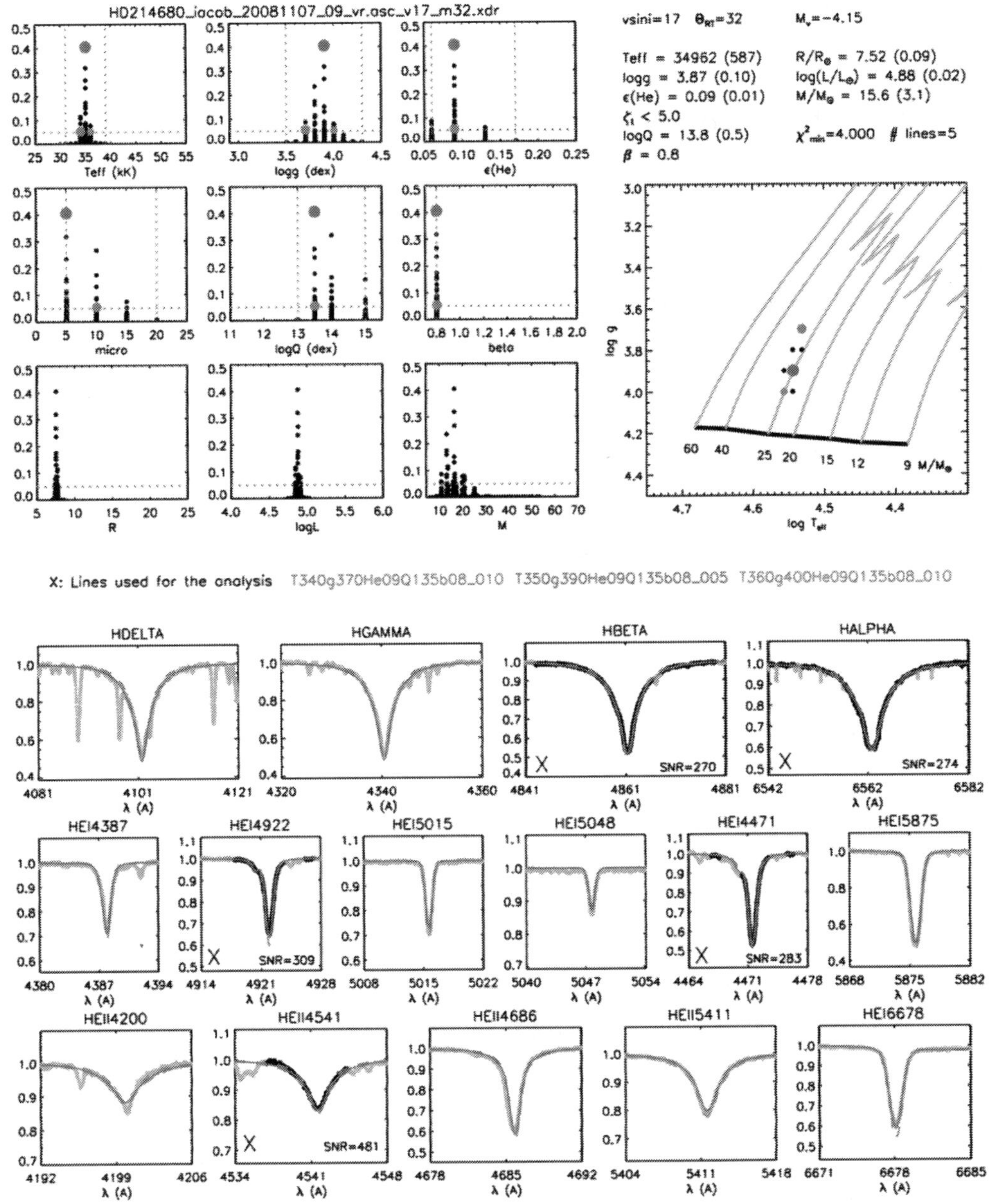

Figure 2. Example of output provided by the IACOB grid-based tool, for the quantitative spectroscopic analysis of the Galactic O9 V star HD 214680. See text for the explanation of all the information contained in the figure. Observed spectrum from the *IACOB spectroscopic database* ($R = 46\,000$). $v \sin i$ and Θ_{RT} were previously determined using a combined Fourier transform + goodness of fit technique (see e.g. [14]). In this example, only 5 H/He lines were used for the actual analysis. Note that also all other synthetic lines fit the observations perfectly at the derived parameters. The analysis was performed in $\sim$2 min. See [2] for details on the distance, apparent visual magnitude and visual extinction used to determine the absolute visual magnitude (M_V).

 doi:10.1088/1742-6596/328/1/012021

These distributions are then used to compute mean values and uncertainties for each parameter (taking into account that models above a given threshold in X^2_T can be discarded).

The tool also allows to easily investigate possible degeneracies (for example, for stars with weak winds the analysis of the optical H/He lines only provides an upper limit for the Q parameter, and does not allow to constrain β), and the contribution of the various parameters to the final uncertainty (by fixing one of the free parameters to a certain value and recomputing the statistics for the other parameters).

Step 5: Visualization of the results

The last step is the creation of a summary plot for better visualization of the results (see Figure 2 as an example). The present version of the tool includes:

- $p_{X^2_T}$ – distributions for the various parameters involved in the analysis (upper-left panels). In those panels, vertical lines indicate the limiting values adopted for the six free parameters (i.e., the sub-grid of FASTWIND models for which a X^2_T is computed), and horizontal lines indicate the $p_{X^2_T}$ value corresponding to X^2_T(threshold). The user can select three of the models that will be indicated as red, blue, and pink dots (the same colors are also used in other parts of the figure).
- a summary of the input parameters ($v \sin i$, $\Theta_{\rm RT}$, and $M_{\rm v}$) and the result from the statistics for each parameter resulting from the analysis (upper-right part of the figure).
- a $\log T_{\rm eff}$ —$\log g$ diagram including evolutionary tracks (e.g., from [11], without rotation, in the figure) and the position of all models with $X^2_T \leq X^2_T$(threshold).
- a set of panels (lower part of the figure) were the observed and synthetic H/He lines indicated by the user are compared. The spectral regions that are used to compute the X^2_T for each line are indicated in black, while the clipped (or not used) regions are presented in green. Note that the plotted lines are not limited to those used in the analysis process (the latter are marked with an 'X').

2.3. Problems that can be investigated with the IACOB grid-based tool

The fast performance and versatility of the IACOB grid-based tool not only allows to analyze large samples of O-type stars in a reasonable time, but also to easily investigate the effects of

- *the assumed line-broadening*: The common strategy in previous spectroscopic analyses of O-stars was to assume pure rotational line-broadening (in addition to natural, thermal, Stark/collisional and microturbulent broadening included in the synthetic spectra). However, recent studies have shown that an important extra-broadening contribution (commonly quoted as *macroturbulent broadening*) affects the shape of the line-profiles of this type of stars (see e.g. [15], and references therein). How much are the derived parameters affected when this extra-broadening is neglected? Can both broadening contributions be represented by one 'fake' rotational profile (v_f sin$i$), without disturbing the resulting parameters? What is the effect of the uncertainty in the derived/assumed broadening? Some results from a preliminary investigation of these questions can be found in [16].
- *the placement of the continuum*: The normalization of the spectra of O-type stars is sometimes complicated, especially in the case of echelle spectra. It is commonly argued that the derived gravity can be severely affected by the assumed normalization, but, to which extent? To investigate this effect one could use the IACOB grid-based tool, applying small modifications to the continuum placement. An example of this type of analysis for the case of low resolution spectra of B-supergiants can be found in [6].

IOP Publishing
doi:10.1088/1742-6596/328/1/012021

- *neglecting/including a variety of different commonly used diagnostic lines*: Before computing the global X_T^2 (see equation 3), the resulting X_L^2 for each line are stored. This way the user can easily recompute X_T^2, discarding some of the initially considered lines. This option allows, for example, to investigate the change in the wind-strength parameter when both Hα and He II 4686 lines or only one of them are included in the analysis.
- *clipping part of the lines*: Most of the O-stars are associated to H II regions. In some cases, the stellar spectra can be heavily contaminated by nebular emission lines (mainly the cores of the hydrogen lines, but also the He I lines). This contamination must be identified and eliminated from the stellar spectrum to obtain meaningful results from the spectroscopic analysis. When the resolution of the spectra is high, the clipped region is small in comparison with the total line width; however, even for a moderate resolution an important region of the line needs to be eliminated. What is the effect of such a strong clipping in the Hα line on the determination of the mass-loss rate? How large is the effect of clipping the core of the He I lines in the $T_{\rm eff}$ determination?
- *the way the statistics is derived from the X_T^2 – distributions*: As commented in Section 2.2, this is a critical point in the determination of final best values and uncertainties.

These are only some examples of things can be investigated with the IACOB grid-based tool. Many other tests can be easily performed. In addition, we expect the tool to be of great benefit for the analysis of the O-star samples included in, e.g., the ESO-Gaia, VFTS, OWN, IACOB and other similar large surveys.

Acknowledgments

Financial support by the Spanish Ministerio de Ciencia e Innovación under projects AYA2008-06166-C03-01, AYA2010-21697-C05-04, and by the Gobierno de Canarias under project PID2010119. This work has also been partially funded by the Spanish MICINN under the Consolider-Ingenio 2010 Program grant CSD2006-00070: First Science with the GTC (`http://www.iac.es/consolider-ingenio-gtc`).

References

[1] Simón-Díaz S, Castro N, Garcia M, Herrero A and Markova N 2011a *BSRS Liege* **80** 514
[2] Simón-Díaz S, Garcia M, Herrero A, Maíz-Apellániz J and Negueruela I 2011b *Proc. Conf. Star clusters and Associations: A RIA Workshop on Gaia Preprint* arXiv1109.2665S
[3] Evans C J, *et al* 2011 *A&A* **530** A108
[4] Mokiem M R, de Koter A, Puls J, Herrero A, Najarro F and Villamariz M R 2005 *A&A* **441** 711
[5] Urbaneja M A, Kudritzki R P, Bresolin F, Przybilla N, Gieren W and Pietrzyński G 2008 *ApJ* **684** 118
[6] Castro N, Herrero A, Urbaneja M A, *et al* (*A&A* submitted)
[7] Puls J, Urbaneja M A, Venero R, Repolust T, Springmann U, Jokuthy A and Mokiem M R 2005 *A&A* **435** 669
[8] Martins F, Schaerer D and Hillier D J 2005 *A&A* **436** 1049
[9] Kudritzki R P and Puls J 2000 *A&A* **38** 613
[10] Asplund M, Grevesse N, Sauval A J and Scott P 2009 *ARA&A* **47** 481
[11] Schaller G, Schaerer D, Meynet G and Maeder A 1992 *A&AS* **96** 269
[12] Herrero A, Puls J and Najarro F 2002 *A&A* **396** 949
[13] Repolust T, Puls J and Herrero A 2004 *A&A* **415** 349
[14] Simón-Díaz S, Herrero A, Uytterhoeven K, Castro N, Aerts C and Puls J 2010 *ApJL* **720** L174
[15] Simón-Díaz S 2011 *BRSR Liege* **80** 86
[16] Sabín-Sanjulian C, Simón-Díaz S, Garcia M, Herrero A, Puls J and Castro N 2011 *Proc. Conf. in Honour of A. Moffat*

Journal of Physics: Conference Series **328** (2011) 012022 doi:10.1088/1742-6596/328/1/012022

The O stars in the VLT-FLAMES Tarantula Survey

A de Koter[1,2], H Sana[1], C J Evans[3], T Bagnoli[1], N Bastian[4], J M Bestenlehner[5], A Bonanos[6], E Bressert[7,8], I Brott[9], M Cantiello[10,11], G Carraro[8], S Clark[12], P Crowther[13], S E de Mink[14,15], E Doran[13], P Dufton[16], P Dunstall[16], M Garcia[17,18], G Gräfener[5], V Hénault-Brunet[19], A Herrero[17,18], I Howarth[20], R Izzard[10], K Köhler[10], N Langer[10], D Lennon[21], J Maíz Apellániz[22], N Markova[23], P Najarro[24], J Puls[25], O Ramirez[1], C Sabín-Sanjulián[17,18], S Simón-Díaz[17,18], S Smartt[16], V Stroud[12,26], J Th van Loon[27], W Taylor[19] and J S Vink[5]

[1] Astronomical Institute Anton Pannekoek, University of Amsterdam, P.O. Box 94249, 1090 GE Amsterdam, The Netherlands
[2] Astronomical Institute, Utrecht University, Princetonplein 5, 3584 CC, Utrecht, The Netherlands
[3] UK Astronomy Technology Centre, Royal Observatory Edinburgh, Blackford Hill, Edinburgh, EH9 3HJ, UK
[4] Excellence Cluster Universe, Garching, [5] Armagh Observatory, [6] National Observatory of Athens, [7] University of Exeter, [8] ESO, [9] University of Vienna, [10] Argelander-Institut für Astronomie der Universität Bonn, [11] Kavli Institute for Theoretical Physics, University of California, [12] The Open University, Milton Keynes, [13] University of Sheffield, [14] Space Telescope Science Institute, [15] Hubble Fellow, [16] Queen's University Belfast, [17] Instituto de Astrofísica de Canarias, La Laguna, [18] Universidad de La Laguna, [19] University of Edinburgh, [20] University College London, [21] ESA Space Telescope Science Institute, [22] Instituto de Astrofísica de Andalucía-CSIC, [23] Institute of Astronomy with NAO, Smoljan, [24] CSIC-INTA, Madrid, [25] Universitäts-Sternwarte München, [26] University of Glamorgan, [27] Keele University

E-mail: `A.deKoter@uva.nl`

Abstract. The VLT-FLAMES Tarantula Survey (VFTS) has secured mid-resolution spectra of over 300 O-type stars in the 30 Doradus region of the Large Magellanic Cloud. A homogeneous analysis of such a large sample requires automated techniques, an approach that will also be needed for the upcoming analysis of the Gaia surveys of the Northern and Southern Hemisphere supplementing the Gaia measurements. We point out the importance of Gaia for the study of O stars, summarize the O star science case of VFTS and present a test of the automated modeling technique using synthetically generated data. This method employs a genetic algorithm based optimization technique in combination with FASTWIND model atmospheres. The method is found to be robust and able to recover the main photospheric parameters accurately. Precise wind parameters can be obtained as well, however, as expected, for dwarf stars the rate of acceleration of the flow is poorly constrained.

1. Introduction

It has been estimated that the disk of the Milky Way should contain about 100 open clusters as massive as R136 ($> 10^4\ M_\odot$) in the 30 Doradus region in the Large Magellanic Cloud [1, 2], some eight of which have been found to date – Westerlund 1 being so far the most massive one

($\sim 5 \times 10^4\, M_\odot$). Though searches for these massive clusters rely mostly on infrared imaging surveys, the Gaia mission may contribute to the identification of these objects as, for instance, current searches tend to be less sensitive for clusters with rather low stellar densities. Phase space correlations, for which Gaia has been designed, of all objects down to $V = 20$ mag may thus help to identify them.

Of the anticipated number of $\sim 10^4$ O-type stars in our Galaxy, so far only a small fraction has been found, though the census of O- (and early B-type) stars within the first 2.5 kpc is thought to be fairly complete ($\sim$ 85%, [3]). For the nearby O-star population, the main importance of Gaia therefore lies in *i)* the determination of distances accurate to within 1-2 percent; *ii)* the identification of spectroscopic and/or astrometric binary systems, *iii)* potentially the measurement of projected rotational velocities, and *iv)* identifying their cluster membership through phase space correlations. The latter item(s) may, for instance, reveal (spun-up) run-away sources [4]. At larger distances the discovery potential of new O stars and young open clusters will be larger, though *i)* through *iv)* remain fundamental objectives albeit the uncertainties involved will increase. All in all, some 150 000 O and B-type stars are expected in the Gaia sample [5], including peculiar objects such as B[e], Luminous Blue Variable, Of/WN, WNh and evolved Wolf-Rayet stars. The legacy of Gaia for these most massive stars will encompass break-throughs in our understanding of single and binary star evolution, cluster dynamics and evolution and the role of massive stars in the formation and evolution of the Milky Way galaxy.

The wavelength range of 847 to 874 nm covered by the medium-resolution Gaia Radial Velocity Spectrograph (RVS) may help to identify O stars, but is not sufficient to characterize their stellar and stellar wind properties [5, 6]. To achieve this, spectroscopy of the hot stars in the FLAMES/VLT Gaia-ESO Survey for the Southern Hemisphere (Blomme, this Volume) and the potential WEAVE/WHT Gaia-Survey for the Northern Hemisphere [7] will be vitally important. The large number of massive objects targeted in these projects warrants the development of automatic spectral analysis techniques. Here we report on one of the two automated approaches developed in the context of our VLT Large Programmes on massive stars [8, 9]. For a discussion of the second method we refer to Simón-Díaz (this Volume). In particular, we will discuss this method in the context of our on-going spectral census of $\sim$ 800 O and early-B stars in and near the central cluster R136 using the Medusa/FLAMES instruments on the VLT, i.e. our VLT FLAMES Tarantula Survey (VFTS) program.

2. The FLAMES Tarantula Survey science rationale

The VFTS targets over 300 O-type stars, including 20 Of/WN and Wolf-Rayet emission line stars, and over 500 early B-type stars, covering the full luminosity range from dwarfs to bright supergiants. An overview of the project is given in [10]. The observational sample is presented by [9]. Early science results are reported in [11, 12]. An image showing the location of the O stars and Wolf-Rayet stars is presented in Figure 1.

Though a main focus of the project is to address outstanding questions in how massive stars evolve from the early main sequence to their deaths as core-collapse supernovae, the unprecedented sample size also allows to understand massive stars *as a population*. The latter includes the possibility to study evolutionary connections between the morphological sub-groups in the field, integrated cluster properties – important for understanding distant, unresolved star clusters, and cluster dynamics – valuable in the context of cluster formation and evolution. Here we focus on the science related to stellar evolution.

The evolution of massive stars is inextricably connected to the physics of mass loss, rotational mixing, core overshooting and binary interaction. Unfortunately, there are many gaps in our knowledge of these physical processes.

The VFTS permits a thorough investigation into how progression of increasing wind strength

Figure 1. The 30 Doradus region in the Large Magellanic Cloud centered on its young (1–2 Myr; [13]) ionizing cluster R136. Overlaid on a $YJKs$-band of the VISTA Magellanic Cloud Survey [14] the location of the O-type (red circles) and Wolf-Rayet stars (blue diamonds) observed by the VFTS are shown. From [10] with permission.

from O to Of to Of/WN to WN spectral type may be coupled to stellar evolution. Recent theoretical results [15–17] agree on the dominant role of the Eddington factor Γ for the mass loss properties of the most massive. By determining the mass-loss rates $\dot{M}$ of a sizable number of Of, Of/WN, and WN objects, we aim to establish the empirical relation $\dot{M}(\Gamma)$ in the regime where it is most pronounced and most pivotal for stellar evolution, such that these predictions may be tested.

The first statistically significant test of rotationally-induced mixing in massive stars by [18, 19] showed that only some 60 percent of a sample of apparently single LMC B-type stars concords with theoretical expectations. Two groups of stars appear in conflict with the predictions: a group of late-MS fast rotators with little mixing, and a group of slow rotators which show up to 0.7 dex increase in the surface nitrogen abundance. The former group may be populated by close binaries, however, it is highly unlikely that binary effects may explain the second group [20]. The effects of mixing are expected to be more significant in O stars but, to date, abundance analyses have been limited by the models available. New theoretical developments [21] now provide the tools by which we can potentially investigate the effects of stellar rotation at the high-mass end of the Hertzsprung-Russell diagram, and establish whether the O star population also show groups similar to those identified among the B stars.

A long-standing and disquieting problem in massive star evolution is the large number of B supergiants [22, 23]. At present it is unclear what is the nature of these objects, i.e. whether they are core hydrogen or helium burning. A striking difference between O dwarfs and B supergiants is that while the former are the most rapid rotators known, the B supergiants rotate much more slowly ($v_{\rm R} \sin i \lesssim 50$ km s^{-1}). This has been ascribed to the expansion of the star after leaving the main sequence. [18, 24] showed that the preference for slow rotation occurs at gravities $g \lesssim 10^{3.2}$ cm s^{-2} and suggested the B supergiants to be post main-sequence. [20] subsequently used this empirical finding to constrain the core overshooting in massive star models, that sensitively controls the position of the cool edge of the main-sequence band. [25] proposed an alternative explanation for the difference between the rotational properties of O V and B I stars, at least for stars initially more massive than $\sim 15-20\, M_{\odot}$, suggesting this behavior to be related to a very substantial increase in the mass-loss rate once stars evolve to past spectral type B1 ([26]). If the overshooting is large enough for this to occur on the main-sequence, it may be the enhanced loss of angular momentum associated with the high mass loss that may spin the stars down. The Tarantula Survey aims to investigate the nature of B supergiants.

One of the key ingredients missing from current theories of both star formation and cluster evolution is an empirical relation for the binary fraction of massive stars [27], and the distribution of the mass ratios and periods of these systems (but see [28] for recent progress). The Tarantula Survey was therefore conceived as a multi-epoch program, with the aim of obtaining clear indications of binarity in the large majority of the targets. This adds a valuable parameter to the subsequent quantitative analysis and interpretation, enabling tests of the predictions of both single-star and binary evolution models that include all the relevant physical processes ([29]). This will provide important insights in our efforts to disentangle the roles of rotation and binarity for the first time, and to draw conclusions for the evolutionary end-points of O stars, including long-duration gamma-ray bursts and magnetars.

3. Spectral analysis by means of a genetic algorithm

VFTS observations were primarily done using the Medusa fibre-feed to the Giraffe spectrograph, yielding a spectral resolving power $R \sim 7\,000$ and $8\,500$ in the LR02 (for a total of 6 epochs) and LR03 (3 epochs) settings. The region containing Hα was observed in the HR15N setting (2 epochs), providing $R \sim 15\,000$. For five pointings of R136 and its direct surroundings the ARGUS integral field unit + Giraffe was used ($R \sim 10\,500$). In parallel to the ARGUS observations, red-arm UVES spectra were taken of a small sample of stars in the inner part of 30 Dor ($R \sim 53\,000$). This yields a grand total of 22 000 spectra with typical signal-to-noise ratios of 70 – 100, reaching 500 for the brightest targets.

Here we discuss the spectral analysis technique of the O stars for which model atmospheres accounting for the outflow of gas are required. To assure a homogeneous approach capable of dealing with large samples, we build on the automatic fitting method developed by [30], based on a genetic algorithm (GA) utilizing the optimization routine PIKAIA by [31] and employing the stellar atmosphere code FASTWIND by [32, 33]. FASTWIND solves the radiative transfer equations for a medium in which the state of the gas is described by statistical equilibrium, subject to the constraint of radiative equilibrium. A radial stellar wind is assumed to stream out from the photosphere with a prescribed velocity structure $v(r) = v_{\infty}\,(1 - b \times r_{\circ}/r)^{\beta}$, where b is chosen such that at $r_{\circ}$ the flow has a prescribed minimum velocity $v(r_{\circ}) = v_{\circ}$. The maximum or terminal

Journal of Physics: Conference Series **328** (2011) 012022 doi:10.1088/1742-6596/328/1/012022

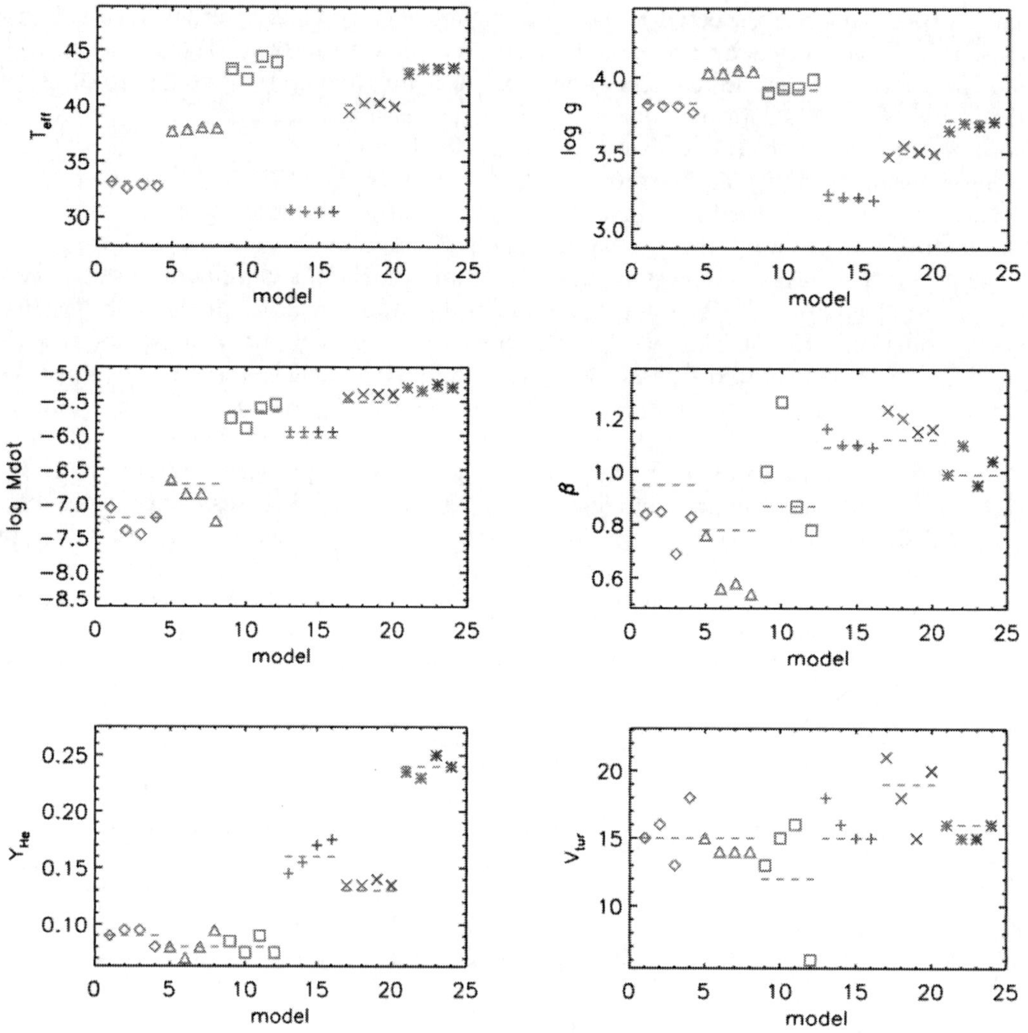

Figure 2. A comparison of the best fit results of the GA fitting method with simulated observations. The horizontal axis provides the model number with the first 12 models representing main sequence (MS) stars and the last 12 models supergiant (SG) stars. The meaning of the symbols is as follows: early-MS (□); mid-MS (△); late-MS (◇); early-SG (X); late-SG (+) and helium rich, high mass-loss SG ($*$). The magenta horizontal dashes indicate the correct model parameters. The fitting of the wind acceleration β is, as expected, poor for the mid- and late-MS stars. The mass-loss determinations for these objects suffer from this.

outflow velocity is v_∞ and β describes the rate of acceleration of the outflow.

In the setup discussed here the method is used to constrain the effective temperature $T_{\rm eff}$, gravity g, luminosity L, number fraction of helium over hydrogen $Y_{\rm He}$, mean photospheric micro-turbulent velocity $\xi_{\rm micro}$, mass-loss rate $\dot{M}$ and measure of the rate of acceleration of the outflow β. Additional parameters can be added to the GA-method. These include the rotational velocity, radial velocity, photospheric macro-turbulent velocity and terminal flow velocity of the stellar

Table 1. The mean shift and mean dispersion of the fitting results (Fit) relative to the synthetic data (Syn), using all 24 test cases. The * superscript denotes values that are based on supergiant models only.

Parameter	Unit	<Fit−Syn>	Dispersion
$T_{\rm eff}$	[kK]	−0.09	0.40
$\log g$	[cm s^{-2}]	0.00	0.03
$\log \dot{M}^\star$	[$\dot{M}$ yr^{-1}]	0.05	0.09
$\beta^\star$		0.04	0.05
Y		0.00	0.01
$\xi_{\rm micro}$	[km s^{-1}]	0.0	2.2

wind. So far, no elements other than hydrogen and helium have been considered, however, we plan to include nitrogen for the analyses of VFTS objects building on the theoretical groundwork by [21].

To test our method we have performed a GA-analysis on a set of simulated stellar spectra. These have been computed using FASTWIND spectra broadened with a range of rotational velocities and degraded to the spectral resolution and typical S/N of our observations. Before we discuss the results it should be pointed out that in several respects this is an idealized test yielding error-bars that should be viewed as lower limits. First, the spectra were normalized to the actual continuum. This essentially implies that the stellar radius is known. Second, the models do not account for macro-turbulent broadening of the lines. Compared to rotational broadening this effect is usually negligible in O (and B) dwarfs but is typically substantial in supergiants [34], though see [35, 36]. Third, the rotational broadening is assumed to be known. This mimics the approach that will be taken in the VFTS analysis, as $v_{\rm rot} \sin i$ values will be determined separately using techniques described by [34]. Fourth, the terminal velocity of the flow is adopted. Fifth, radial velocity shifts are non-existent. Sixth, the test implies that collisional broadening profiles are 'known'. Seventh, the simulated spectra do not suffer from nebular contamination. Eighth, the stars are 'known to be single'.

Figure 2 shows an overview of the results of the test. In six panels we compare the best fit parameters with their true values. The luminosity has been excluded from the comparison as constraining this parameter only requires to perform an absolute scaling of the spectra, using information on a photometric magnitude (for instance V or K) and interstellar extinction properties. The correct model parameters have been indicated using (magenta) horizontal dashed lines. Colors indicate different projected rotational velocities: $v_{\rm R} \sin i = 50$ (blue), 100 (green) and 300 (red) km s^{-1}. The horizontal axis provides the model number of each of the 24 test models. Models 1−12 represent dwarf stars while 13−24 are supergiants. Different symbols denote early- (□), mid- (△) and late- (◇) O dwarfs; early- (X) and late- (+) O supergiants, and helium enriched, high mass-loss O supergiants ($*$).

The mean shift and mean dispersion of the fitting results relative to the actual values are given in Table 1. One can immediately conclude that the fitting method does not suffer from systematic sources of uncertainty and that the synthetic spectra are fitted (very) accurately. Several aspects of Figure 2, however, warrant discussion. The mass-loss rates are rather poorly constrained for late- and mid-O dwarfs and β for all O dwarfs. To constrain β one needs knowledge of how rapid the flow is accelerating, therefore one can only determine robust values for winds that are somewhat substantive. For supergiants relative to dwarfs the photospheric scale height is larger, implying the build-up of line optical depth starts further out in the flow.

For this reason, the β values of supergiant models for $\dot{M} \sim 10^{-6} M_{\odot}\mathrm{yr}^{-1}$ can be constrained, while this is not the case for the dwarf models with similar mass loss. A mitigation strategy here is to adopt theoretical predictions of β for O dwarfs, for instance by [37]. These predictions seem relatively secure, therefore they will likely help to reduce the uncertainty in the mass-loss rates of mid- and late-O dwarfs.

The helium abundances are well reproduced. For the late-O supergiants (+ symbols), which have weak He II lines, the helium abundance appears to anti-correlate with ξ_{micro}. Though the uncertainties in the determination of the micro-turbulent velocity are quite modest in this test, fitting of this parameter in the real data is expected to result in substantially larger errorbars. This may have a negative effect on the error in Y of late supergiants.

4. Conclusions

We have presented a test of our automated genetic algorithm based fitting technique of O stars using synthetic observations. This method is one of two that is to be applied to analyze the spectra of $\sim$300 O stars secured within the context of the VFTS project. An automated techniques is a prerequisite for the analysis of such a large dataset, to be surpassed even by the sub-set of O and early-B stars expected in the FLAMES/VLT and WEAVE/WHT surveys planned to support the Gaia observations themselves. The results are extremely promising, yielding only small error bars and no systematic offsets. The rate of acceleration of the outflow could not be measured accurately for mid- and late-O stars. This is expected when using optical spectra only, but can be remedied by adopting theoretical predictions.

References

[1] van den Bergh S and Lafontaine A 1984 AJ **89** 1822–1824

[2] Hanson M M and Popescu B 2008 One Hundred 30 Dors? *IAU Symposium* (*IAU Symposium* vol 250) ed F Bresolin, P A Crowther, & J Puls pp 307–312

[3] Reed B C 2003 AJ **125** 2531–2533

[4] Blaauw A 1993 *Massive Stars: Their Lives in the Interstellar Medium* (*Astronomical Society of the Pacific Conference Series* vol 35) ed J P Cassinelli & E B Churchwell p 207

[5] Bouret J C, Lanz T, Frémat Y, Martins F and Lefever K *et al* 2008 *Revista Mexicana de Astronomia y Astrofisica Conference Series* (*Revista Mexicana de Astronomia y Astrofisica, vol. 27* vol 33) pp 50–50

[6] Simón-Díaz S, Garcia M, Herrero A, Maíz Apellániz J and Negueruela I 2011 *ArXiv e-prints* (*Preprint* 1109.2665)

[7] Balcells M, Benn C R, Carter D, Dalton G B and Trager S C *et al* 2010 *Society of Photo-Optical Instrumentation Engineers (SPIE) Conference Series* (*Society of Photo-Optical Instrumentation Engineers (SPIE) Conference Series* vol 7735)

[8] Evans C J, Smartt S J, Lee J K, Lennon D J and Kaufer A *et al* 2005 A&A **437** 467–482

[9] Evans C J, Taylor W D, Hénault-Brunet V, Sana H, de Koter A and *et al* 2011 A&A **530** A108

[10] Evans C, Taylor W, Sana H, Hénault-Brunet V, Bagnoli T, Bastian N, Bestenlehner J, Bonanos A, Bressert E, Brott I, Campbell M, Cantiello M, Carraro G, Clark S, Costa E, Crowther P, de Koter A, de Mink S, Doran E, Dufton P, Dunstall P, Garcia M, Gieles M, Gräfener G, Herrero A, Howarth I, Izzard R, Köhler K, Langer N, Lennon D, Apellániz J M, Markova N, Najarro P, Puls J, Ramirez O, Sabín-Sanjulián C, Simón-Díaz S, Smartt S, Stroud V, van Loon J, Vink J S and Walborn N 2011 *The Messenger* **145** 33–38

[11] Taylor W D, Evans C J, Sana H, Walborn N R, de Mink S E, Stroud V E, Alvarez-Candal A, Barbá R H, Bestenlehner J M, Bonanos A Z, Brott I, Crowther P A, de Koter A, Friedrich K, Gräfener G, Hénault-Brunet V, Herrero A, Kaper L, Langer N, Lennon D J, Maíz Apellániz J, Markova N, Morrell N, Monaco L and Vink J S 2011 A&A **530** L10

[12] Bestenlehner J M, Vink J S, Gräfener G, Najarro F, Evans C J, Bastian N, Bonanos A Z, Bressert E, Crowther P A, Doran E, Friedrich K, Hénault-Brunet V, Herrero A, de Koter A, Langer N, Lennon D J, Maíz Apellániz J, Sana H, Soszynski I and Taylor W D 2011 A&A **530** L14

[13] de Koter A, Heap S R and Hubeny I 1998 ApJ **509** 879–896

[14] Cioni M R L, Clementini G, Girardi L, Guandalini R and Gullieuszik M *et al* 2011 A&A **527** A116

[15] Vink J S 2006 *Stellar Evolution at Low Metallicity: Mass Loss, Explosions, Cosmology* (*Astronomical Society of the Pacific Conference Series* vol 353) ed H J G L M Lamers, N Langer, T Nugis, & K Annuk p 113

[16] Gräfener G and Hamann W R 2008 A&A **482** 945–960

[17] Vink J S, Muijres L E, Anthonisse B, de Koter A, Gräfener G and Langer N 2011 A&A **531** A132

[18] Hunter I, Brott I, Lennon D J, Langer N, Dufton P L, Trundle C, Smartt S J, de Koter A, Evans C J and Ryans R S I 2008 ApJ **676** L29–L32

[19] Hunter I, Brott I, Langer N, Lennon D J, Dufton P L, Howarth I D, Ryans R S I, Trundle C, Evans C J, de Koter A and Smartt S J 2009 A&A **496** 841–853

[20] Brott I, Evans C J, Hunter I, de Koter A, Langer N, Dufton P L, Cantiello M, Trundle C, Lennon D J, de Mink S E, Yoon S C and Anders P 2011 A&A **530** A116

[21] Rivero Gonzalez J G, Puls J and Najarro F 2011 *ArXiv e-prints* (*Preprint* 1109.3595)

[22] Fitzpatrick E L and Garmany C D 1990 ApJ **363** 119–130

[23] Evans C J, Lennon D J, Smartt S J and Trundle C 2007 A&A **464** 289–289

[24] Hunter I, Lennon D J, Dufton P L, Trundle C, Simón-Díaz S, Smartt S J, Ryans R S I and Evans C J 2008 A&A **479** 541–555

[25] Vink J S, Brott I, Gräfener G, Langer N, de Koter A and Lennon D J 2010 A&A **512** L7

[26] Vink J S, de Koter A and Lamers H J G L M 1999 A&A **350** 181–196

[27] Zinnecker H and Yorke H W 2007 ARA&A **45** 481–563

[28] Sana H and Evans C J 2011 *IAU Symposium* (*IAU Symposium* vol 272) ed C Neiner, G Wade, G Meynet, & G Peters pp 474–485

[29] Langer N, Cantiello M, Yoon S C, Hunter I, Brott I, Lennon D, de Mink S and Verheijdt M 2008 *IAU Symposium* (*IAU Symposium* vol 250) ed F Bresolin, P A Crowther, & J Puls pp 167–178

[30] Mokiem M R, de Koter A, Puls J, Herrero A, Najarro F and Villamariz M R 2005 A&A **441** 711–733

[31] Charbonneau P 1995 ApJS **101** 309

[32] Santolaya-Rey A E, Puls J and Herrero A 1997 A&A **323** 488–512

[33] Puls J, Urbaneja M A, Venero R, Repolust T, Springmann U, Jokuthy A and Mokiem M R 2005 A&A **435** 669–698

[34] Simón-Díaz S and Herrero A 2007 A&A **468** 1063–1073

[35] Simón-Díaz S 2011 *Bulletin de la Societe Royale des Sciences de Liege* **80** 86–91

[36] Markova N, Markov H, Puls J, Simón-Díaz S and Herrero A 2011 *BigAJ* **17**, *in press*

[37] Muijres L E, Vink J S, de Koter A, Müller P E and Langer N 2012 A&A, *submitted*

Journal of Physics: Conference Series **328** (2011) 012023 doi:10.1088/1742-6596/328/1/012023

XSL: The X-Shooter Spectral Library

Yanping Chen[1], Scott Trager[1], Reynier Peletier[1] and Ariane Lançon[2]

[1]Kapteyn Astronomical Institute, University of Groningen, Postbus 800, 9700 AV, Groningen, Netherlands

[2]Observatoire Astronomique, 11, rue de l'Université, F-67000 Strasbourg, France

E-mail: yanping@astro.rug.nl

Abstract. We are building a new spectral library with the X-Shooter instrument on ESO's VLT: XSL, the X-Shooter Spectral Library. We present our progress in building XSL, which covers the wavelength range from the near-UV to the near-IR with a resolution of $R{\sim}10\,000$. At the time of writing we have collected spectra for nearly 240 stars. An important feature of XSL is that we have already collected spectra of more than 100 Asymptotic Giant Branch stars in the Galaxy and the Magellanic Clouds.

1. Introduction

Stellar population models are powerful tools which are widely used to study galaxy evolution. Using these models, one can determine galaxy ages, metallicities and abundances. Spectral libraries are an integral component of stellar population models (e.g., [1]). A spectral library gives the behavior of individual stellar spectra as function of effective temperature ($T_{\rm eff}$), gravity ($\log g$) and metallicity ([Fe/H]). A stellar population model integrates these spectra together with a set of stellar isochrones and an initial mass function to produce a model spectrum of an entire population. In order to reproduce galaxy spectra as precisely as possible, one requires a comprehensive stellar spectral library that covers the entire desired parameter space of $T_{\rm eff}$, $\log g$, and [Fe/H]. Moreover, extended wavelength coverage is strongly desirable, because different stellar phases contribute their light in different bands. For instance, bright giants contribute more light in the near-infrared than the faint giant and subgiant stars, while in the optical, the situation is reversed [2]. These Asymptotic Giant Branch (AGB) stars dominate the light of intermediate-aged and old stellar populations in the near-infrared but are unimportant in the optical (e.g., [3; 4]). Detecting their presence requires *broad wavelength coverage* in both the target and model spectra.

Stellar spectral libraries can be classified into empirical and theoretical libraries, depending on how the library is obtained. Both theoretical and empirical libraries have improved in recent years. The most widely used theoretical libraries in stellar population models are those of [5], [6], [7], [8], [9], and [10]. Theoretical libraries have the advantage of (nearly) unlimited resolution and selectable abundance patterns – not only scaled-solar abundances but also non-solar patterns. Unfortunately, theoretical libraries suffer from systematic uncertainties, as they rely on model atmospheres and require a reliable list of atomic and molecular line opacities [8]. Empirical stellar libraries, on the other hand, have the advantage of being drawn from real, observed stars and therefore do not suffer this limitation; however they frequently have relatively low resolution (with a few exceptions; see below) and are unable to reproduce the indices measured in giant

Table 1: Some previous stellar libraries.

Library	Resolution $R=\lambda/\Delta\lambda$	Spectral range (nm)	Number of stars	Reference
STELIB	2000	320–930	249	[14]
ELODIE	10 000	390–680	1300	[15–17]
INDO-US	5000	346–946	1237	[18]
MILES	2000	352–750	985	[19]
IRTF-SpeX	2000	800–2500	210	[20]
NGSL	1000	167–1025	374	[21]
UVES-POP	80 000	307–1030	300	[22]
LW2000	1100	500–2500	100	[23]

elliptical galaxies [11; 12], because they are based on local stars with typical Milky Way disk abundance patterns.

In this proceeding we present the X-Shooter Stellar Library, XSL. XSL is being obtained using the new X-Shooter three-arm spectrograph on ESO's VLT [13]. XSL has the unique advantage of simultaneously acquiring spectra covering the near-ultraviolet up to the near-infrared. Furthermore, the fact that X-Shooter is mounted on an 8.2 m telescope allows us to include faint objects in stars in the Galactic bulge and the Magellanic Clouds for the first time, along with stars in the Galactic disk and halo, at moderate spectral resolution ($R\sim10\,000$).

1.1. Previous stellar libraries

We begin with a review of several previous empirical stellar libraries and their principal features, listed in Table 1. In the optical, we have (among others) Lick/IDS [24], MILES [19], ELODIE [15–17], STELIB [14], NGSL [21], and the Pickles library [25]. Libraries in the near-IR are a challenging task, but pioneering work has been done by Lançon and Wood [23] (LW2000), Rayner et al. [26] (IRTF-SpeX), and Mármol-Queraltó et al. [27]. However, extended-wavelength-coverage spectral libraries at moderate resolution are still largely missing.

STELIB [14] has been used to construct the widely used stellar population synthesis models of Bruzual & Charlot 2003 ([1], here after BC03). This library has a wide wavelength range, 3200–9300 Å, at modest resolution ($R\sim2000$). The spectra of local galaxies have been reproduced by the models of BC03 through this library [28]. However, stars with low and high metallicites are sparse in this library, leading to problems in some regimes, like modeling old, metal-poor globular cluster spectra [29].

The ELODIE library [15–17] is a moderate-resolution spectral library with $R\sim10\,000$. It has been applied in the PÉGASE-HR synthetic model [30]. The atmosphere parameter coverage of this library had been improved through its updated version, but the wavelength range is still limited in the optical region (4100–6800 Å).

The INDO-US library [18] includes a large number of stars (1273), covers a fair range in atmosphere parameters with moderate-resolution ($R\sim5000$) . Atmospheric parameters of its stars are given in Wu et al. [31]. The data have been used for automated spectral classification of high-resolution spectra over a wide wavelength range [32]. The problem of this library is the lack of accurate spectrophotometry, making inclusion in stellar population models problematic.

The MILES library [19] is widely used in stellar population models. This library profits from its good physical parameter coverage, careful flux calibration, and a large number of stars (985),

Journal of Physics: Conference Series **328** (2011) 012023 doi:10.1088/1742-6596/328/1/012023

enabling stellar population models to predict metal-poor or metal-rich systems. The MILES library's wavelength range covers 3500–7500 Å at modest resolution (R~2000) [33].

The Next General Spectral Library (NGSL[1], [21]) is a low resolution (R~1000) stellar library obtained with STIS on HST. It contains ~400 stars having a wide range in metallicity and age. Since its spectra were observed in space, this library does not suffer from telluric absorption or seeing variations, and hence has excellent absolute spectrophotometry. Other features of NGSL are its broad wavelength coverage, from the space UV to 1 μm, and high signal-to-noise. There is also a high-resolution (R~40 000, covering 3600–11 000 Å) extension of NGSL's southern stars taken with the UVES spectrograph of ESO's VLT (Hanuschik et al., in prep.).

The UVES Paranal Observatory Project (UVES-POP, [22]) is a library of spectra of ~300 nearby bright stars taken with UVES. The spectra cover the optical region at R~80 000 and have a typical S/N ratio is 300–500 in the V band. The library has been flux calibrated but has not been corrected for the severe telluric absorption in the red. For stellar population models the sample is incomplete, since it only contains bright stars around solar metallicity.

In the near-infrared range, the IRTF Spectral Library (IRTF-SpeX, [20]) contains 200 stars observed with the cross-dispersed infrared SpeX spectrograph on IRTF at a resolving power of R~2000. The largest spectral library of very cool supergiants and giants over the wavelength range λ=0.5–2.5 μm was published by Lançon & Wood (LW2000, [23]), which contains ~100 stars taken at a resolution of R=1100, including observations over multiple phases. Unfortunately, cool giants are variable in optical and NIR, so simultaneous observations are required. Libraries with more limited spectral coverage in the near-infrared, like the 2.3 μm library of Marmól-Queraltó et al. [34], also exist, but are limited to a small number of spectral features (like the CO bandhead).

2. XSL

X-Shooter [13] was built by a consortium of 11 institutes in Denmark, France, Italy and the Netherlands, together with ESO. It is currently mounted on UT2 on ESO's VLT. A unique capability of X-Shooter is that it collects spectra in the wavelength range from the near-ultraviolet to the near-infrared – 300–2500 nm – through its three arms *simultaneously*. This character is extremely useful for observing variable stars, especially very cool stars – like stars – whose spectra vary substantially during their pulsation cycles.

2.1. Sample selection

XSL targets were selected from many of the above libraries as well as supplementary literature sources. We took stars from Lick/IDS, MILES, and NGSL to cover $T_{\rm eff}$, $\log g$, and [Fe/H] as uniformly as possible. However, these libraries mostly lack the cool, bright stars so important in the near-infrared. For this purpose we selected AGB and LPV stars from LW2000 and IRTF-SpeX with declination < 35° marked M, C or S-stars. Long-period variable (LPV) stars were also collected from the LMC [35] and SMC [36]. Red supergiant stars were taken from the lists of LW2000 and [37; 38]. To cover metal-rich stars with abundances similar to giant elliptical galaxies, we also included Galactic Bulge giants from the samples of [39; 40]. Our intention is that XSL will contain about 600 stars at the end of the survey.

2.2. Observations and data reduction

To date, 251 observations of 236 unique stars from the XSL input catalog have been made. Figure 1 shows the distribution of our sample stars with different stellar types. We have collected a large sample of M, C, and LPV stars, interesting for many different cool star studies. In Fig. 2 we show those sample stars with known stellar parameters in an HR diagram.

[1] See `http://archive.stsci.edu/prepds/stisngsl/`

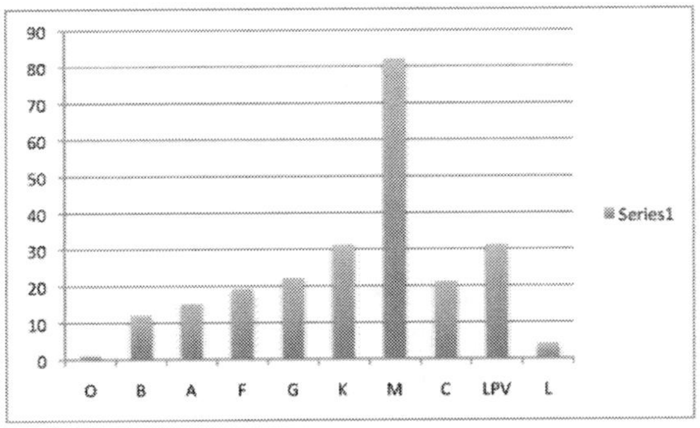

Figure 1: Distribution of spectral types of the main XSL samples (excluding telluric calibrators), retrieved from SIMBAD or based on educated guesses from the source library or atmospheric parameters.

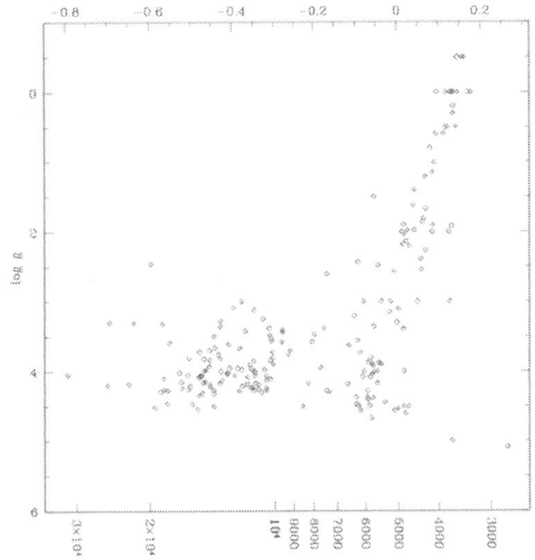

Figure 2: HR diagram of the 114 X-Shooter library stars with known $T_{\rm eff}$ and $\log g$. Bottom axis is $T_{\rm eff}$; top axis is log θ, where $\theta = 5040/T_{\rm eff}$.

The targets were observed with X-Shooter with a $0.5'' \times 11''$ slit in the UVB arm (3000–6000 Å, R=9100), a $0.7'' \times 11''$ slit in the VIS arm (6000–10200 Å, R=11 000), and a $0.6'' \times 11''$ slit in the NIR arm (1–2.5 μm, R=8100). Stars were nodded along the slit to remove the sky background, a serious problem in the NIR arm even for very short exposures. In addition, for every star a wide (5.0″) slit spectrum was taken to allow for excellent flux calibration (see, e.g., [19]).

The basic data reduction was performed with the public release of the X-Shooter pipeline version 1.1.0, following its standard steps up to the production of 2-D spectra, including bias and/or dark correction, flat-fielding, wavelength calibration, sky subtraction. We then used IRAF's `twodspec.apextract` routine to extract 1-D spectra from the 2-D spectra. We show typical X-Shooter library spectra in Fig. 3, which contains several spectral types from B2 to M3. In this figure flux calibration has been applied to all the three arms. The UVB arm is shown in blue, the VIS arm in green and the NIR arm in red. Telluric corrections have been applied to the VIS and NIR arms. Grey bars demonstrate the telluric regions in the NIR arm, in which the light grey bars cover the areas of severe telluric absorption while the dark grey ones cover the areas nearly complete telluric absorption.

2.3. Telluric correction

Since X-Shooter is a ground based instrument, correction for telluric absorption in the VIS and NIR arm spectra is very important. For this reason a hot star (typically a mid-B dwarf) was observed after every science object at a similar airmass. Unfortunately, the telluric absorption lines change strength on timescales shorter than the 'long' exposure times ($\geq$ 90 seconds) of faint XSL stars and the total observational overhead time of ~900 seconds, resulting in a noisy telluric correction. To optimize the telluric correction, we built a telluric stellar spectrum library, in which the hot stars were carefully wavelength calibrated. First we masked out strong H and other lines from the telluric region, since we know that B and other hot stars usually have strong H lines which must be corrected before the spectra can be used as telluric templates. We then

Journal of Physics: Conference Series **328** (2011) 012023 doi:10.1088/1742-6596/328/1/012023

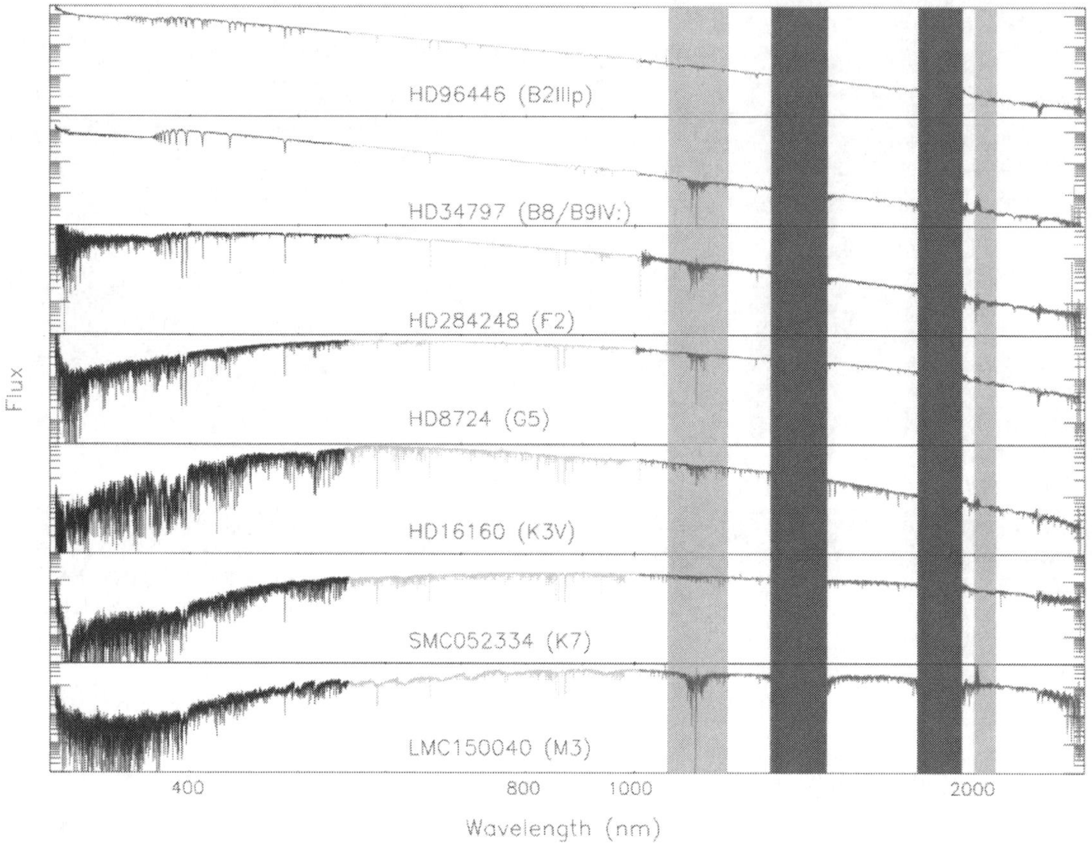

Figure 3: A sample of XSL stellar spectra sorted by temperature. Grey bars cover areas of severe (light grey) or nearly complete (dark grey) telluric absorption in the NIR arm.

normalized the masked spectra to create our telluric standard library.

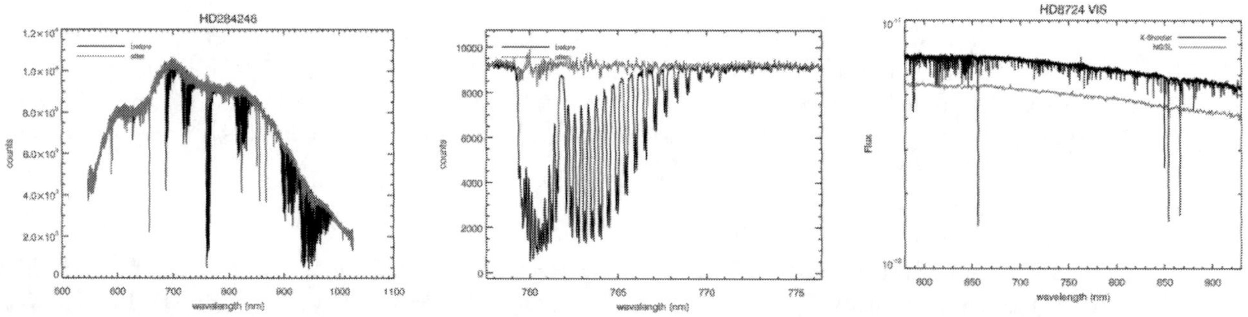

(a) (b) (c)

Figure 4: Telluric correction and flux calibration of XSL spectra. (a): Telluric corrected version (red) of HD 284248 together with the original spectrum with different telluric absorption bands (black). (b): Same as in Fig. 4a, zoomed in wavelength range $\lambda\lambda 758 - 776\,\mathrm{nm}$. (c): Flux-calibrated spectrum of HD 8724. The X-Shooter spectrum is shown in black and the NGSL spectrum is shown in red.

For each science object, we defined several telluric absorption regions and ran the spectral-fitting program pPXF [41] to find a 'best fit' telluric correction spectrum. We fitted the science spectrum with a linear combination of the spectra in the (normalized) telluric library, fitting at the same time the line-broadening of the science spectra, and multiplying the result with a higher order polynomial. This step gave a 'best fit' for the telluric absorption features to every science object. We then divided the science spectrum by this 'best fit', resulting in a telluric-corrected science spectrum. Figure 4a shows an example of HD 284248 in the VIS arm (without flux calibration), where the black line is the original reduced spectrum before telluric correction and the red line is after correction. To show the quality of the correction in more detail, we zoom into the telluric-corrected spectrum in atmospheric 'A' band between 758 and 776 nm in Fig. 4b.

2.4. Flux calibration

To perform a reliable flux calibration, we observed several spectrophotometric standards (BD+17 4708, GD71, LTT1020, GD153, EG274 and EG21) with a wide (5.0″) slit in 'stare' mode at different airmasses. These flux standard star were then reduced and extracted through the procedures mentioned above. The spectra were compared with the flux tables of the appropriate stars from the CALSPEC HST database [42][2] and averaged to produce the response function. The Paranal extinction curve [43] was applied to perform the atmosphere extinction correction. Figure 4c shows the flux-calibrated VIS arm spectrum of HD 8724 from X-Shooter (black line) and NGSL (red line). The UVB arm spectra are corrected in a similar matter. The NIR arm spectra will be telluric- and flux-calibrated simultaneously using the hot star telluric correction spectra in combination with the flux standards.

3. Summary

XSL, the X-Shooter Stellar Library, is intended to be the largest stellar library with complete wavelength coverage from 320–2480 nm and covering as large a range in stellar atmospheric parameters as possible. We will soon have a first version of XSL with spectra of ~240 stars. The final library will contain ~600 stars at moderate resolution (R~10 000). The stellar parameter coverage will be enforced through carefully selected samples, so that not only stars with solar metallicities and abundances are included but also metal-poor or metal-rich stars from the Bulge and Magellanic Clouds. With X-Shooter's unique capability, the variable stars will be observed consistently, which will yield more reliable stellar population models.

Acknowledgments

We thank our collaborators on the first phase of XSL, D. Silva and P. Prugniel, and B. Davies and M. Koleva for useful discussions. We would also like to extend our great thanks to V. Manieri, A. Modigliani, J. Vernet, and the ESO staff for their help during the XSL observations and reduction process.

References

[1] Bruzual A G and Charlot S 2003 *MNRAS* **344** 1000
[2] Frogel J A 1988 *ARA&A* **26** 51
[3] Maraston C 2005 *MNRAS* **362** 799
[4] Conroy C and Gunn J E 2010 *ApJ* **712** 833
[5] Kurucz R 1993 *ATLAS9 Stellar Atmosphere Programs and 2 km/s grid* Kurucz CD-ROM No. 13 (Cambridge, Mass.: Smithsonian Astrophysical Observatory)

[2] `http://www.stsci.edu/hst/observatory/cdbs/calspec.html`

Journal of Physics: Conference Series **328** (2011) 012023 doi:10.1088/1742-6596/328/1/012023

[6] Munari U, Sordo R, Castelli F and Zwitter T 2005 *A&A* **442** 1127

[7] Gustafsson B, Edvardsson B, Eriksson K, Jørgensen U G, Nordlund Å and Plez B 2008 *A&A* **486** 951

[8] Coelho P, Barbuy B, Meléndez J, Schiavon R P and Castilho B V 2005 *A&A* **443** 735

[9] Coelho P, Bruzual G, Charlot S, Weiss A, Barbuy B and Ferguson J W 2007 *MNRAS* **382** 498

[10] Martins L P, González Delgado R M, Leitherer C, Cerviño M and Hauschildt P 2005 *MNRAS* **358** 49

[11] Peletier R F 1989 *Ph.D. thesis* University of Groningen

[12] Worthey G, Faber S M and Gonzalez J J 1992 *ApJ* **398** 69

[13] Vernet J, Dekker H, D'Odorico S, Masonand E, di Marcantonio P, Downing M, Elswijk E, Finger G, Fischer G, Kerber F, Kern L, Lizon J L, Lucuix C, Mainieri V, Modigliani A, Patat F, Ramsay S, Santin P, Vidali M, Groot P, Guinouard I, Hammer F, Kaper L, Kjærgaard-Rasmussen P, Navarro R, Randich S and Zerbi F 2010 *Society of Photo-Optical Instrumentation Engineers (SPIE) Conference Series* (*Society of Photo-Optical Instrumentation Engineers (SPIE) Conference Series* vol 7735) p 50

[14] Le Borgne J F, Bruzual G, Pelló R, Lançon A, Rocca-Volmerange B, Sanahuja B, Schaerer D, Soubiran C and Vílchez-Gómez R 2003 *A&A* **402** 433

[15] Prugniel P and Soubiran C 2001 *A&A* **369** 1048

[16] Prugniel P and Soubiran C 2004 (*Preprint* astro-ph/0409214)

[17] Prugniel P, Soubiran C, Koleva M and Borgne D L 2007 (*Preprint* astro-ph/0703658)

[18] Valdes F, Gupta R, Rose J A, Singh H P and Bell D J 2004 *ApJS* **152** 251

[19] Sáncez-Blázquez P, Peletier R F, Jiménez-Vicente J, Cardiel N, Cenarro A J, Falcón-Barroso J, Gorgas J, Selam S and Vazdekis A 2006 *MNRAS* **371** 703

[20] Rayner J T, Cushing M C and Vacca W D 2009 *ApJS* **185** 289

[21] Gregg M D *et al.* 2006 (*The 2005 HST calibration workshop* vol NASA/CP2006-214134) ed Koekemoer A M, Goudfrooij P and Dressel L L (Greenbelt MD: NASA) p 209

[22] Bagnuolo S, Jehin E, Ledoux C, Cabanac R, Melo C and Gilmozzi R 2003 *ESO Messenger* **114** 10

[23] Lançon A and Wood P R 2000 *A&AS* **146** 217

[24] Worthey G and Ottaviani D L 1997 *ApJS* **111** 377

[25] Pickles A J 1985 *ApJS* **59** 33

[26] Rayner J T, Cushing M C and Vacca W D 2009 *ApJS* **185** 289

[27] Mármol-Queraltó E, Cardiel N, Cenarro A J, Vazdekis A, Gorgas J, Pedraz S, Peletier R F and Sáncez-Blázquez P 2008 *A&A* **489** 885

[28] Gallazzi A, Charlot S, Brinchmann J, White S D M and Tremonti C A 2005 *MNRAS* **362** 41

[29] Koleva M, Prugniel P, Ocvirk P, Le Borgne D and Soubiran C 2008 *MNRAS* **385** 1998

[30] Le Borgne D, Rocca-Volmerange B, Prugniel P, Lançon A, Fioc M and Soubiran C 2004 *A&A* **425** 881

[31] Wu Y, Singh H P, Prugniel P, Gupta R and Koleva M 2011 *A&A* **525** 71

[32] Pickles A J 2007 *IAUS* **241** 82

[33] Falcón-Barroso J, Sáncez-Blázquez P, Vazdekis A, Ricciardelli E, Cardiel N, Cenarro A J, Gorgas J and Peletier R F 2011 *A&A* **532** 95

[34] Mármol-Queraltó E, Cardiel N, Cenarro A J, Vazdekis A, Gorgas J, Pedraz S, Peletier R F and Sánchez-Blázquez P 2008 *A&A* **489** 885

[35] Hughes S M G and Wood P R 1990 *AJ* **99** 784

[36] Cioni M R L, Blommaert J A D L, Groenewegen M A T, Habing H J, Hron J, Kerschbaum F, Loup C, Omont A, van Loon J T, Whitelock P A and Zijlstra A A 2003 *A&A* **406** 51

[37] Levesque E M, Massey P, Olsen K A G, Plez B, Josselin E, Maeder A and Meynet G 2005 *ApJ* **628** 973

[38] Levesque E M, Massey P, Olsen K A G and Plez B 2007 *ApJ* **667** 202

[39] Blanco V M, McCarthy M F and Blanco B M 1984 *AJ* **89** 636

[40] Groenewegen M A T and Blommaert J A D L 2005 *A&A* **443** 143

[41] Cappellari M and Emsellem E 2004 *PASP* **116** 138

[42] Bohlin R C 2007 *ASPC* **364** 315

[43] Patat F, Moehler S, O'Brien K, Pompei E, Bensby T, Carraro G, de Ugarte Postigo A, Fox A, Gavignaud I, James G, Korhonen H, Ledoux C, Randall S, Sana H, Smoker J, Stefl S and Szeifert T 2011 *A&A* **527** 91

Journal of Physics: Conference Series **328** (2011) 012024 doi:10.1088/1742-6596/328/1/012024

A quantitative study of the O stars in NGC 2244

L Mahy[1], F Martins[2], D J Hillier[3] and G Rauw[1]

[1] Institut d'Astrophysique et de Géophysique, Allée du 6 Août, 17, Bât B5C, B-4000 Liège, Belgium

[2] LUPM-UMR 5299, CNRS & Université Montpellier II, Place Eugène Bataillon, F-34095, Montpellier Cedex 05, France

[3] Department of Physics and Astronomy, University of Pittsburgh, 3941 OHara street, Pittsburgh, PA 15260, USA

E-mail: mahy@astro.ulg.ac.be

Abstract. NGC 2244 located in the Rosette Nebula is a young open cluster composed of seven O-type stars. A first paper focused on the multiplicity of these stars, revealed only one binary system out of the seven studied stars. The minimum binary fraction of this cluster ($\sim$ 14%) differs to the average fraction measured on the nearby clusters ($\sim$ 44%). In order to better constrain this discrepancy, an analysis based on the determination of the stellar and wind parameters of these stars with the CMFGEN atmosphere code was performed. The main results confirm that all the stars have an age between 0 and 5 Myr, and that the N surface abundance appears to be consistent with the evolutionary models for a population of stars of the same age. Moreover, this investigation exhibits the existence of dynamical interactions inside this young open cluster sufficiently strong to eject the hottest component from its centre.

1. Introduction

Massive stars play a key role in the ecology of the galaxies mainly because of their strong ionizing fluxes, their powerful winds and the heavy chemical elements synthesized during their spectacular death as supernovae. However, these stars still hide a part of mystery. The earliest stages of their formation but also some of their physical properties are indeed poorly known. As their evolution is governed by several main parameters such as the binarity, their mass, their mass-loss rate or their rotation, a better knowledge of their multiplicity and their fundamental parameters can provide us additional constraints to improve our understanding of these stars.

In this context, several investigations were made on the northern Monoceros star-forming region and in particular the young open cluster NGC 2244 located at the core of the Rosette nebula. This region gathers seven O-type stars. The known age of about 1–6 Myrs (Bonatto & Bica [1]) and distance of about 1.4–1.7 kpc (Hensberge et al. [2]) also make the early-type stars of this cluster suitable candidates for asteroseismology. Four of these stars and two other massive stars in Mon OB2 association were thus recently observed by the CoRoT satellite (Baglin et al. [3]) and the data were analysed in a series of papers (Blomme et al. [4], Briquet et al. [5], Degroote et al. [6], Mahy et al. [7]). However, the comparison between the theoretical models and the observations is still challenging and needs a good knowledge of the fundamental parameters of these stars.

Therefore, the requirement of reliable stellar and wind parameters for such stars is strong. For that purpose, we thus performed an analysis of these objects by means of atmosphere models.

Journal of Physics: Conference Series **328** (2011) 012024 doi:10.1088/1742-6596/328/1/012024

We describe in the present paper the main results of this investigation but a more detailed study of these results and of, notably, the wind properties, the N enrichment and the stellar masses is reported in Martins et al. [8]. We organized this paper as following. Section 2 summarizes the observations and the results of the spectroscopic investigation of the multiplicity of O-type stars in NGC 2244. Section 3 describes the analysis of the atmosphere models whilst Sect. 4 presents the results. Finally, our conclusions are provided in Sect. 5.

2. Observations and spectroscopic binary fraction

The data were collected in order to determine the binary fraction of the O-type stars in NGC 2244. A total of 207 spectra were gathered with several telescopes. The majority of our data were obtained with the 1.52m telescope equipped with the Aurélie spectrograph at the Observatoire de Haute Provence (OHP, France). We also retrieved several spectra taken with the Elodie spectrograph at the 1.93m OHP telescope.

In April 2007 and March 2008, another set of data was obtained at Observatorio Astronómico Nacional of San Pedro Màrtir (SPM) in Mexico with the Espresso spectrograph mounted on the 2.10m telescope. This échelle spectrograph covers the wavelength domain [3780–6950] Å. Also in the 'échelle spectra category', we retrieved spectra taken with FEROS spectrograph mounted on the 2.20m telescope at La Silla (Chile).

Among this dataset, we selected, for the atmosphere models analysis, the spectra with the widest wavelength coverage, highest spectral resolution and highest signal-to-noise ratio.

These data, spread over nine years, already allowed us to determine the binary fraction of the O-type stars in NGC 2244. The details of this investigation were reported in Mahy et al. [9] but we summarize in the present paper the main results. Among the seven O stars in the young cluster, one, HD 258691, was however not studied because of its faintness. We detected for the first time the SB2 signature in the spectrum of HD 46149 and classified its components as an O 8.5V and an B 0–1V for the primary and the secondary, respectively. Located at the center of the cluster, HD 46150 (O 5.5V((f))) was reported as a binary candidate. Indeed, small variations were detected in its line profiles but we were not able to attribute them to binarity or to atmosphere motion. HD 46223, the hottest member in NGC 2244, was classified as an O 4V((f^+)), this star does not present any variations in its radial velocity nor in its line profiles. The same conclusion was obtained for HD 46202, the latest O-type star in the cluster. Finally, the two remaining O stars, HD 46056 and HD 46485, are rapid rotators with projected rotational velocities estimated to about 330 and 300 km s^{-1}, respectively. Both stars were classified as presumably single stars even though line profiles variations were observed. The latter are probably due to non-radial pulsations rather than a secondary component. Therefore, the spectroscopic binary fraction in NGC 2244 was established between 14% and 43%.

3. Modeling

In order to better know the stellar and wind parameters of these O stars, we performed a quantitative study of the optical and UV spectra on the basis of the CMFGEN atmosphere code (Hillier & Miller [10]). CMFGEN provides non-LTE atmosphere models including winds and line-blanketing. This code needs as input an estimate of the hydrodynamical structure that we constructed from TLUSTY models (taken from the OSTAR2002 grid of Lanz & Hubeny [11]) connected to a β velocity law of the form $v = v_{\infty}(1 - R/r)^{\beta}$ where v_{∞} is the wind terminal velocity. We adopted $\beta = 0.8$ since this is the typical value for O dwarfs (e.g., Repolust et al. [12]). Our models include H, He, C, N, O, Ne, Mg, Si, S, Fe with the solar composition of Grevesse et al. [13] unless otherwise stated. The super-level approach is used to reduce the amount of memory requirements. We have included X-ray emission in the wind since this can affect the ionization balance and the strength of key UV diagnostic lines. For the formal solution of the radiative transfer equation leading to the emergent spectrum, a microturbulent velocity varying linearly (with velocity) from 10 km s^{-1} to 0.1 $\times$ v_{∞} was used. In practice we use the same diagnostic lines as in Martins et al. [8] to derive the stellar and wind parameters.

The obtained synthetic spectra were convolved a first time with a rotational profile. The $v\sin i$ were determined from the Fourier transform method of Simón-Díaz & Herrero [14]. For slowly and moderately rotating objects, we convolved the resulting spectra with a Gaussian profile to introduce the macroturbulence in our profiles. We used He I 4713 as the main indicator of macroturbulence since it is present with sufficient SNR in all sample stars. Secondary indicators were the C IV 5801–5812 doublet and He I 5876.

We also determined the N content of stars from the N III absorption lines between 4500 and 4520 Å. They are not affected by winds and are strong enough for abundance determination. The uncertainties are of the order of 50%. The errors do not take into account any systematics related to atomic data.

Clumping is implemented in CMFGEN by means of a volume filling factor f following the law $f = f_{\infty} + (1 - f_{\infty})e^{-v/v_{cl}}$ where f_{∞} is the maximum clumping factor at the top of the atmosphere and v_{cl} a parameter indicating the position where the wind starts to be significantly clumped. As shown by Bouret et al. [15], O V 1371 and N IV 1720 are two UV features especially sensitive to wind inhomogeneities in early O stars. For mid to late O stars, these UV clumping diagnostic are not identified, as we confirmed by our study. Therefore, we decided to adopt $f_{\infty} = 1$ (i.e., homogeneous model) for these stars.

4. Results

The main stellar and wind parameters are listed in Table 1 whilst Figure 1 compares each synthetic spectrum to the observed spectrum in the UV and the optical domains. For all the presumably single stars, the best-fit CMFGEN models are rather of good quality. For the binary system HD 46149, we applied a disentangling program based on the method of González & Levato [16] to separate the contributions of the primary component from the secondary star. This method provides mean spectra of both components. The resulting spectra are then corrected from the light factor which is deduced from the brightness ratio (here $l_1/l_2 = 2.3$). To derive this brightness ratio, we compute the equivalent widths of several lines on the observed spectra and compare these values to those given by Conti et al. [17] and Conti [18] as well as those obtained from synthetic spectra which have same spectral classifications as our two components. However, to obtain reliable spectra, three conditions have to be met:

- the sampling of the data has to be as uniform as possible (in Doppler shift or in orbital phase) over the orbital cycle,
- the spectral separation between both components has to be large enough to sample the full width of the lines, especially for the Balmer lines,
- the normalization has to be as perfect as possible to avoid oscillations in the continuum of the resulting spectra.

However, in the case of HD 46149, these conditions are not completely respected. Indeed, the orbital period of the system is estimated to be about 800 days but the collected data are not sufficiently well sampled to allow us to be more accurate on this value and compute its orbital solution. Indeed, the spectra were mainly taken at three different phases: when the difference between the radial velocities of the secondary and the primary ($RV_S - RV_P$) are equal to -30, ~ 0 and $+140$ km s^{-1}. Moreover, the large eccentricity and the orientation of the orbit imply an asymmetric excursion in radial velocity preventing the sampling of the full width of the Balmer lines. Therefore, the derived parameters and notably the $\log g$ are rather uncertain, especially for the secondary component.

The O-type stars of NGC 2244 form a homogeneous sample of stars with the same age, distance, and chemical composition. The effective temperatures and the luminosities are

Journal of Physics: Conference Series **328** (2011) 012024 doi:10.1088/1742-6596/328/1/012024

sufficiently accurate to report the positions of the O-type stars on the Hertzsprung-Russell diagram (Fig. 4). These results reveal that all the O stars in NGC 2244 have an inferred age between 0 and 5 Myrs. However, we clearly detect a trend between the two most massive stars (HD 46150 and HD 46223) and the other ones. Indeed, these two stars appear to be younger than the lower-mass O stars. The origin of this trend is unclear and could come either from real phenomena or from artefacts related to the comparison method. The former would assume that either the early-type stars emerge from their parental cloud faster than the later-type stars or that they from after these later-type stars. The latter would imply that the utilization of a same initial rotational velocity for the evolutionary tracks of all stars is unadapted. Unfortunately, in this case, we do not know whether the initial rotational velocity of more massive stars is larger than that of lower-mass O stars. If this trend was real, that would therefore confirm that dynamical interactions exist inside NGC 2244. The study made by Wang et al. [19] in the X-ray domain indeed revealed the presence of X-ray sources, associated to pre main-sequence (PMS) stars, around HD 46150 and not around HD 46223 which is however the most massive star of the cluster. The authors explained these observations by the fact that HD 46223 is either younger than HD 46150 or that it was ejected from the center of the cluster due to dynamical interactions. We can now assume that the second possibility seems to be the most probable one.

The determination of the $\dot{M}$ systematically reveals that the mass-loss rates derived on UV P-Cygni profiles are smaller than those estimated from Hα (see Table 1). Our estimates could however be biased because of the few available features to constrain the models, because of the CMFGEN code itself which requires an accurate wind ionization structure or because of the porosity (or vorosity) which could alter the shape of diagnostic lines, for example. This vorosity (i.e., the porosity in velocity space) could partly explain this discrepancy. Sundqvist et al. [20] have indeed shown that accounting for vorosity, the UV line profiles can be strongly reduced for a given mass-loss rate whilst Hα mass-loss rate is less affected. Their simulations have notably emphasized the importance of vorosity on the formation of strong UV lines. However, as this property is observed in all the stars, the assumption of vorosity would mean that the winds of all the stars in our sample are clumped, even for the weak wind stars. Therefore, the Hα mass-loss rates derived by using homogeneous models should be considered as upper limits.

In addition to the stellar and wind parameters, we have also derived the N abundances for all the stars in NGC 2244. These values are particularly interesting to compare the theoretical predictions with the observations. Indeed, rotation is supposed to affect the angular momentum and the chemical elements transport through mixing processes (Meynet & Maeder [21]). Therefore, we have put in perspective the N content with the projected rotational velocities (Fig. 3, left-hand panel). In this context, we observed a clear trend of larger N/H with larger $v \sin i$, except for the two rapid rotators (HD 46056 and HD 46485). When we compare the N content with the stars' luminosity (Fig. 3, right-hand panel), a clear trend of larger enrichment in higher mass stars is observed. We do not see any outliers, thereby meaning that these fast rotators are too young to have experienced a strong enrichment in spite of their large $v \sin i$. The binary system HD 46149 does not show any evidence of interaction between its components. That suggests that the components of this system have not been affected by a mass transfer or too severe tidal effects. They are still supposed to evolve as if they were single stars.

Finally, our investigation shows a discrepancy between the evolutionary masses and the spectroscopic masses (i.e., the masses derived from the radii and the gravities of the stars). The determination of both masses are given in Table 1.

Journal of Physics: Conference Series **328** (2011) 012024 doi:10.1088/1742-6596/328/1/012024

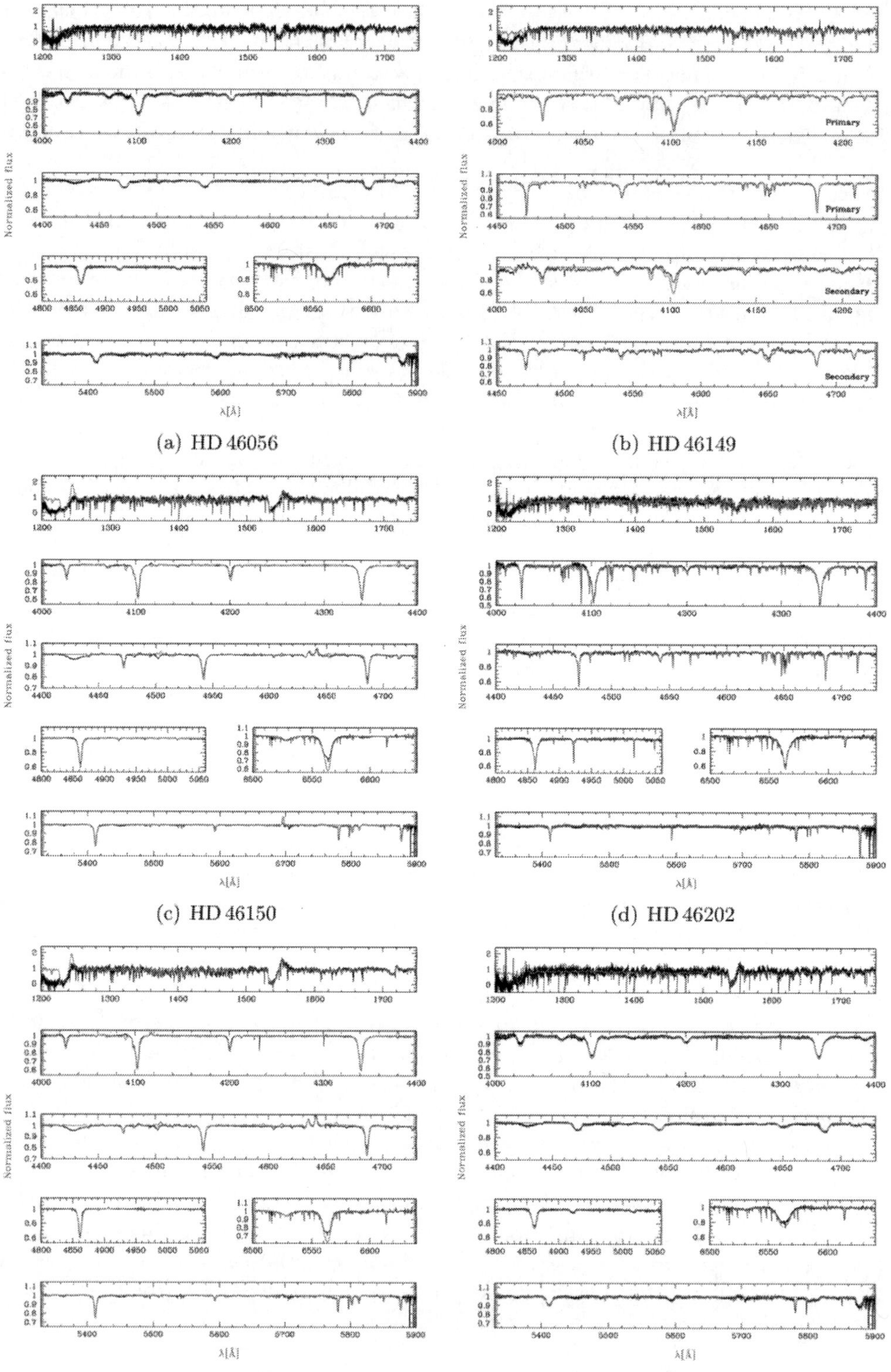

(a) HD 46056 (b) HD 46149

(c) HD 46150 (d) HD 46202

(e) HD 46223 (f) HD 46485

Figure 1. Best-fit CMFGEN models for the O-type stars in NGC 2244.

Journal of Physics: Conference Series **328** (2011) 012024 doi:10.1088/1742-6596/328/1/012024

Table 1. Derived stellar properties of the sample stars: name, spectral type, effective temperature (uncertainty 1000 K, except for HD 46149, 2000 K), luminosity, effective gravity (uncertainty 0.1 dex, except for HD 46149, 0.2 dex), UV mass-loss rate, terminal velocity, projected rotational velocity, macroturbulent velocity, evolutionary mass, spectroscopic mass, nitrogen content, ionizing flux. Spectroscopic masses are computed from the true gravity (i.e., $\log g$ corrected from the effect of centrifugal force).

Source	$T_{\rm eff}$ [kK]	$\log \frac{L}{L_\odot}$	$\log g$	$\log(\dot{M}_{UV})$	$\log(\dot{M}_{\rm H\alpha})$	f_∞	v_∞ [km s^{-1}]	$v\sin i$ [km s^{-1}]	$v_{\rm mac}$ [km s^{-1}]	$M_{\rm evol}$ [$M_\odot$]	$M_{\rm spec}$ [$M_\odot$]	N/H [$\times 10^4$]
NGC 2244												
HD 46223	43.0	5.60±0.11	4.01	−7.17	−6.20	0.1	2800	100	32	$52.1^{+6.2}_{-5.9}$	48.3 ± 19.3	7.0±2.0
HD 46150	42.0	<5.65	4.01	−7.30	−6.40	0.1	2800	100	37	< 53.6	59.5 ± 23.7	3.0±2.0
HD 46485	36.0	5.05±0.11	3.85	−7.80	−6.45	1.0	1850	300	-	$28.6^{+3.3}_{-4.3}$	19.5 ± 7.9	≤1.2
HD 46056	34.5	4.85±0.12	3.89	−8.50	-	1.0	1500	330	-	$23.1^{+2.4}_{-1.9}$	15.8 ± 6.7	≤0.6
HD 46149-1	37.0	4.90±0.12	4.25	<-9.0	-		1300	∼0	24	$25.9^{+3.1}_{-3.5}$	30.9 ± 21.6	0.8±0.5
HD 46149-2	33.0	4.60±0.16	3.53	-	-	-	-	100	27	$19.2^{+2.9}_{-2.5}$	4.7 ± 3.6	0.6±0.5
HD 46202	33.5	4.85±0.12	4.10	-9.0	−7.10	1.0	1200	20	17	$22.7^{+2.0}_{-2.4}$	29.0 ± 12.4	1.0±0.5

Journal of Physics: Conference Series **328** (2011) 012024 doi:10.1088/1742-6596/328/1/012024

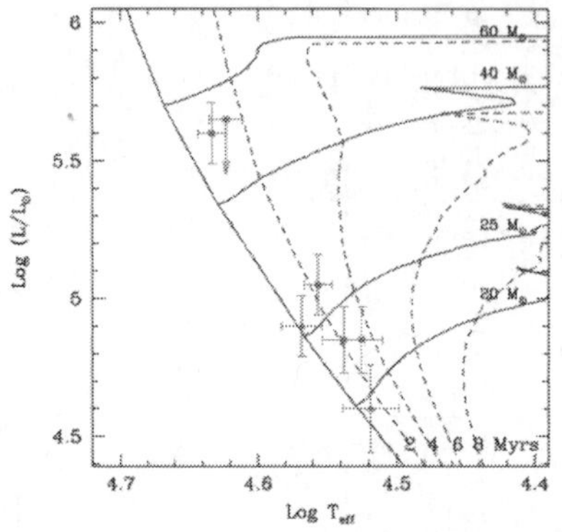

Figure 2. HR diagram of NGC 2244. Curves, computed from an initial rotational velocity of 300 km s^{-1}, are from Meynet & Maeder [22].

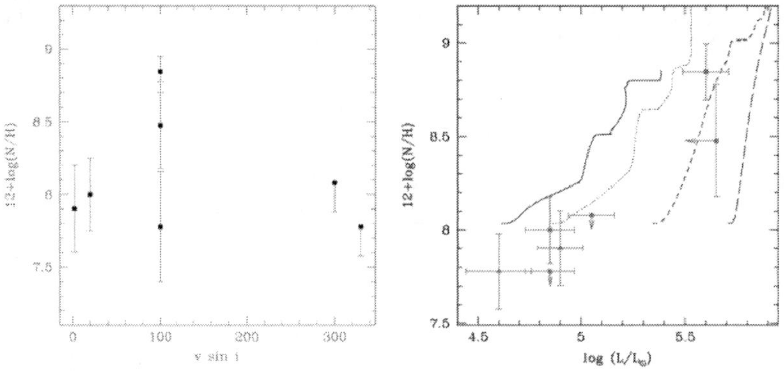

Figure 3. *Left-hand panel*: Nitrogen surface abundance (in units of $12 + \log(N/H)$) as a function of projected rotational velocity. *Right-hand panel*: Nitrogen surface abundance (in units of $12 + \log(N/H)$) as a function of luminosity. Evolutionary tracks (shown by the solid, dot-long dashed, dot-dashed, dashed lines for M=20, 25, 40 and 60 $M_\odot$) are from Meynet & Maeder [22].

5. Conclusions

We have analyzed six O-type stars belonging to NGC 2244 and among them HD 46149 which is a binary system. Optical and UV spectroscopy have been obtained. Atmosphere models computed with the code CMFGEN have been used to derive the main stellar and wind parameters of the target stars.

All the stars have an age between 0 and 5 Myr, with a trend of lower age for the most massive stars. Moreover, we confirm the existence of weak winds in the latest type stars of our sample and, for all the stars, we see a clear discrepancy between the mass-loss rates computed from the UV domain and those derived from Hα in the optical wavelengths. Finally, a discrepancy between the evolutionary masses and the spectroscopic masses (estimated from the radii and the gravities of the stars) is observed for the latest type O stars. However, we need to analyze more stars of the same age to confirm the trends observed from a point of view of their ages, their masses and their mass-loss rates.

Finally, this investigation improves our knowledge of these stars and paves the way for the computations of excited modes and pulsations which could be compared to the results obtained from the CoRoT photometry.

Acknowledgments

This research was supported by the PRODEX XMM/Integral contract (Belspo), the Fonds National de la Recherche Scientifique (F.N.R.S.) and the Communauté française de Belgique – Action de recherche concertée – A.R.C. – Académie Wallonie-Europe.

References

[1] Bonatto C and Bica E 2009 *MNRAS* **394** 2127–2140 (*Preprint* 0901.0833)
[2] Hensberge H, Pavlovski K and Verschueren W 2000 *A&A* **358** 553–571
[3] Baglin A, Auvergne M, Barge P, Deleuil M, Catala C, Michel E, Weiss W and The COROT Team 2006 *ESA Special Publication* (*ESA Special Publication* vol 1306) ed M Fridlund, A Baglin, J Lochard, & L Conroy pp 33–+
[4] Blomme R, Mahy L, Catala C, Cuypers J, Gosset E, Godart M, Montalban J, Ventura P, Rauw G, Morel T, Degroote P, Aerts C, Noels A, Michel E, Baudin F, Baglin A, Auvergne M and Samadi R 2011 *A&A* **533** A4+ (*Preprint* 1107.0267)
[5] Briquet M, Aerts C, Baglin A, Nieva M F, Degroote P, Przybilla N, Noels A, Schiller F, Vučković M, Oreiro R, Smolders K, Auvergne M, Baudin F, Catala C, Michel E and Samadi R 2011 *A&A* **527** A112+ (*Preprint* 1101.3162)
[6] Degroote P, Briquet M, Auvergne M, Simón-Díaz S, Aerts C, Noels A, Rainer M, Hareter M, Poretti E, Mahy L, Oreiro R, Vučković M, Smolders K, Baglin A, Baudin F, Catala C, Michel E and Samadi R 2010 *A&A* **519** A38+ (*Preprint* 1006.3139)
[7] Mahy L, Gosset E, Baudin F, Rauw G, Godart M, Morel T, Degroote P, Aerts C, Blomme R, Cuypers J, Noels A, Michel E, Baglin A, Auvergne M, Catala C and Samadi R 2011 *A&A* **525** A101+ (*Preprint* 1010.4959)
[8] Martins F, Mahy L, Rauw G and Hillier D J 2011 *ArXiv e-prints* (*Preprint* 1110.4509)
[9] Mahy L, Nazé Y, Rauw G, Gosset E, De Becker M, Sana H and Eenens P 2009 *A&A* **502** 937–950 (*Preprint* 0905.1592)
[10] Hillier D J and Miller D L 1998 *ApJ* **496** 407–+
[11] Lanz T and Hubeny I 2003 *ApJS* **146** 417–441 (*Preprint* arXiv:astro-ph/0210157)
[12] Repolust T, Puls J and Herrero A 2004 *A&A* **415** 349–376
[13] Grevesse N, Asplund M and Sauval A J 2007 *Space Science Reviews* **130** 105–114
[14] Simón-Díaz S and Herrero A 2007 *A&A* **468** 1063–1073 (*Preprint* arXiv:astro-ph/0703216)
[15] Bouret J C, Lanz T and Hillier D J 2005 *A&A* **438** 301–316 (*Preprint* arXiv:astro-ph/0412346)
[16] González J F and Levato H 2006 *A&A* **448** 283–292
[17] Conti P S and Alschuler W R 1971 *ApJ* **170** 325–+
[18] Conti P S 1973 *ApJ* **179** 161–179
[19] Wang J, Townsley L K, Feigelson E D, Broos P S, Getman K V, Román-Zúñiga C G and Lada E 2008 *ApJ* **675** 464–490 (*Preprint* 0711.2024)
[20] Sundqvist J O, Puls J and Feldmeier A 2010 *A&A* **510** A11+ (*Preprint* 0911.3361)
[21] Meynet G and Maeder A 2000 *A&A* **361** 101–120 (*Preprint* arXiv:astro-ph/0006404)
[22] Meynet G and Maeder A 2003 *A&A* **404** 975–990 (*Preprint* arXiv:astro-ph/0304069)

Journal of Physics: Conference Series **328** (2011) 012026 doi:10.1088/1742-6596/328/1/012026

Hα and Gaia–RVS domain spectroscopy of Be stars and interacting binaries with Ondřejov 2m telescope

P Koubský, L Kotková and V Votruba

Astronomical Institute, Academy of Sciences of CR, 251 65 Ondřejov, Czech Republic

E-mail: koubsky@sunstel.asu.cas.cz

Abstract. A long term project to investigate the spectral appearance over the Gaia RVS domain of a large sample of Be stars and interacting binaries has been undertaken. The aim of the Ondřejov project is to create sufficient amounts of training data in the RVS wavelength domain to complement the Bp/Rp classification of Be stars which may be observed with Gaia. The project's current status is described and sample spectra in both the Hα and RVS wavelength domains are presented and discussed.

1. Introduction

Be stars are usually defined as early type stars of luminosity classes III–V which once showed emission in at least one Balmer line. Emission lines are usually not limited to the Balmer series, but they can be present in the Paschen, Brackett, Pfund, and Humphreys series as well, and also in lines of other ions such as Fe II, Ca II, or He I. The emission is attributed to a circumstellar envelope/disc. Despite the large efforts in both observational and theoretical fields, the emission line formation is not yet well explained. The emission lines can disappear and possibly reappear with non-regular periods of several years. Moreover, Be stars are variable on other time-scales from hours to months, but the long-term variations (usually related to the building up and decline of a disc) are the most important ones. In case of interacting binaries, which mimic some properties of Be stars, the line variations are related to the orbital period. Be stars are not rare objects, they actually represent 17 % of all B-type stars (Zorec and Briot [1]).

Gaia has aboard three instruments ASTRO (for astrometry), the Bp/Rp photometer, and the RVS (Radial Velocity Spectrometer) which will provide fundamental information about the still unexplained Be phenomenon. Lobel et al. [2] processed ground-based spectra of emission line stars with the Gaia Object Generator to analyze simulated Bp/Rp photometer fluxes for objects to G=15. They showed that Be stars can be identified using an Hα/Hβ-index diagram. In this paper we investigate RVS data to provide a more precise description of the behaviour of Be stars observed by Gaia. We also discuss Ca II near-IR triplet emission lines in Be stars.

The Gaia RVS wavelength region will cover Paschen, Ca II, and N I lines, which in Be stars can display complicated blends of both absorption and emission components (Munari and Tomasella [3]). Various descriptions of near-IR spectra of Be stars are available in the literature; an atlas of O-G0 stars in the IR is provided in Andrillat et al. [4], and a rather brief description of the Paschen and near-IR Ca II lines is given in Jaschek et al. [5], and in Tomasella et al.

Journal of Physics: Conference Series **328** (2011) 012026 doi:10.1088/1742-6596/328/1/012026

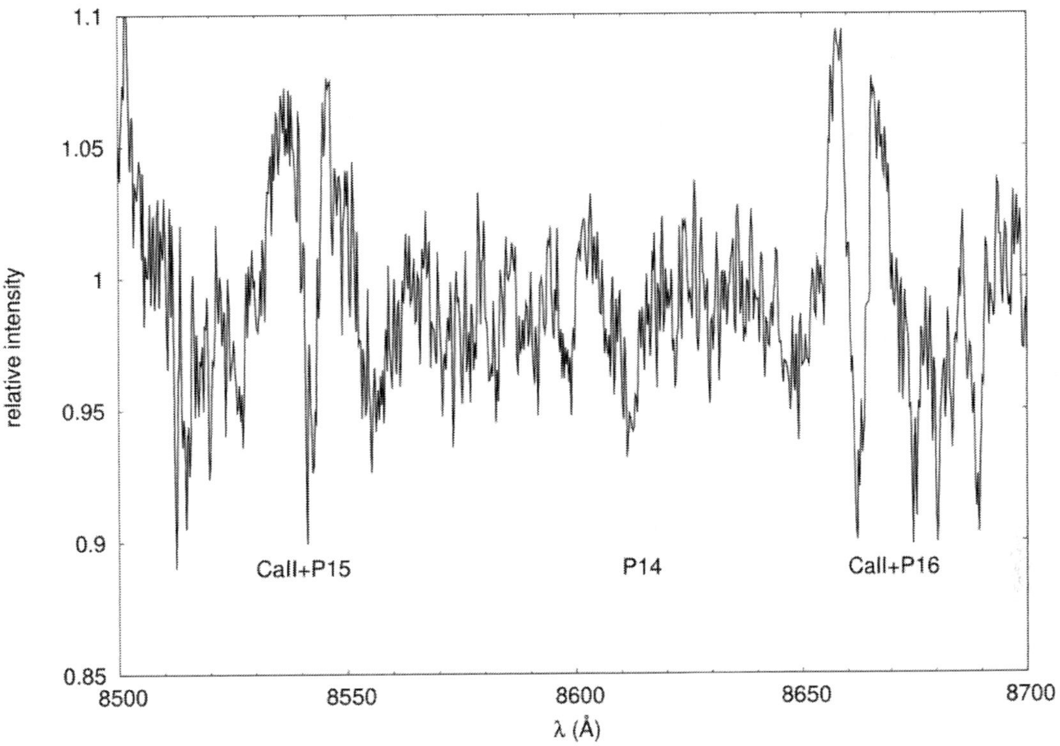

Figure 1. Spectrum of Be star SAO 349 observed with the Ondřejov coudé spectrograph.

[6]. Many IR Be star spectra are available in digital archives; e.g., low-resolution IR spectra (∼1000) of Be stars in the RVS wavelength region are provided in the MAST HPOL archive, and high-resolution spectra of Be stars to λ=8881 Å are available from the ESO FEROS archive, and from the VLT-UVES archive (note with a wavelength gap from 8540 Å to 8660 Å). No systematic attempt has been undertaken so far to investigate the spectra of Be stars in the Gaia RVS domain. We therefore decided to perform a large-scale documentation project of Be stars with the coudé spectrograph of the Ondřejov 2 m telescope. We compare Hα and Paschen spectra of Be stars and exploit the possibility to investigate the properties and the behaviour of the disc from observations of the higher members of the Paschen series. In addition, we also investigate the near-IR Ca II emission lines in Be stars and interacting binaries.

It is expected that the RVS spectrometer aboard Gaia will observe Be stars brighter than V=10 mag., in the wavelength range from 8470 Å to 8740 Å, with a spectral resolution of R=10500 (while R∼6000 for fainter objects). In Fig. 1 we show an example of a Be star spectrum observed in the framework of the Ondřejov project in the Gaia RVS domain. The signal-to-noise ratio of the spectrum is ∼30 which should become available with Gaia data of stars to V∼13 mag. (e.g., combined over the full mission). The spectral resolution is comparable to synthetic RVS spectra (see Katz [7] for details).

2. Data collection and reduction

The spectra were observed with the Ondřejov single-order spectrograph and a SITe 2000x800 chip at the focus of a 702-mm camera (R=12 000, λ=6250–6750 and 8392–8900 Å). In all cases, calibration arc frames were observed before and after each stellar object frame. During each

Journal of Physics: Conference Series **328** (2011) 012026 doi:10.1088/1742-6596/328/1/012026

night, a series of flat-field and bias exposures was obtained. The stellar target and calibration spectra (using Th-Ar and tungsten lamps) were reduced with IRAF. The spectra of ~150 Be stars were observed in both the Hα and Paschen wavelength regions.

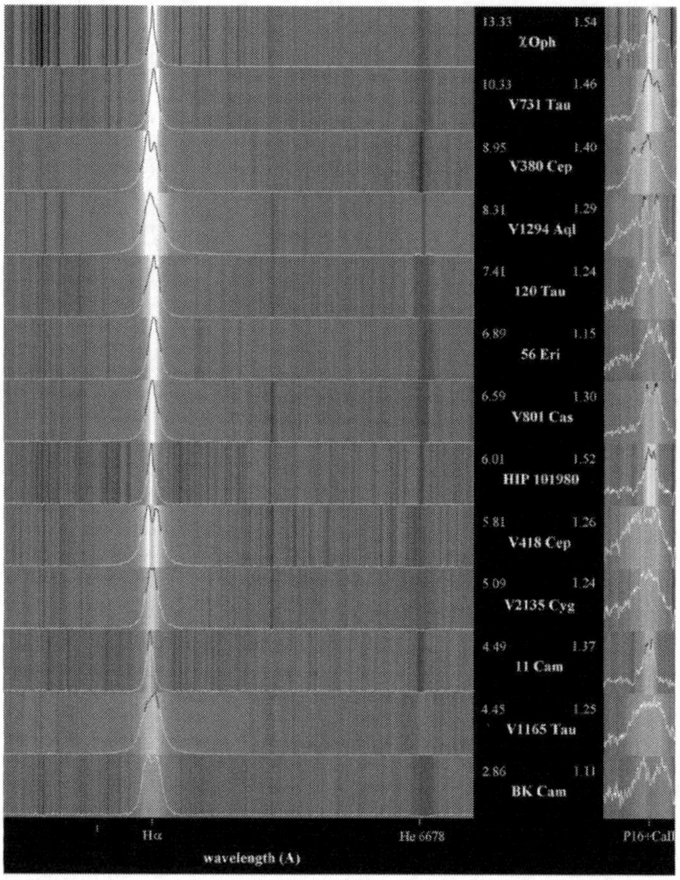

Figure 2. Spectra of Be stars in the Hα and Gaia RVS wavelength regions ordered according to the normalized Hα emission flux (see the numbers in the left-hand corner of the middle panel; the normalized emission fluxes of P16 are given in the right-hand corner above the name label of every star). All spectra are displayed both in the classical (continuum normalized) way, combined with a gray-scale representation. Spectra of Be stars with strong Hα emission are shown.

3. Results

3.1. Be stars

A straightforward relation between the properties of Hα and Paschen line profiles in Be stars is summarized in Fig. 2. The spectra are ordered according to the normalized flux of Hα emission with an approximate range from 13 to 3. Note that the Hα line profiles do not show important central absorption components, i.e., the discs are observed for small inclination angles (Silaj et al. [3]). In all cases the Paschen lines are also observed in emission. Thus, for Be stars with strong Hα emission, the Gaia RVS spectra, which will cover the Paschen P16 – P13 lines, can provide independent information about the Hα line.

Journal of Physics: Conference Series **328** (2011) 012026 doi:10.1088/1742-6596/328/1/012026

Figure 3 presents Be stars with more complicated Hα line profiles. They include the narrow profiles observed in FW CMa, or a combination of emission and absorption line components in ϕ And. They cover a smaller range of normalized emission fluxes of 6.3 – 1.1. We find that all Hα lines with normalized emission fluxes above 2.5 correspond to emission in the Paschen lines. The presence of Hα emission in o Aqr or V1507 Cyg (see Section 3.2) might therefore be deduced from near-IR spectra after careful analysis of the Paschen lines.

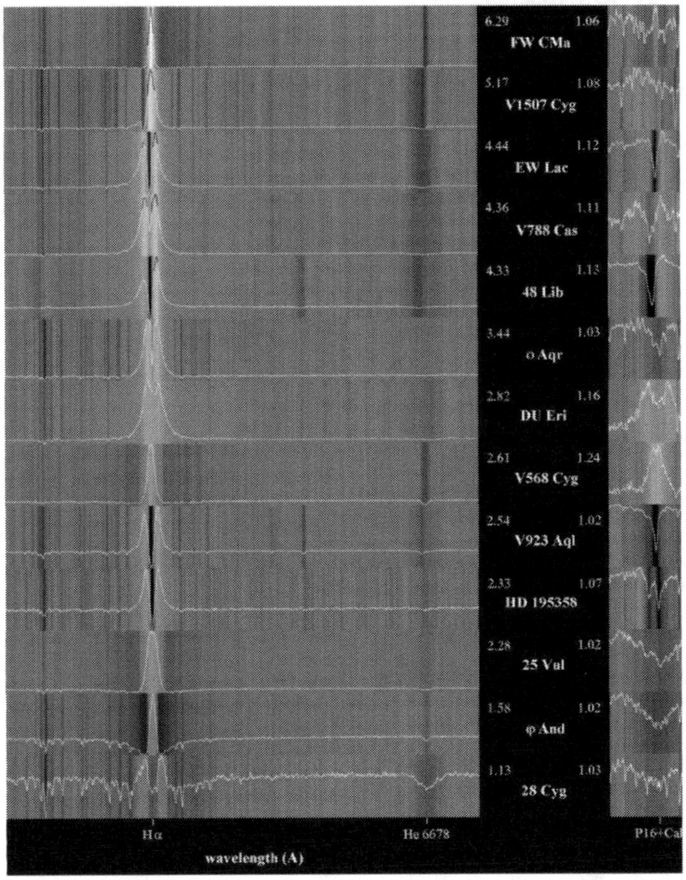

Figure 3. As in Fig. 2, but for Be stars with moderate Hα emission.

3.2. Infrared calcium triplet emission

Very interesting spectral features in the Gaia RVS wavelength region are the Ca II triplet lines (λλ 8498, 8542, and 8662) which nearly coincide with the wavelengths of the Paschen P16, P15, and P13 lines. In many Be stars both these Paschen and Ca II triplet lines are observed in emission. In these cases, the P14 emission line (which does not blend with Ca II lines) is weaker than P15 and P13. However, some peculiar objects such as υ Sgr (a hydrogen deficient binary), Mira Ceti stars, and T Tau (a pre-main sequence star) show only Ca II triplet emission, without emission of P14. The spectra of Be stars with Ca II triplet emission are shown in Figs. 4 and 5. Comparing Figs. 2 and 3 with Fig. 4 we find that Ca II emission very often enhances the Paschen emission, possibly in all cases where the Hα disc is adequately strong. We assemble in Fig. 5 examples of Be star spectra in which the Ca II triplet line emission dominates. Some stars

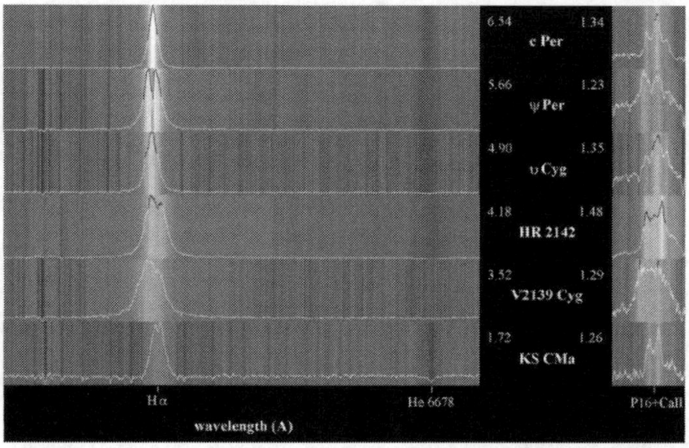

Figure 4. As in Fig. 2, but for Be stars showing emission in the Paschen and Ca II lines.

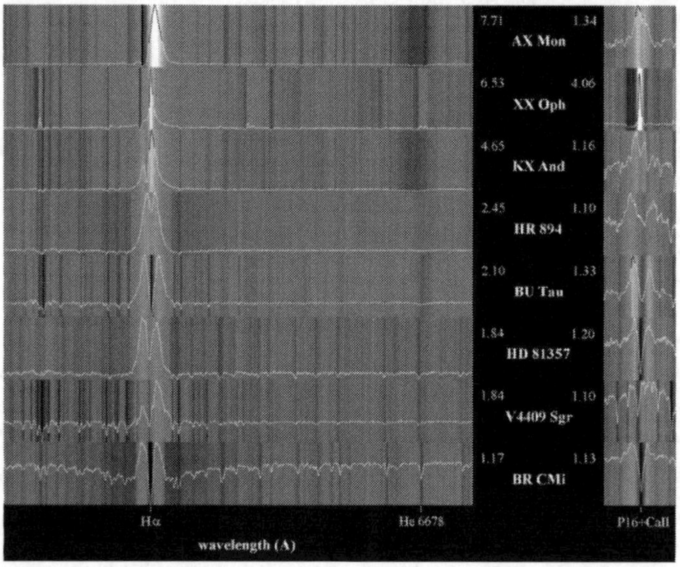

Figure 5. As in Fig. 2, but for Be stars with dominating Ca II emission.

are proved interacting binaries (AX Mon, KX And, BR CMi). HD 81357 was recently detected to be a 33 d binary with a spectrum that mimics BR CMi. The spectrum of V1507 Cyg, shown in Fig. 3, may also belong to this group. This 28 d binary system has a very complex nature (Hutchings and Redman [9]) consisting of a visible component that loses mass to a more massive, but observationally undetectable, star due to its huge circumstellar envelope. On the other hand, there is no evidence of a Roche-lobe filling secondary in BU Tau, and the binarity of XX Oph, HR 894, or V4409 Sgr has not been confirmed. It leaves the suggestion of Polidan [10] still open that there is a possible connection between the presence of Ca II triplet emission and the binary nature of Be stars. To our best knowledge, Polidan's suggestion has so far not been studied in further detail, and we hope that the Gaia RVS data will shed more light on this problem as well.

Journal of Physics: Conference Series **328** (2011) 012026 doi:10.1088/1742-6596/328/1/012026

4. Conclusions

We have shown that data from the Gaia RVS instrument can be combined with Bp/Rp data for detecting Be stars with normalized Hα emission exceeding 2.5. The high incidence of interacting binary stars among those stars with Ca II triplet emission lines is also worthy of further study. We hope that the Gaia RVS instrument will provide new and interesting data for the reliable modeling of circumstellar discs in both single and binary stars.

Acknowledgments

We thank the anonymous referee for helpful comments and for constructive criticism to an earlier version of the paper. The authors would like to thank Dr. Y. Frémat for insightful comments which helped to improve the paper. We would like to acknowledge the assistance of Drs. Jiří Kubát and Marek Wolf for obtaining observations used in this paper. The research in this study was supported by ESA PECS grant 98058.

References

[1] Zorec J and Briot D 1997 *Astron. Astroph.* **318**, 443
[2] Lobel A, Frémat Y and Blomme R 2010 Gaia report GAIA-C8-TN-ROB-AL-001-1, Gaia Livelink
[3] Munari U and Tomasella L 1999 *Astron. Astroph. Suppl* **137**, 521
[4] Andrillat Y, Jaschek C and Jaschek M 1995 *Astron. Astroph.* **112** 475
[5] Jaschek C, Jaschek M, Egret D and Andrillat Y 1988 *Astron. Astroph.* **192**, 285
[6] Tomasella L, Munari U and Zwitter T 2010 *Astron. Journal* **140**, 1758
[7] Katz D 2009 *Proceedings of the Annual Meeting of the French Society of Astronomy and Astrophysics* eds M Heydari-Malayeri C Reylé and R Samadi p 57
[8] Silaj J, Jones C E, Tycner C, Sigut T A A and Smith A D 2010 *Astrop. Journal Suppl.* **187** 228
[9] Hutchings J B and Redman R O 1973 *Montly Notices of Royal Astron. Soc.* **163**, 209
[10] Polidan R S 1976 *Be and Shell Stars, IAU Symp. 70* ed A Slettebak (Dordrecht,Holland; Boston: D.Reidel Pub.Co.) p 401

Journal of Physics: Conference Series **328** (2011) 012027 doi:10.1088/1742-6596/328/1/012027

Stellar atmospheres in the Gaia era

Alex Lobel
Royal Observatory of Belgium, Ringlaan 3, 1180 Brussels, Belgium
E-mail: alobel@sdf.lonestar.org; Alex.Lobel@oma.be

Abstract. Highlights of the meeting on *Stellar Atmospheres in the Gaia Era: Quantitative Spectroscopy and Comparative Spectrum Modeling* (http://great-esf.oma.be) held on 23-24 June 2011 in Brussels, Belgium are emphasized. New research results are summarized and a record of the scientific discussions during the meeting is provided, as well as important open questions for future research.

1. Introduction

About two years ago in November 2009 the Working Group B4 for Stellar Atmospheres was formed during the third GREAT Plenary meeting in Nice, France. At the meeting I was asked to become co-facilitator for WGB4, but rather hesitantly accepted. The GREAT Programme (Gaia Research for European Astronomy Training[1]) is a pan European science driven research infrastructure which facilitates, through focused interaction, the fullest exploitation of the ESA-Gaia 'cornerstone' astronomy mission. It brings together relevant scientific expertise by promoting topical workshops, training events, exchange visits, conferences and so forth with the aim of addressing the major scientific issues that the Gaia satellite will impact upon.

Since its kick-off meeting in Nice the Stellar Atmospheres Working Group has steadily grown, currently involving over 30 scientists from 20 research institutions. My initial hesitation stemmed from the fact that contemporary stellar atmospheres research covers a very wide variety of science topics that are often difficult to unite under the same banner. Perhaps a more appropriate common denominator is "Advanced Analyses and Interpretation of Light from Stellar Objects", but that sounds somewhat awkward.

Stellar atmospheres research is likely one of the oldest and richest fields of modern astrophysics. 150 years have passed since Gustav Kirchhoff produced the first detailed map of the solar spectrum, thereby introducing stellar physics research. Wonderful discoveries came our way since that time (the Sun is a nuclear fusion reactor, the Andromeda nebula is an independent Galaxy, radio waves from stars, the Universe expands, space and time were born together, and many more), and yet much remains unknown about the Sun and stars. The Gaia (expected launch May 2013) census of the Milky Way will observe over one billion stars with three instruments to collect astrometric, photometric, and spectroscopic data down to the $20^{\rm th}$ visual magnitude. The satellite will observe ~150 million spectra of stars brighter than the $17^{\rm th}$ visual magnitude (but fainter than $V{\sim}6^{\rm m}.0$).

The goal of the Gaia mission is ambitious: to unravel the structure and formation history of the Galaxy with a 6-D (or 5-D) position-velocity map. The astrometric accuracies are a

[1] http://www.ast.cam.ac.uk/ioa/GREAT/

hundred times more precise than the Hipparcos catalog, while the spectra will provide reliable radial velocity values of a staggering amount of stars. It is therefore safe to say that we should expect the unexpected astrophysics with Gaia data, although the current preparations for the data processing and analyses are quite a challenge. The confrontation of these data with theoretical models will drastically advance our understanding of the physics of stellar atmospheres. For example, new stellar populations such as previously unknown emission line stars will be discovered, and fundamental questions such as the basic scenarios of stellar evolution will be addressed with Gaia data.

The meeting is summarized here in five major topics emphasizing important points and new research results presented in the 23 oral and 10 poster presentations. I only mention the name of the speaker and of the first author of the poster, as the complete list of authors of the contributions can be consulted in this Volume. A record of the four main Discussion sessions following the Morning and Afternoon Sessions is added to the relevant topic. The record is probably incomplete but provides a concise overview of interactions between the participants that stimulated further debates. I also add a number of personal comments and views about important open questions for future stellar atmospheres research.

2. Radiation transport codes

The presentations and discussions at the meeting identified key areas for the improvement of model spectra from various modern spectral synthesis codes required to determine reliable physical parameters of hot and cool stars. The presentations of the first day highlighted new results with the PHOENIX, MARCS, and MOOG spectral synthesis codes for cool stars, but also included a presentation about advanced future spectroscopic instrumentation. The second day focused on new results with the ATLAS, TLUSTY, and CMFGEN synthesis codes applied for modeling the spectra of hot stars. A substantial number of presentations also addressed very recent research initiatives, showing some preliminary results, for large ground-based spectroscopic sky surveys to support and/or complement future Gaia data analysis.

2.1. Cool stars

After an introduction about the Gaia satellite design and current development status, Ulrike Heiter presented results of the GREAT-ESF Stellar Atmospheres Workshop on Comparative Spectrum Modeling in Vienna of August 2010. Observed and simulated spectra of cool giants were modeled by 14 research groups with different atmospheric models and analysis approaches. For example, for the observed spectrum of α Tau the resulting spread of computed atmospheric parameters ($T_{\rm eff}$, $\log g$, and [Fe/H]) is larger than the estimated error bars, although most of the parameter values agree within 2-σ. The differences in the synthetic spectra are largest in the blue wavelength region and decrease towards the near-IR. The normalized central absorption line depths can differ by 0.4 around 5000 Å, while averages around 0.1 are measured around 6600 Å [1]. There do not appear to be clear trends in the computed parameter values that depend on specific assumptions of the modeling method. However, the main cause for the differences in atmospheric parameter values can be attributed to the atomic input data (adopted oscillator strengths and damping constant), and to the choice of the modeled wavelength regions. I conclude that the 'comparative modeling experiment' for cool giants should be repeated for other types of stars, including hot stars for which atomic input data of the higher ions is less well-tested compared to cool star spectra. The results for cool giants also show that one should first homogenize a number of assumptions in the modeling method for pinpointing a specific cause of the different parameter values. For example, utilize the same 1-D atmosphere model grid for exposing possible differences in the methods and data to compute the stellar continuum fluxes, or for example, caused by missing/incomplete molecular opacity (for cool stars) and differences in the adopted equation of state.

Journal of Physics: Conference Series **328** (2011) 012027 doi:10.1088/1742-6596/328/1/012027

Of central importance for the ongoing preparations of Gaia data processing is the development of codes that properly classify all observed objects (stars should be separated from galaxies, quasars, solar system bodies) and determine the astrophysical parameters (APs), which also include reddening, α-element enrichment, radial- and projected rotation velocity, activity. Rosanna Sordo presented an overview of high-resolution synthetic spectral libraries of model spectra computed with a large variety of spectrum synthesis codes [2]. The model girds will be utilized to compute stellar parameters with the Discrete Source Classifier Algorithms currently under development in Gaia DPAC-Coordination Unit 8 (CU8). They are implemented in the Stellar Population Code to train the codes for the classification of many different types of hot and cool stars; FGK & M stars, OB stars, white dwarfs, red supergiants, peculiar A and B stars, carbon stars, ultra cool dwarfs, various types of emission line stars, Be stars, Wolf-Rayet stars, etc. An important point is that more sophisticated spectral grids computed with 3-D atmosphere models will unlikely tackle the basic problems that such an extensive compilation of training data must deal with. Apart from the large computational expenses in 3-D to cover the full ($T_{\rm eff}$, log g, [Fe/H])-range for hot and cool stars, the model 'gridpoints' have to be interpolated for continuous coverage of the APs space. The interpolation procedures yield errors of $\sim$10 % in the Gaia RVS region, while for libraries with 0.1 nm wavelength sampling they can amount to $\sim$4 % for $T_{\rm eff}$=5000 K.

The application of input libraries from different spectral synthesis codes for training the DPAC-CU8 codes introduces in the computed APs errors that should be quantified at some point. Detailed comparisons of MARCS, PHOENIX, and ATLAS spectra for $T_{\rm eff}$= 4000 K, log g=4.0, and [Fe/H]=0.0 reveal mean flux differences to $\sim$10 % below 5000 Å. For $T_{\rm eff}$=8000 K they decrease and overall agreement improves, although with rather different spectral energy distribution (SED) slopes in the blue. Bertrand Plez very eloquently pointed out major flux differences between the three synthesis codes by comparing $T_{\rm eff}$ and $V-K$ color, and color-color diagrams for 2500 K $\leq T_{\rm eff} \leq$ 9000 K, for log g=0.5 (giants) and 4.5 (dwarfs). The codes agree well above 4000 K because the opacity sources are almost identical. Below 4000 K ATLAS dwarf models are too blue in the near-IR for currently unknown reasons. For $T_{\rm eff}$ below 4000 K considerable differences in the blue between MARCS and PHOENIX fluxes should also be investigated.

An important conclusion of the meeting is that SED differences between the synthesis codes for cool stars ($T_{\rm eff}\leq$10 kK) are not attributable to (small) differences in the 1-D atmosphere model structures. For example, differences in log($\tau_{\rm Ross}$) versus gas kinetic temperature (T) computed with ATLAS9 and ATLAS12 are limited, although the opacity treatment (ODFs versus opacity sampling) is rather different [3]. Comparisons of spherical ATLAS models and PHOENIX NEXTGEN models for cool giants [4] reveal almost identical T-structures that cannot account for the large SED differences in the emergent flux spectra. Plez shows similar good agreement between models computed with MARCS and ATLAS (using new ODFs) for dwarfs, giants, and supergiants. This is also the case by comparing with (spherical) PHOENIX-ACES models (using an extended EOS), except perhaps for the limited differences that occur at small optical depths of log($\tau_{\rm Ross}$)≤ -2 for $T_{\rm eff}\leq$3000 K. The flux differences in the spectra are therefore likely the result of appreciable differences of the input background opacity sources, for example due to different strengths of broad TiO and FeH molecular bands. I also add that the partition functions of the most important hydrides and oxides require further improvements.

More detailed comparisons of the shapes of the Ca II near-IR triplet lines (in the Gaia RVS wavelength range) reveal that the line wings predicted with MARCS are somewhat broader than with PHOENIX, which can be reconciled by adopting the same line damping and broadening constants in both codes. I remark that detailed line profile comparisons with observed spectra of large signal-to-noise ratio also require a proper treatment of isotopic line splitting (in log(gf)-components) and, if possible, hyperfine splitting for a number of important elements such as Mn

Journal of Physics: Conference Series **328** (2011) 012027 doi:10.1088/1742-6596/328/1/012027

and Co. Hyperfine splitting broadens the wings of absorption lines and causes remarkably flat central line cores. Much progress is likely to come from detailed comparisons with high-resolution observed spectra of stars having model-independent $T_{\rm eff}$- and log g-values, for example obtained from accurate angular diameter and bolometric flux measurements, or from comparisons with seismologic surface gravity values (see Section 4.1).

New results with PHOENIX for brown dwarfs ($T_{\rm eff}$=1800 K and 2000 K) were presented by Peter Hauschildt. The new models include the formation/destruction of dust grains depending on the location (depth) of the convection zone. They incorporate the feedback of dust nucleation & growth and dust gravitational settling on the overall atmospheric structure PHOENIX computes (in dust-drift mode) for a stationary solution. The theoretical SEDs very closely fit the observed SEDs of brown dwarfs (log g=5.5) and of metal-poor cool giants ($T_{\rm eff}$=1600 K, log g=3.0, [Fe/H]=−1.0). The PHOENIX (v16) package is a versatile suite of codes for computing 1-D or 3-D atmosphere models and the emergent flux spectra in non-LTE. It can include velocity fields (of winds, supernovae) and chemical equilibrium calculations. PHOENIX has been optimized for massively parallel processing. For example, ongoing tests show overall good agreement for fully line blanketed 1-D synthesis calculations compared to 3-D (using spherical coordinates) for UV and optical spectra of stars with $T_{\rm eff}$=3000 K, 5700 K (solar type), and 9000 K. 3-D spectrum synthesis calculations are currently computationally rather expensive. However, they will become very important for testing a large variety of physical processes in cool star atmospheres with the fast increases of compute power.

2.2. Hot stars

The determination of accurate APs of OB stars with Gaia photometry and RVS spectra will be challenging. In hot stars the small RVS wavelength region is dominated by a handful of broad H Paschen series lines from P13 to P16. Ground-based spectroscopic surveys of hot stars will therefore become very important to test and verify the APs obtained with Gaia data using various other wavelength intervals that can also yield reliable elemental abundance values for these stars. Norbert Przybilla presented an overview of quantitative spectrum modeling of hot dwarfs and giants from B4 to O9 (15 kK $\leq T_{\rm eff} \leq$ 35 kK). He compares the results of three spectrum synthesis codes: ATLAS/SYNTHE in LTE, TLUSTY/SYNSPEC in non-LTE, and a hybrid approach that combines LTE atmosphere models (from ATLAS9) with non-LTE line transfer calculations using DETAIL/SURFACE, called 'ADS' [5]. For $T_{\rm eff}$ below 22 kK 1-D atmospheric structures and SEDs computed in LTE and non-LTE are very similar ($\Delta T/T \leq 1$ %). Above 22 kK non-LTE effects become more important in the cores and wings of H Balmer lines and He I/II absorption lines. In detailed line profile modeling the non-LTE effects generally become stronger towards larger $T_{\rm eff}$, and differences of line fluxes computed with the three codes increase towards smaller log g-values. Non-LTE line spectra computed with relatively simple atomic structures (e.g., H I, He I/II, and Mg II) yield reliable APs, but which is not valid for more complicated species. A remarkable conclusion of Przybilla's code comparisons is that advanced non-LTE computations which employ simplified model atoms (but in fact too simplified or outright inappropriate) often yield wrong results that can largely exceed the systematic errors of (simple) LTE line approximations. He finds that the (carbon, nitrogen, silicon, iron) model atoms of TLUSTY require extensive revision and updates for comparison to observed OB spectra. It is my opinion that the study profoundly and thoroughly demonstrates the enormous impact of missing or inadequate atomic input data on quantitative stellar spectroscopy. Garbage in, garbage out, whatever the amount of megaflops and number of processor threads used. It underscores the demand for new and *tested* atomic input data (both laboratory and astrophysical), and the great care we must take when using them.

In case the energy density in the spectral lines formation region of a stellar atmosphere is dominated by the ambient radiation field the detailed atomic balance (e.g., ionization and excitation rates and the resulting population fractions) can very far depart from local TE conditions. The local gas state (T, ρ, $P_{\rm el}$) must be coupled to the atomic rate equations for computing the line opacity and emergent fluxes. These overall non-LTE conditions occur in the atmospheres of stars with $T_{\rm eff}\geq$ 30 kK. In massive O-type stars the stellar atmosphere also expands due to radiative forces. The radiation-driven winds of O stars rapidly accelerate to highly supersonic velocities within one stellar radius above the stellar surface. It determines the transport of radiation because the line formation physics (line opacity and scattering) in the expanding wind is influenced by the Doppler effect, requiring a rather sophisticated calculation of the line source function. The spectral lines adopt a characteristic horn shape or P-Cygni profile. These 'wind' lines dominate the UV spectra of massive hot stars with many resonance transitions of N, C, and Si ions. They are however also observed in the optical spectra of several types of OB supergiants, such as Luminous Blue Variables (LBV) and Wolf-Rayet (WR) stars.

Jose Groh discussed the determination of APs with the CMFGEN code for these stars [6]. CMFGEN can solve the transfer equation in spherical symmetry for a user-specified radial monotonic velocity field using the co-moving frame prescription (e.g., opacity and emissivity are isotropic). It can compute a hydrostatic atmosphere structure to just below the sonic point in the wind, which incorporates full line blanketing due to metals with Z up to 30. Back-warming is mainly caused by iron. The $T_{\rm eff}$ is classically computed from the ionization balance with detailed fits to diagnostic spectral absorption lines such as He I λ4471 & He II λ4542 for O stars, and He I λ4471, λ4713/5876/6678 & He II λ4686 or optical Si II/III lines in LBVs. Important stellar wind parameters such as the mass-loss rate ($\dot{M}$ or $\dot{M}/\sqrt{f}$) and the volume-filling factor (f) for modeling the inhomogeneity or (micro-)clumping of the wind can be obtained from fits to Hα and He II λ4686, while UV wind lines or strong H Balmer lines provide the terminal wind velocity (v_∞). The effective atmospheric acceleration of OB dwarfs with log $g\geq$3.0 can be constrained from the pressure-broadened Hδ and Hγ absorption lines. In LBVs and WR stars, however, the H Balmer and Paschen lines are filled in by wind emission and adopt P Cygni profiles, generally preventing an accurate determination of log g-values for these stars. Interestingly, Groh also showed that surface abundances of nitrogen and carbon can be determined from optical NIII/IV and C III/IV lines. Diagnostic lines of the same species are less affected by model details, although they may require simultaneous fits to absorption and emission profiles (i.e., N III). New versions of CMFGEN are currently tested with time-dependent co-moving frame transfer solutions for monotonic spherical outflow. Relativistic effects are included and one can hence expect exciting new results with CMFGEN by comparing with the SNe atmosphere models of PHOENIX.

2.3. Other radiative transfer codes

For completeness I also mention the MOOG, MAFAGS, and STEPAR LTE spectrum synthesis codes that were employed for computing APs of FGK stars in a number of presentations at the meeting. Non-LTE abundance determinations in cool stars were also presented using the MULTI2.3 code, while quantitative analyses with the FASTWIND code were presented for O-type stars (see Section 5). All radiative transfer codes can be separated depending on the numerical method for solving the radiative transfer equation; either a discrete ordinate method (short, long, or mixed characterisitcs) or a Monte Carlo method is implemented. A plethora of codes have been developed for modeling radiative transfer in Earth's atmosphere. A small literature search reveals the names of various other spectrum synthesis and line modeling codes actively used in the stellar physics community such as CLOUDY, SPECTRUM, MALI, ATA, COSSAM, XSPEC, and many more. Most of them solve the transfer equation, while some employ approximative methods such as photon escape probability. I apologize to everyone whose pet code is not listed or was not discussed at the meeting (but am delighted to learn about them, so contact me). There are also codes for solving radiative transport in dusty envelopes and discs around stars,

Journal of Physics: Conference Series **328** (2011) 012027 doi:10.1088/1742-6596/328/1/012027

such as DUSTY, 2-DUST, RADCM, to name a few.

The longevity of transfer codes depends on much needed resources (financial and personnel) for new science applications. It requires new code updates and a proper understanding of how a code works internally. The development of a code is different from its application, but experienced users usually become developers provided they are helped (e.g., document your code) and are given sufficient time to become familiar with them. For example, as an undergraduate I worked with a transfer code called SCAN originating from Kees de Jager and co-workers at Utrecht, The Netherlands, for studying microturbulent broadening in the spectra of cool super- and hypergiants. Almost two decades later I still employ a more advanced version (SCANSPEC) for computing the detailed spectra of cool stars and for testing the quality of the atomic input data. Astrophysical $\log(gf)$-values are collected in an on-line spectral database called SPECTROWEB[2] [7]. Later I adapted spherical and molecular versions of the well-known MULTI code for investigating self-reversed emission lines formed in the chromospheres of cool supergiants such as Betelgeuse, including the effects of partial frequency redistribution in the lines. More recently I developed a transfer code (WIND3D) for modeling internal wind structures of hot stars using the dynamic spectra of wind lines. My point with these examples is that many transfer codes are initially developed by one (or a small group) of person(s) for modeling a given type of feature or property observed in stellar spectra. It is reassuring that these long-term radiative transfer code developments by many research groups worldwide will now benefit the entire stellar astrophysics community interested in the Gaia mission and its data products.

3. Radiation hydrodynamics codes

3.1. 3-D stellar atmosphere models

Two presentations at the meeting addressed very recent developments in 3-D hydrodynamic modeling of the atmospheres of late-type stars. Remo Collet showed that the convective pattern of large granules in STAGGER-CODE simulations of red giants yield density and temperature inhomogeneities (with correlations to macroscopic velocity fields) at the stellar surface that can profoundly alter the profiles and strengths of spectral lines compared to 1-D atmosphere models [8]. Interestingly, he pointed out that differences between the temperature stratifications in the upper atmosphere ($\log \tau_{5000} \leq -2$) of 1-D and 3-D models can amount to $\sim$1000 K, which severely affect the excitation, ionization, and molecular equilibria in those layers. In 3-D hydro models of metal-poor stars the physical cause is related to a reduced efficiency of radiative heating due to line opacity for balancing the adiabatic cooling rate compared to 1-D models. LTE transfer calculations of the detailed profiles and strengths of temperature-sensitive Fe I lines (hence with a large fraction of their line contribution function below $\log \tau_{5000} = -2$) in cool giants yield abundance differences of -0.5 dex or more, compared to the 1-D line transfer. The computed 3D$-$1D LTE abundance differences increase (they become more negative) with decreasing [Fe/H]- and $\log g$-values. The grid resolution and non-grey radiative transfer (i.e., the opacity wavelength resolution or binning) may also have profound effects on these differences, but the importance of 3-D hydro and transfer modeling for reliable determinations of the APs of cool-type spectroscopic (metal-poor) standard stars is clear. Andrea Chiavassa investigated surface convection in K-giants by combining the STAGGER-CODE with the radiative transfer code OPTIM3D. He finds convective Doppler shifts of up to 0.75 $\mathrm{km\,s^{-1}}$ for spectral lines in the Gaia RVS wavelength range. For a *star-in-a-box* model of red supergiant Betelgeuse from the CO^{5}BOLD code, the convective blue-shifts of the Ca II near-IR triplet lines can increase to $\sim -8\ \mathrm{km\,s^{-1}}$ [9]. He also finds that significant excursions of the Gaia G-band photocenter of RSGs can amount to $\sim$3 % of the optical stellar radius during the 5 years of the Gaia mission. It results from brightness fluctuations, with spots of up to 50 times brighter than dark surface

[2] `http://spectra.freeshell.org`

regions on the RSG, over time-scales of weeks.

3.2. Astrophysical microturbulence

Current *box-in-a-star* hydro-simulation research ought to provide more profound and valuable information for unraveling the physics of astrophysical 'microturbulence'. It is not sufficient to point out that spectral synthesis using 3-D atmosphere models does not require any microturbulence broadening without addressing the basic physical properties (or underlying mechanisms) of the atmospheric velocity fields that increase the line EWs. It is paramount for stellar astrophysics because it is impossible by means of classical spectroscopic methods to reliably determine APs (such as [Fe/H]) without considering microturbulence broadening. Astrophysical microturbulence is therefore often called a 'fudge-factor', an ad-hoc individual-to-the-star, free parameter for non-thermal line broadening that applies to spectral lines formation in hot stars as well. It is an AP that urgently requires a firm physical foundation. 3-D atmosphere models can address a number of long-standing pertinent astrophysical questions. What is the energy spectrum (e.g., the typical length scales of the velocity fluctuations with energy) of local atmospheric movements in 3-D hydro models that match traditional 1-D microturbulence velocities (ζ_μ)? Mean bulk velocities, or the velocity dispersion, in the lines formation regions of 3-D models of spectroscopic standard stars (Procyon, Arcturus, Canopus, etc.) can be compared to the 'observed' 1-D values, including depth-dependent models. An informed guess is that the energy spectrum adopts a (Kolmogoroff) power-law known for hydrodynamical turbulence, but advanced 3-D atmosphere models (e.g., properly resolved hydro-simulations) ought to provide a quantitative answer as to how it scales with the local thermodynamic conditions in stellar atmospheres. Why are microturbulence velocities in supergiants observed to exceed the local speed of sound? Is supersonic microturbulence signaling shock wave propagation or do the large values result from the 1-D thermodynamic model structure in the lines formation region? Do microturbulence velocities depend on the local gas kinetic temperature, in other words, do the non-thermal velocity fluctuations result from the balancing of local atmospheric heating and cooling mechanisms? The physical properties of astrophysical microturbulence are presumably different from plasma microturbulence modeled and observed at much larger gas kinetic temperatures in tokamaks, but how different?

Discussion session I

Workshop Question 1: *Do you think it is possible to define a standard format for (exchange of) model atmosphere data? How could we start to define such a standard? This would make comparisons and tests of different implementations of model physics and numerical approaches easier.*

Answer by Ulrike Heiter: I think it is worth a try to define a "stellar atmosphere data model". With that I mean a definition of physical and other quantities, with detailed description, units and keywords, that should be included in every model atmosphere file. One could start by doing a "model atmosphere user survey", asking users what they need for their calculations. These could be spectrum synthesis calculations, but also limb darkening (e.g., for interferometric angular diameters), boundary conditions for stellar evolution models, or non-LTE calculations.

Answer by Remo Collet: On the practical side, one possibility would be to share the data using "self-describing" file formats (i.e., NetCDF), which allow to easily add documentation and description of variables to the files. A more fundamental issue to consider when exchanging model atmospheres or synthetic spectra is the one of consistency. One should bear in mind that, in general, erroneous results and conclusions may be derived when spectra are computed with an equation-of-state and opacities that are inconsistent with the ones assumed for the model atmospheres. When exchanging model atmosphere data, one should consider using variables that are less sensitive to this problem.

Answer by Maria Bergemann: It is necessary to provide more data relevant to the equation-of-state (EOS) used to compute model atmospheres. Common experience says that the equation-of-state used in the latter and in spectrum synthesis packages adopted by a user are inconsistent. In the case of 1-D versus 3-D this inconsistency may be fatal. Individual partial pressures for as many elements as possible can be used to cross-check for this. Besides, some spectrum synthesis codes demand partial pressures as input.
Answer by Leo Houziaux: Is it workable to define a standard format? Letting aside the usual trinity $T_{\rm eff}$, $\log g$, [M/H], it seems necessary to decide on 1-D or multi-D, (N)LTE, various velocity fields, $v\sin i$, dust around the object, to say nothing about the influence of magnetic fields on the structure of the atmosphere that nobody mentioned. Inevitably, I think we will have to cope with various codes.
Answer by Bertrand Plez: In Montpellier we have set up the Pollux database of theoretical spectra. It contains spectra for O to M stars calculated for CMFGEN, ATLAS, and MARCS models (SED and high-resolution spectra). The database if fully Virtual Observatory (VO) -compliant and can be addressed through tools such as VO-Spec. It uses the VO format and data descriptors defined by VO. More spectra will be added in future releases.

Workshop Question 2: *We computed 1-D* MARCS *models for AGB stars. Although we made a lot of assumptions (such as LTE, hydrostatic, homogeneous layers, etc.) far from reality, the synthetic spectra fit the observed spectra fairly well, and computed colors match very good to observed colors (both in the optical and infrared). Is this just a coincidence, or does this show that 1-D models suffice for the moment to compute atmospheric parameters for AGB stars?*
Answer by Bertrand Plez: It is true that 1-D LTE models often fit observations pretty well, although not fully (the more and better observations we get, the more "details" don't fit). The only way to know if 1-D hydrostatic LTE models can be used to analyze observations is to try with non-LTE, 3-D, and look at the differences. If it turns out that we can use simple models and methods combined with tabulated corrections, that is very well! Otherwise we will have to do it the hard way!
Answer by Leo Houziaux: In this respect, Robert Kurucz showed about forty years ago that in the case of the solar spectrum most of the features, even very tiny ones, could be reproduced with a sort of "stone age" theory of line formation: hydrostatic equilibrium, plane-parallel atmosphere, LTE, radiative equilibrium. We know now that all these hypotheses are subjects of discussion.
Answer by Remo Collet: I think it is definitively important to pursue 3-D modeling: current 3-D models of late-type stellar atmospheres actually show that there are large systematic differences between, e.g., abundances derived with 3-D and with 1-D models, especially at low metallicities. Furthermore, predictions concerning other observables (e.g., center-to-limb variations) also differ between 1-D and 3-D calculations.
Answer by Maria Bergemann: Matching observed colors of a star to a model is very simple, and it doesn't tell much about the physics of stellar atmospheres. Models are usually computed to cover a broad range of observational data (colors or SEDs), so it is no surprise that one or a few from your grid accidentally match observed colors. The problem here is deeper: the model is far too simple to describe complex phenomena in stellar atmospheres. Especially for low-gravity cool stars, systematic errors caused by extension/LTE/1-D approximations are large. There is a large amount of studies describing these problems.

Workshop Question 3: *In coming years new 3-D hydrodynamic atmosphere model grids will become available. For example, in metal-poor cool stars oxygen abundances can be substantially overestimated with 1-D models. Should we decide when to use 3-D instead of 1-D modeling for determining atmospheric parameters, also for Gaia data? Is it feasible for very large datasets,*

Journal of Physics: Conference Series **328** (2011) 012027 doi:10.1088/1742-6596/328/1/012027

e.g. survey data?
Answer by Andreas Korn: That has to be the ambition! One has to realize the following: 3-D and non-LTE are the *current challenges* in cool stars modeling. Only by going to this new level of modeling sophistication can we hope to uncover and address *new challenges* in the physics of stars (e.g., atomic diffusion, rotation and mixing) or in the modeling (fully self-consistent 3-D non-LTE, with non-LTE feedback on the model; non-steady-state statistical equilibrium). We are still far from exploiting the full information encoded in stellar spectra!
Answer by Remo Collet: Once the grid of 3-D models will be complete, one possibility is that we directly provide the 3-D – 1-D corrections to oxygen abundances computed for different indicators and for all 3-D models and make them available to the community for direct application to abundances derived with classical 1-D methods.
Another possibility would be to directly use the average 3-D models which we will make available to the community as a substitute for classical 1-D models. This may yield results in closer agreement with "full 3-D" calculations, but at a lower computational cost and easier implementation for users of classical 1-D models.
Answer by Leo Houziaux: For Gaia spectral data, which are very limited both in wavelength range and in signal-to-noise ratio, 1-D models including microturbulence seem appropriate. 3-D models are certainly more realistic, provided the necessary atomic and molecular parameters are accurately known, especially the electronic collision cross-sections. On the other hand, oxygen abundance is very delicate to determine, not only because it is present in several molecules, but even as an atomic species, it presents well-known fluorescence mechanisms with H(Lyβ) and even with H(Lyα). In addition, care should be taken to use an appropriate Stark broadening theory for the higher lines of the H Paschen series, as the energy levels get very close to each other.

Discussion session II
Question by Remo Collet to Ulrike Heiter and Bertrand Plez: *You have reported that the GREAT-ESF workshop on comparative spectrum modeling (for cool giants in Vienna of Aug. 2010), for some of the test cases, the various participating groups derived significantly different stellar parameters from the same spectrum. Did you investigate the reasons why the different analysis strategies led to different results? Would it be possible to identify an optimal strategy using the ones that were considered?*
Answer by Ulrike Heiter: No. We did no see any clear relation between parameter differences and type of model or analysis method. We did a comparison of spectra computed with different codes for the same set of atmospheric parameters and found differences of the same order of magnitude as the differences between best-fit model and observations. This means that the modeling differences play a certain role. It would be good to repeat the experiment, letting every participant use the same strategy for parameter determination.
Answer by Bertrand Plez: One way to identify the source of errors/differences is to use (i) a single parameter determination, with spectra from the different groups, that will give us the impact of differences in modeling (due to line lists, equation of state, etc.), and (ii) use spectra from a single model to test the various fitting approaches.

Question by Bertrand Plez to Remo Collet: *What is the meaning of depth-dependent microturbulence? Why do you use the velocity dispersion from 3-D simulations as an estimate of it? Isn't microturbulence related to turbulence at microscopic (inter-atomic) scales?*
Answer by Remo Collet: In late-type stars, the so-called "microturbulent" broadening adopted in classical 1-D calculations is most likely due (mainly) to turbulence occurring on microscopic scales comparable with the inter-atomic distances, as the name seems to suggest. The microturbulence parameter in late-type stellar atmospheres is actually related more to the distribution of relative bulk velocities in the line formation region at scales of the order of the mean free

path of photons which can be quite large at the optical surface and above it. The value of the (depth-dependent) micro-turbulence for a given stellar atmosphere could be estimated by looking at the dispersion of the bulk velocities at various optical depths as predicted by 3-D hydrodynamic simulations (discussed in recent publications). As Maria Bergemann showed in her talk, such a choice of micro-turbulence removes the trends of derived iron abundances with equivalent widths of Fe I and Fe II lines in non-LTE calculations using average 3-D models. This suggests that present simulations are already resolving the important scales and velocity fields that contribute to micro-turbulence broadening.

Question by Ulrike Heiter to Remo Collet: *On the question of 1-D vs. 3-D hydrodynamic models for cool stars: 1-D models can be calculated for spherical geometry, while 3-D "box-in-the-star" models are plane-parallel. What are the prospects for 3-D spherical models and is this important?*
Answer by Remo Collet: Based on the experience from 1-D models, I think we can say that the "box-in-the-star" approximation is safe for models with log $g >\sim 1.5$ (cgs units) and that sphericity effects can be neglected to first order. For lower gravities, it would be in principle better to switch to the "star-in-the-box" mode.

4. Atmospherical parameters, abundance, metallicity, and chemical tagging studies

The workshop provided an overview of research activities for determining astrophysical parameters; the effective temperature, surface gravity, metallicity (e.g., surface abundances), mean microturbulence velocity, α-enrichment, radial velocity, projected rotation velocity, and macroturbulence velocity, from the spectra of cool and hot stars. Other parameters such as the stellar mass-loss rate, wind acceleration (β), terminal wind velocity, wind clumping factor, and the number fraction of helium over hydrogen, were also discussed for luminous hot stars.

4.1. Cool star parameters

The three APs $T_{\rm eff}$, log g, and [Fe/H] are traditionally determined from 1-D LTE spectrum synthesis calculations in FGK stars. Maria Bergemann described results of non-LTE transfer calculations with DETAIL and MULTI2.3 of Fe I and Fe II lines in FGK dwarfs and giants [10]. The iron model atom is very advanced with 296 energy levels of Fe I and 112 levels for Fe II, where 16207 permitted bound-bound transitions are considered, and assuming complete frequency redistribution in the lines. She finds lower number densities of Fe I compared to LTE, whereas Fe II lines are hardly affected by NLTE. It yields the weakening of Fe I lines compared to LTE, therefore requiring a larger iron abundance for fitting observed spectral lines. The computed "NLTE abundance corrections" for Fe I lines do not exceed $\sim$0.05 dex for solar metallicity dwarfs, but they can increase to 0.6 dex in metal-poor giants. The Fe I NLTE corrections increase with decreasing [Fe/H]- and log g-values, thus the assumed LTE ionization balance yields progressively underestimated surface gravities and metallicities. A database is under construction to provide the NLTE abundance corrections online.

Nadya Gorlova determined the APs of post-AGB F-type giant BD+46 442 with a dust disc, and finds $T_{\rm eff}$=6250±250 K, log g=1.5±0.5, [Fe/H]=−0.7±0.2 dex (slightly metal-poor), and ζ_μ=4.0±0.5 km s^{-1} [11]. She compares individual line abundances computed with the LTE spectrum synthesis codes WIDTH9 and MOOG, using EW-values of weak and medium-strong lines observed in high-resolution Mercator-HERMES spectra. She concludes that the dominant source of the abundance errors is the uncertainty in the model $T_{\rm eff}$ and log g, followed by the uncertainty in the EW measurements, and the employed line oscillator strengths. Both transfer codes yield very similar metal abundance values (with differences smaller than $\sim$0.03 dex). A comparison between the original ATLAS9 atmosphere models and the new ODF ATLAS9 models (from Fiorella Castelli) yields, however, abundance differences to −0.16 dex for 7 N I lines with

Journal of Physics: Conference Series **328** (2011) 012027 doi:10.1088/1742-6596/328/1/012027

35 mÅ<EW<110 mÅ, and to −0.11 dex for Fe I and Fe II lines. Effects of the adopted 1-D atmosphere models may therefore be non-negligible for element abundance measurements, provided the LTE assumption is appropriate. A study of [Fe/H] in a sample of FGK stars with and without dusty debris discs by Jesús Maldonado reveals (using WIDTH9 and ATLAS9) a lack of correlation between the presence of debris discs and large metal abundances [12]. These metal abundance studies can provide important clues for planet formation because the formation of planetary cores is favored in discs with high-metal content. Gaia is expected to detect 10 000 exoplanets. Giancarlo Pace presented an interesting literature study to assess abundance differences between dwarfs and giants in open clusters [13]. He warns that reported [Fe/H] differences should be closely scrutinized because they can result from the adopted measurement procedures (i.e., LTE versus NLTE modeling) and spectral line lists. For example, is the lack of giants with [Fe/H]>0.2 dex compared to dwarfs in the solar neighborhood the result of comparing two different classes of objects? Actual abundance differences between dwarfs and giant have profound theoretical implications for planet formation rates and the chemical evolution of the Galaxy.

David Montes presented a study of APs and chemical tagging of nearby FGK stars. It is based on high-resolution spectra observed with a large number of spectrographs, and also collected from various telescope archives. $T_{\rm eff}$, log g, [Fe/H], and ζ_μ are computed with the STEPAR code using EW-values of some 300 Fe I and Fe II lines. STEPAR (a combination of MOOG and ATLAS9) determines the excitation equilibrium, while the ionization equilibrium is imposed by log(ϵ(Fe I)) = log(ϵ(Fe II)). Differential abundances (Δ[Fe/H]) are then used to study candidate membership of stars to i.e., the Hyades Supercluster, with respect to [Fe/H] of a known cluster member. This permits to investigate the stellar kinematics of various groups. For example, field-like stars are associated with the Galaxy spiral structure or with dynamical resonances (bar), whereas young co-eval stars with debris of star-forming aggregates in the disc. Chemical tagging studies (the analysis of Fe, Na, Mg, Al, Si, Ca, Sc, Ti, V, Cr, Mn, Co, and Ni) of possible Hyades members show 41% homogeneity of the abundances. An investigation of the metallicity and Li abundance of the Hyades Stream compared to Hyades Cluster was presented by Alain Jorissen. He finds that the Hyades Stream has a metallicity excess of 0.2 dex compared to the thin disc, and that the Stream is possibly a 4:1 resonance of the spiral pattern.

In a study of S stars Sophie Van Eck presented a grid of 3522 synthetic spectra computed with TURBOSPECTRUM for 2700 K $\leq T_{\rm eff} \leq$ 4000 K, C/O = 0.5, 0.750, 0.899, 0.925, 0.951, 0.971, 0.991, [s/Fe] = 0.0, +1.0, +2.0 dex, [Fe/H]=0.5 and 0.0 dex, and log g = 0, 1, 2, 3, 4, 5. From a comparison of color-color diagrams of observed and synthetic spectra she finds that for a given $V-K$-value, the $T_{\rm eff}$-range of models with different C/O ratios can be as large as 400 K. The application of the usual $T_{\rm eff}$-scale for M-stars (traditionally based on the $V-K$-color index) can therefore yield $T_{\rm eff}$ errors for S stars of up to 400 K [14]. The Tc, Zr, and Fe abundances of S stars are determined with the grid by Pieter Neyskens. He compared the Tc/Zr abundances (as a function of [Zr/Fe] overabundances) computed with the STAREVOL code for AGB stellar evolution models of different stellar masses and sizes of the partial mixing zone, with Tc abundances from optical Tc I lines [15]. The computed [Zr/Fe] overabundances are in good agreement with the AGB stellar evolution model predictions, while the Tc/Zr abundances are slightly over-predicted. The study helps to constrain nucleosynthesis and mixing mechanisms in AGB stars.

Alejandra Recio-Blanco presented an overview of research results for classifying FGK stars in the Gaia Data Processing and Analysis Consortium (DPAC-CU8). The Generalized Stellar Parametrizer spectroscopy group develops and tests various algorithms for automated AP determinations with Gaia RVS spectra. She compared internal errors of APs computed with the MATISSE (projection method) & DEGAS (oblique decision tree method), Nelder-Mead, and artificial neural network classification codes using spectral signal-to-noise ratios S/N≤150. She

Journal of Physics: Conference Series **328** (2011) 012027 doi:10.1088/1742-6596/328/1/012027

finds for S/N=50 that $\Delta T_{\rm eff}\sim$100 K, Δlog $g\sim$0.2–0.3 dex, Δ[M/H]$\sim$0.15 dex, and $\Delta[\alpha$/H]$\sim$0.1 dex. For cool dwarfs the Ca II near-IR triplet lines in the RVS wavelength domain do not provide sufficiently accurate log g-values. The development and (combined) application of these codes will also be important for the planned Gaia-ESO large public survey with the Giraffe-UVES/FLAMES instrument of VLT. Thierry Morel discussed an interesting study for improving log g-determinations of late-type stars, and hence the accuracy of elemental abundance values, based on a sample to $\sim$40 G-type dwarfs and red giants. The sample provides accurate *seismic* log g-values computed from $T_{\rm eff}$ and the frequency of the maximum oscillation power obtained from ground-based observations, and the *CoRoT* and *Kepler* missions. The log g-values computed from traditional spectroscopic methods agree remarkably well with the seismic values, showing systematic differences below 0.04 dex with 1-$\sigma\leq$0.19 dex [16]. The model-independent methods are important for parameterizing benchmark stars that will be utilized to calibrate the automated AP algorithms currently under development for Gaia and large spectroscopic surveys.

4.2. Hot star parameters

A very interesting study of CNO abundances in OB main sequence stars was presented by Maria-Fernanda Nieva with remarkable new results. She finds from spectral synthesis calculations with the SURFACE code, combined with NLTE level populations from DETAIL and atmosphere models from ATLAS9, a much larger degree of chemical homogeneity of C/O and N/O abundances with respect to the O/H abundance for a sample of 64 B-type stars, compared to previous analyses. In the past the enormous scatter in B-star abundances was difficult to reconcile with data from solar-type stars. The presented analysis accounts for all spectroscopic indicators, e.g., Balmer, helium lines, and metal ionization equilibria, and yields substantially smaller systematic errors for the APs and chemical abundances [17]. The large increase of accuracy and precision of this method for determining the APs of early-type stars is important for future analyses of the Galactic abundance gradient, i.e., with high-resolution and large S/N ratio spectra of the planned Gaia-ESO public survey.

Laurent Mahy modeled the spectra of O-type stars in young open cluster NGC 2244 of the Rosette Nebula. He determined stellar and wind parameters with NLTE synthetic spectrum fits of CMFGEN to UV and optical spectra. The atmosphere models from TLUSTY are combined with a β wind velocity law for O-type dwarfs. The analysis yields systematically smaller mass-loss rates derived from UV P Cyg profiles compared to Hα, possibly due to wind 'vorocity', or the micro-clumping of the expanding wind that can be modeled with a volume filling factor. He also finds a clear trend of N enrichment (from optical N III absorption lines) with larger stellar luminosity or mass [18].

Thomas Dall presented a nice study of the effects of non-spherical and non-homogeneous stellar surfaces on synthetic spectrum calculations. The visible stellar surface is divided into (nearly) equal geometrical sectors ('tiles') for which the emergent spectra are calculated with classical atmosphere models (ATLAS, MARCS, etc.) and summed up to a theoretical spectrum. Preliminary results for FK Com (and Altair) show rather small differences with traditional spherical synthetic spectrum calculations. The new code is however very flexible for assessing spectroscopic effects of various stellar properties, such as star-spots, fast rotation, variable temperature, gravity, and abundances across the stellar surface [19].

Discussion session III

Workshop Question 4: *How to define accuracy with which we determine atmospheric parameters? Should we make a clear distinction between internal and systematic errors? Systematic errors can be due to different modeling methods, LTE vs. non-LTE, 1-D vs. multi-D, inaccurate or incomplete line lists, etc. Internal errors can result from differences in best fit*

methods, for example spectral synthesis fits vs. classical equivalent line width (EW) methods.
Answer by Ulrike Heiter: Yes, internal and systematic errors should be analyzed and estimated separately. Nowadays, authors are doing a good job citing internal errors resulting from the numerical accuracy of their method, but systematic errors are usually poorly characterized. Total uncertainties in atmospheric parameters and abundances seem to be underestimated.
Answer by Alex de Koter: In the hot stars community, we indeed try to tackle systematic errors due to different fitting approaches. In particular, we compare a grid based method with a genetic algorithm method. The methods show very good agreement. In those remaining cases where significant discrepancies occur we can explain why these occur and have ideas how to resolve them.
Answer by Maria Bergemann: It is absolutely necessary to make a careful assessment of errors associated with missing/unrealistic physics in the models used for the Gaia stellar parameter pipeline. Furthermore, systematic errors must be accommodated in the input parameters characterizing synthetic spectra, which are used in the numerical algorithms ("nearest neighbor", "neural network", etc.) developed for Gaia Radial Velocity Spectrometer (RVS) spectra.
Answer by Leo Houziaux: The main question is that we do not really know the noise generated by the location of the Gaia instrument in the vicinity of the Earth. It is quite likely that cosmic rays etc. will lower the signal-to-noise ratio for a lot of objects. What is important is that error limits (both internal and systematic) be indicated by the members of the various consortia.
Answer by Remo Collet: I think the only feasible approach is to use a set of standard stars for which high-quality spectra and photometric data as well as measurements of parallaxes, radii, etc., are available to validate theoretical model atmospheres and synthetic spectra. One should try to reproduce with the same suite of models as many observables as possible for standard stars, in a consistent way (i.e., making sure that different indicators of effective temperature, surface gravity, etc., yield the same results).

5. Large spectroscopic surveys

A large number of presentations at the meeting discussed several techniques for measuring the abundances of various chemical elements from stellar spectra. The methods are important for ongoing developments and testing of automated & supervised algorithms for determining detailed chemical composition and tagging studies in large (chemo-dynamical) spectroscopic surveys planned to complement the Gaia (astrometric and kinematic) census of the Galaxy. The Gaia-ESO public survey (GES) will observe $\sim$100 000 stars, and the LAMOST survey possibly millions of stars. Smaller spectroscopic surveys of hot stars (Tarantula Survey, VLT-X-Shooter, VLT-CRIRES, IACOB project, Be-star survey, Low Dispersion Sky plate surveys) were also extensively discussed, including quantitative analyses for determining APs and advanced spectral classification methods.

Ronny Blomme provided an overview of the preparations for observing O, B, and A-stars in GES (PIs are Gerry Gilmore and Sofia Randich). It is currently estimates that $\sim$1500 hot star spectra will be observed with Giraffe ($R\sim$15 000), and about 600 with UVES (R=47 000) using VLT FLAMES over 300 nights during 5 years (2011–2016). The number of spectra of FGK stars is estimated more than an order of magnitude larger (10^5 Giraffe and $\sim$6000 UVES spectra). The definition of GES work packages is ongoing and the hot stars work package (GES-WP13) involves cluster selection and the coordination of spectrum analyses with a variety of transfer codes discussed at the meeting [20]. About 60 old star clusters will be observed as part of GES (these are defined as having an age larger than 100 Myr), and 40 young clusters. All stellar populations in the clusters will be observed by GES. Among the young clusters, 13 were selected for their massive-star content. A large variety of science topics will be addressed with GES data of hot stars. To name a few; constraining the upper part of the Initial Mass Function, the possible correlation of N abundance with $v\sin i$, modeling of micro-clumping in massive hot star winds, and the flattening of the Galactic abundance gradient for B-stars in clusters for studying thin disc formation. Note that the first GES observations are planned for December 2011.

Alex de Koter presented preliminary results for (automated) analyses of VLT FLAMES Tarantula Survey spectra. The survey is possibly the largest homogeneous study of O and early-B stars to date and consists of multi-epoch spectroscopy of ~1000 massive stars in 30 Dor [21]. It will allow for comprehensive tests of the theory of massive star evolution. Long-standing problems such as the discrepancy between spectroscopic and evolutionary masses of O stars will be addressed. He described test results with a genetic algorithm (GA) fit method using simulated spectra of the NLTE FASTWIND radiative transfer code for atmospheric hydrogen & helium composition $Y_{\rm He}$ (N abundance is to be included). The analysis method for O-type stars incorporates spherically outflowing wind models. The fit results for a sample of 24 O-type dwarfs and supergiants are very accurate, and appear free of systematic offsets in $T_{\rm eff}$, log g, $\dot{M}$, β, $Y_{\rm He}$, and ζ_{μ}. The analysis results will establish empirical relations between the mass-loss rate and the Eddington factor (Γ) in massive stars, investigate rotationally induced mixing (i.e., N surface abundance versus $v\sin i$), the effects of angular momentum loss through wind mass-loss, and important aspects of the binary fraction of massive stars, the binary period distribution and evolution. Frans Arjen Stap showed results of the automated GA fitting method to hydrogen and helium line profiles in the J, H, K, and L-band of 5 bright O-type dwarfs. The near-IR high-resolution profiles are observed with VLT-CRIRES (R~50 000). He compares the best fit parameters with the results of optical spectrum analyses, and finds $T_{\rm eff}$ differences within a spectral subtype and Δlog g<0.2 dex [22]. The near-IR and optical spectra yield differences of computed mass-loss rates of 0.2–1.0 dex, that may possibly be reconciled using differential micro-clumping in the wind models.

Sergio Simon-Diaz presented results of the IACOB project, a grid-based automatic tool for determining stellar and wind parameters of O-type stars. The grid consists of ~192 000 models computed with FASTWIND for $T_{\rm eff}$>25 kK and 2.6 dex $\leq$ log g $\leq$ 4.3 dex. It employs a χ^2-minimization method and uses the detailed profiles of Balmer H and He I/II (absorption) lines and the line equivalent widths as input values [23]. The stellar evolutionary mass can also be computed from evolutionary tracks. The tool is very fast and provides graphics output. It will certainly benefit the analysis of large O star samples observed in current and future ground-based surveys. Yanping Chen discussed a new library of ~600 stars that will be observed with VLT X-Shooter for all spectral types. The spectra cover the complete wavelength range from 300 nm to 2500 nm with R≃10 000. Since the library will cover a wide range of $T_{\rm eff}$-, log g-, and [Fe/H]-values of stars observed in the LMC, SMC, and the Galactic disc and bulge, it will provide reliable stellar population models for studying galaxy formation and evolution. The X-Shooter spectra will also be calibrated in absolute fluxes and corrected for absorption in Earth's atmosphere [24]. A study of ~150 Be star spectra observed with the Coudé spectrograph (R=12 000) of the Ondřejov 2 m telescope was presented by Pavel Koubsky. He showed that the P13 to P16 Paschen emission lines are proxies for normalized Hα emission line fluxes exceeding 2.5. The Gaia RVS spectra, which will include these Paschen lines, will hence provide independent information about the Hα line [25]. Several thousands of Be stars will be observed by Gaia, and the study of their emission lines provides important information about the physical properties of Be-star discs.

Rene Hudec provided an overview of various low-dispersion spectroscopic sky surveys. The data have been observed with objective prisms on photographic plates and are compared to simulated low-dispersion observations of the Gaia BP/RP photometers. He develops algorithms for the automated classification of digitized spectra from the photographic plates. It includes, among other things, searches for spectroscopically variable objects, strong and wide spectral emission features in cataclysmic variables, PNe, white dwarfs, carbon stars, QSOs, AGN and peculiar galaxies, and studies of the continuum flux distribution of objects at high redshift [26].

Journal of Physics: Conference Series **328** (2011) 012027 doi:10.1088/1742-6596/328/1/012027

Teresa Aparicio Villegas investigated the capabilities of the ALHAMBRA photometric system for estimating APs and stellar classification. She presented a fit method based on BaSeL models for reliably estimating $T_{\rm eff}$-, log g-, [Fe/H]-, and E($B-V$)-values of a wide variety of spectral types (O to M stars) and luminosity classes, including hot white dwarfs and chemically peculiar stars. Three extinction laws were investigated yielding differences that are not significant to decide if one law should be preferred over the other [27].

Jiannan Zhang and Ali Luo presented the comparison of synthetic LTE spectra computed with the ATLAS9 (new ODFs) and MAFAGS-OS 1-D atmosphere models. The latter models are computed with opacity sampling and convection using a convective efficiency parameter. The ATLAS9 model grid is calculated with opacity distribution functions and mixing-length theory. The comparison of both synthetic spectrum grids for dwarf and giants of 5000 K $\leq T_{\rm eff} \leq$ 15 kK, based on photometric color indexes, absorption line indexes, and the complete spectra, shows appreciable differences for $T_{\rm eff}$<6000 K [28]. The reported differences can systematically influence planned AP determinations of large spectroscopic sky surveys such as LAMOST and Gaia.

Andreas Quirrenbach discussed concepts and plans for future massively parallel spectroscopic instruments that can provide huge numbers of stellar spectra. It is obviously important to match the properties and capabilities of such instruments to the requirements of quantitative spectroscopic modeling efforts. He discussed the 4MOST project proposed as a large international collaboration to go 2–3 magnitudes fainter than Gaia. It involves multi-fibre spectroscopy (300 fibers for R=20 000, and 1500 fibers for R=3000 (red) to R=5000 (blue)) of millions of stars with S/N=20, using large facilities such as NTT and VISTA. 4MOST is approved for study by ESO. Very large surveys are important for unraveling the role of stellar migration for the formation of the Galactic thick and thin disc.

6. New atomic database

Nicholas Walton presented ongoing developments of the Virtual Atomic & Molecular Data Centre (VAMDC), a new atomic database that will become online available by late 2012. The database will provide advanced and cost-effective access methods to the current 'best' fundamental atomic and molecular data. It will provide access infrastructure to a wide range of users (astronomy, nuclear, climatology, biology) in academia and industry. The new database will be important for astrophysics for improved interpretation of the huge spectral datasets that will be generated by the Gaia mission and complementary large ground-based surveys.

Discussion session IV

Workshop Question 5: *A number of large surveys are under way. What additional surveys would be needed to substantially advance our knowledge of stellar atmospheres (both for cool and hot stars)?*

Answer by Elena Pancino: In the area of surveys, it becomes more and more important to define a set of "calibrators" or "calibrating fields" of stars with known parameters and abundance ratios, well studied with high-resolution and high signal-to-noise ratios, and analyzed with different methods, to serve as a means to compare the results of different surveys and eventually combine them. The Gaia-ESO survey is trying to do this in coordination with Hermes and Apogee, but at some point this should become a community-wide effort.

Answer by Bertrand Plez: When planning surveys, including asteroseismology fields are very important so that we get additional independent constraints on stellar parameters.

Workshop Question 6: *What are the most important atomic and molecular data that should be improved or determined for realistic modeling of hot and cool star spectra? Which species and/or which type of data, e.g. level energies, transition probabilities, line broadening parame-*

Journal of Physics: Conference Series **328** (2011) 012027 doi:10.1088/1742-6596/328/1/012027

ters, etc.; what wavelength region?
Answer by Alex de Koter: One (of the many) problems the analysis of hot star atmospheres may suffer is poorly known line broadening parameters and/or assumptions in the description of the broadening theory. This may have significant impact on the gravity determinations, which may be part of the reason why we find discrepancies between spectroscopically determined masses and those following from comparison to to evolutionary predictions.
Answer by Norbert Przybilla: Near-IR spectroscopy will gain in importance in the future, in particular with the upcoming generation of extremely large telescopes. In order to use their full potential, observations will depend on adaptive optics, which is technologically best feasible in the near-IR domain. Pioneering high-resolution spectroscopic surveys like CRIRES-POP show that there is an enormous lack of atomic data at near-IR wavelengths. Even line-identifications face severe problems, requiring laboratory measurements of energy levels and wavelengths for many transitions, both atomic and molecular. Quantitative analyses will require all kinds of data, ranging from transition probabilities to line-broadening data, which in the latter case are in part lacking even for hydrogen and helium.
What is currently mostly needed for non-LTE modeling are more collisional data for electron-ion impact excitation, for a large number of ions/elements. This is in contrast to the situation of data for radiative processes, where the Opacity Project and the IRON Project have provided a large quantity of data already.

Workshop Question 7: *Do the standard models used for spectroscopic analysis of massive hot stars (1-D, spherical symmetric, stationary wind, full line blanketing) provide sufficiently accurate stellar and wind parameters? Are we neglecting important physical mechanisms, such as micro- and macro-clumping, radiative instabilities, non-spherical winds?*
Answer by Artemio Herrero: We have to implement new physical phenomena both in 1-D and 3-D. As it has been shown in this conference many times, only after a more sophisticated calculation can we say whether a given effect will be important for a given type of star. Therefore we cannot anticipate whether it is more important to consider a 3-D geometry or to include more opacity sources, just to mention two possible improvements. We have to keep improving our models, because we can not be sure which is the reason for a failure in fitting the observations until we calculate it.

Question by Alex Lobel: *We have not heard much about mass-loss properties in cool stars at the meeting? Would somebody want to comment on mass-loss research for cool stars as well?*
Answer by Alex de Koter: Concerning the study of mass-loss properties of cool stars, Asymptotic Giant Branch stars, and Red Supergiants; the HIFISTARS project uses HIFI/Herschel to probe the (sometimes time-dependent) outflows using molecular line diagnostics (e.g., CO, H_2O). The goal is to model different rotational transitions – formed in different parts of the outflow – consistent with the thermal emission of grains formed in the wind. This allows to constrain the gas-to-dust ratio (from the gas mass-loss rate and the dust mass-loss rate), a fundamental quantity if one wants to understand these cool outflows. So, this is a very active field of research!

7. Summary
The GREAT-ESF workshop on Stellar Atmospheres in the Gaia Era held at the Vrije Universiteit Brussel in June 2011 has considerably improved communication and helped to establish new science collaboration between researchers and students of the hot and cool stars communities, and among members of the GREAT WGs and the Gaia DPAC for the development and application of advanced spectrum modeling codes. It addressed and identified key areas for the improvement of model spectra from various modern spectral synthesis codes required to

determine reliable physical parameters of hot and cool stars. Its outcome can form the basis for future improvements in advanced spectrum synthesis calculations. The results represent a new milestone towards accurate stellar parameters of cool and hot stars that Gaia will observe, and for Gaia follow-up programmes.

Acknowledgments
I would like to express my sincere thanks for financial support from the European Science Foundation (GREAT-ESF) and the Research Foundation - Flanders. I specially thank Jean-Pierre De Greve, Walter van Rensbergen, and Merel Fabré of the LOC for help with organizing the meeting at the Vrije Universiteit Brussel, and with editing of the Proceedings volume. The Royal Observatory of Belgium is gratefully acknowledged for help with the conference grants management, transportation, and administration. I would like to thank all workshop participants for their valuable contributions and lively discussion interactions that resulted in a truly successful scientific meeting.

References
[1] Heiter U and Lebzelter T 2011 Gaia spectroscopy overview and comparative spectrum modeling for cool giants *J. Phys.: Conf. Series this Volume*
[2] Sordo R, Vallenari A, Tantalo R, Liu C, Smith K, Allard F, Blomme R, Bouret J-C, Brott I, de Laverny P, *et al* 2011 Stellar libraries for Gaia *J. Phys.: Conf. Series this Volume*
[3] Plez B 2011 Cool star model atmospheres for Gaia : ATLAS, MARCS, and PHOENIX *J. Phys.: Conf. Series this Volume*
[4] Lester J B and Neilson H R 2008 *A&A* **491** 633–641 (*Preprint* 0809.1870)
[5] Przybilla N, Nieva M-F and Butler K 2011 Testing common classical LTE and NLTE model atmosphere and line-formation codes for quantitative spectroscopy of early-type stars *J. Phys.: Conf. Series this Volume*
[6] Groh J H 2011 Modeling the wind and photosphere of massive stars with the radiative transfer code CMFGEN *J. Phys.: Conf. Series this Volume*
[7] Lobel A 2011 *Can. J. Phys.* **89** 395 (*Preprint* arXiv:astro-ph/1012.5243)
[8] Collet R, Magic Z and Asplund M 2011 The StaggerGrid project: a grid of 3-D model atmospheres for high-precision spectroscopy *J. Phys.: Conf. Series this Volume*
[9] Chiavassa A, Bigot L, Thévenin F, Collet R, Jasniewicz G, Magic Z and Asplund M 2011 3-D hydrodynamical model atmospheres: a tool to correct radial velocities and parallaxes for Gaia *J. Phys.: Conf. Series this Volume*
[10] Bergemann M, Lind K, Collet R and Asplund M 2011 NLTE effects on Fe I/II in the atmospheres of FGK stars and application to the abundance analysis of their spectra *J. Phys.: Conf. Series this Volume*
[11] Gorlova N 2011 Abundance analysis of post-AGB stars *J. Phys.: Conf. Series this Volume*
[12] Maldonado J, Mora A, Montesinos B, Villaver E and Eiroa C 2011 Spectroscopic properties of stars with debris discs *J. Phys.: Conf. Series this Volume*
[13] Pace G 2011 The metallicity scale of dwarf and giant stars *J. Phys.: Conf. Series this Volume*
[14] Van Eck S, Neyskens P, Plez B, Jorissen A, Edvardsson B, Eriksson K, Gustafsson B, Jørgensen U G and Nordlund Å 2011 A grid of S stars MARCS model atmospheres *J. Phys.: Conf. Series this Volume*
[15] Neyskens P, Van Eck S, Plez B, Goriely S, Siess L and Jorissen A 2011 Accurate abundance determinations in S stars *J. Phys.: Conf. Series this Volume*
[16] Morel T, Miglio A and Valentini M 2011 Using seismic targets as benchmarks for spectroscopic analyses of cool stars *J. Phys.: Conf. Series this Volume*
[17] Nieva M-F, Przybilla N and Irrgang A 2011 High-precision atmospheric parameter and abundance determination of massive stars, and consequences for stellar and Galactic evolution *J. Phys.: Conf. Series this Volume*
[18] Mahy L, Martins F, Hillier D J and Rauw G 2011 A quantitative study of the O stars in NGC 2244 *J. Phys.: Conf. Series this Volume*
[19] Dall T H and Sbordone L 2011 Visualization and spectral synthesis of rotationally distorted stars *J. Phys.: Conf. Series this Volume*
[20] Blomme R 2011 Hot stars in the Gaia-ESO Survey *J. Phys.: Conf. Series this Volume*
[21] de Koter A, Sana H, Evans C J, Bagnoli T, Bastian N, Bestenlehner J M, Bonanos A, Bressert E, Brott I, Cantiello M, *et al* 2011 The O stars in the VLT-FLAMES Tarantula Survey *J. Phys.: Conf. Series this Volume*

[22] Stap F A, Sana H and de Koter A 2011 Quantitative near infra-red spectroscopy of massive stars *J. Phys.: Conf. Series this Volume*

[23] Simón-Díaz S, Castro N, Herrero A, Puls J, Garcia M and Sabín-Sanjulián C 2011 The IACOB project: A grid-based automatic tool for the quantitative spectroscopic analysis of O-stars *J. Phys.: Conf. Series this Volume*

[24] Chen Y, Trager S, Peletier R and Lançon A 2011 XSL: The X-Shooter Spectral Library *J. Phys.: Conf. Series this Volume*

[25] Koubský P, Kotková L and Votruba V 2011 Hα and Gaia-RVS domain spectroscopy of Be stars and interacting binaries with Ondřejov 2m telescope *J. Phys.: Conf. Series this Volume*

[26] Hudec R and Hudec L 2011 Tests of simulated Gaia BP/RP spectra with LDS (Low Dispersion Spectroscopy) photographic sky surveys *J. Phys.: Conf. Series this Volume*

[27] Aparicio Villegas T, Alfaro E J, Cabrera-Caño J, Moles M, Benítez N, Perea J, del Olmo A, Fernández-Soto A, Cristóbal-Hornillos D, Aguerri J A L, *et al* 2011 Stellar physics with the ALHAMBRA photometric system *J. Phys.: Conf. Series this Volume*

[28] Zhang J, Luo A, Song Y and Zuo F 2011 Comparison between MAFAGS-OS spectra and Kurucz-ODF spectra *J. Phys.: Conf. Series this Volume*

AUTHOR INDEX

AUTHOR INDEX